T0388192

Emerging Nanomedicines for Diabetes Mellitus Theranostics

Emerging Nanomedicines for Diabetes Mellitus Theranostics

MICHAEL K. DANQUAH

JAISON JEEVANANDAM

ELSEVIER

Elsevier
Radarweg 29, PO Box 211, 1000 AE Amsterdam, Netherlands
The Boulevard, Langford Lane, Kidlington, Oxford OX5 1GB, United Kingdom
50 Hampshire Street, 5th Floor, Cambridge, MA 02139, United States

Notices

Knowledge and best practice in this field are constantly changing. As new research and experience broaden our understanding, changes in research methods, professional practices, or medical treatment may become necessary.

Practitioners and researchers must always rely on their own experience and knowledge in evaluating and using any information, methods, compounds, or experiments described herein. In using such information or methods they should be mindful of their own safety and the safety of others, including parties for whom they have a professional responsibility.

To the fullest extent of the law, neither the Publisher nor the authors, contributors, or editors, assume any liability for any injury and/or damage to persons or property as a matter of products liability, negligence or otherwise, or from any use or operation of any methods, products, instructions, or ideas contained in the material herein.

ISBN: 978-0-323-85396-5

For Information on all Elsevier publications
visit our website at https://www.elsevier.com/books-and-journals

Publisher: Matthew Deans
Acquisitions Editor: Sabrina Webber
Editorial Project Manager: Rafael Guilherme Trombaco
Production Project Manager: Anitha Sivaraj
Cover Designer: Greg Harris

Typeset by MPS Limited, Chennai, India

Working together
to grow libraries in
developing countries

www.elsevier.com • www.bookaid.org

Contents

About the authors

Dr. Michael K. Danquah is a UC Foundation and Guerry Professor of Chemical Engineering and the Associate Dean of the College of Engineering and Computer Science, University of Tennessee at Chattanooga, United States. His research interest includes the utilization of biomolecular engineering principles and nanotechnology to develop emerging biopharmaceuticals, nanomedicines, biosensing, and bioremediation systems. His research findings are well published with about 300 publications and presentations. He is a Chartered Engineer (CEng),  a Chartered Scientist (CSci), a Fellow of the Institution of Chemical Engineers (FIChemE), and a Fellow of the Royal Society of Chemistry (FRSC). He was recently ranked in the top 2% of world scientists based on citations by a Stanford University publication.

Dr. Jaison Jeevanandam is a Senior Researcher at Centro de Química da Madeira, Universidade da Madeira, Portugal. He obtained his PhD degree from the Department of Chemical Engineering, Faculty of Engineering and Science, Curtin University, Malaysia. His PhD project covered the development of novel nanomedicines for reversing insulin resistance in Type 2 diabetes. He has experience in nanoparticle synthesis, characterization, nanoformulation, and cytotoxicity analysis. His current research focuses on the application  of bionanotechnological approaches to fabricate novel nanoformulations for the treatment of various diseases. Dr. Jeevanandam has authored about 40 book chapters and 50 journal articles along with some conference papers.

Preface

Diabetes is one of the most common diseases associated with lifestyle choices in the modern world and has a significant mortality rate. Conventional treatment methods for diabetes mainly include insulin-based therapies. However, insulin therapy possesses several limitations, such as weight gain and hypoglycemia. Thus advanced research in nanomedicine is targeting the development of new and improved diagnostics and treatment methods for diabetes. For example, there exists a plethora of literature supporting the development of novel nanomaterials as glucose sensors for rapid diabetes diagnosis as well as nanoformulations for controlled delivery of pharmaceutical ingredients to treat diabetes. This book, *Emerging nanomedicines for diabetes treatment*, is an anthology that provides readers with knowledge and an appreciation of advancements in the development of efficacious nanomedicines and nanotechnologies for diabetes treatment. The book discusses the application of various novel nanomaterials and nanocomposites for targeted delivery of insulin, glucose sensing, including nano-tattoos as glucose monitors, the potential of biosynthesized nanoparticles for diabetes treatment, and preclinical and clinical assays for evaluating nanomedicines for diabetes treatment.

Chapter 1 presents an overview of diabetes with recent diabetes statistics, current insulin therapy, and advancements in the diagnosis and treatment of diabetes. Chapter 2 covers the application of nanomaterials for insulin delivery, focusing on the types of nanoparticles for insulin delivery and the administration routes, and the role of nanoparticles in controlled delivery of insulin. Chapter 3 focuses on the development and application of nanomaterials for in situ glucose sensing and monitoring to control blood glucose levels. Chapter 4 covers the development and application of nano-tattoos as a novel approach for detecting and monitoring glucose levels in diabetic patients. Chapter 5 discusses the synthesis and characteristic properties of metal and metal oxide nanoparticles as potential nanomedicines for diabetes treatment. Chapter 6 discusses various biosynthesis approaches for the development of nanoparticles as potential nanomedicines for diabetes treatment. Chapter 7 provides an overview of the cytotoxicity profiles of various diabetes nanomedicines. Chapter 8 discusses in vivo assessments of diabetes nanomedicines using animal models. Chapter 9 provides an account of the current preclinical and clinical overview on the development of nanomedicines for diabetes treatment as well as the regulatory framework. Chapter 10 provides some future perspectives on the application of nanomaterials and nanomedicines for diabetes treatment applications.

Key features of this book include:

- provision of up-to-date knowledge on advancements in the development of nano-medicines for diabetes treatment;
- discussions on the potential of nanomaterials and nanocomposites for controlled delivery of insulin and effective diagnosis of diabetes;
- discussions on the efficacy of nano-tattoos as an emerging glucose monitoring approach and the potential of biosynthesized nanoparticles as pharmaceutical ingredients for diabetes treatment; and
- discussions on various preclinical and clinical assays for nanomedicines evaluation, as well as methods to mitigate challenges associated with effective diabetes therapy via the use of nanorobots, nanoformulations, and smartphone-based technologies.

Acknowledgments

This book would have not been completed without the support of several people, their help and encouragement, for which we will always be deeply grateful. First, we wish to thank our families and friends for the encouragement to pursue excellence, for which this book is a part. We also wish to thank our respective institutional departments for providing the support and resources for the completion of this book. We wish to extend our special thanks to Prof. João Rodrigues and colleagues from Madeira Chemistry Center (CQM—Centro de Química da Madeira), Universidade da Madeira (University of Madeira), Madeira, Portugal, and Academy of Competitive Examination and Research Training (ACERT), Chennai, India, for their support. We thank everyone in the Elsevier publication team, particularly Rafael G. Trombaco for his guidance in the preparation and completion of the book. Finally, and most importantly, we thank God for this wonderful opportunity to contribute this body of knowledge to promote advances in diabetes treatment. We would like to dedicate this book to all the doctors, physicians, caregivers, patients, and families who are fighting various diabetes complications.

CHAPTER 1

The current state of diabetes treatment

Introduction

The human life faces a number of threats. These threats range from natural calamities to deadly diseases. The humongous developments in science, technology, pharmaceutical, and biomedical fields help to fight, treat, and cure these diseases. Diabetes is one of such diseases that pose a threat to the human life. Sometimes referred to as a deficiency, diabetes can lead to several metabolic complications, making it difficult for diabetic patients to lead a normal life. Generally, diabetes mellitus or diabetes is used to describe the condition of heterogenous metabolic disturbances, especially in cases such as chronic hyperglycemia caused by impaired insulin action, insulin secretion, or both (Kerner & Brückel, 2014). According to the World Health Organization (WHO), diabetes is termed as a multiple metabolic disorder etiology with the characteristics of chronic hyperglycemia that disturbs the metabolism of fat, carbohydrate, and protein, due to defects in insulin action or secretion (World Health Organization, 1999). In common terms, diabetes is a disordered metabolism syndrome that relates to hyperglycemia (high blood glucose level) (McPhee et al., 2010). The impairment in the secretion or action of insulin caused by disordered metabolism leads to hyperglycemic conditions in diabetic patients. The guidelines given by international standardization of measurements indicate that persons with $\geq 6.5\%$ (≥ 48 mmol/mol) of HbA1c (glycated hemoglobin), ≥ 200 mg/dL (≥ 11.1 mmol/L) of random plasma glucose, ≥ 126 mg/dL (≥ 7.0 mmol/dL) of fasting plasma glucose, and ≥ 200 mg/dL (≥ 11.1 mmol/L) of glucose in venous plasma, measured by oral glucose tolerance test are considered as diabetic patients. A person with HbA1c (glycated hemoglobin) level below $<5.7\%$ is not considered as a diabetic patient and a person with HbA1c between 5.7% and 6.4% is considered as prediabetic patient. Thus HbA1c measurement is the primary diagnostic assay to determine the diabetes status of an individual (Kerner & Brückel, 2014).

In the past 20 years, diabetes has emerged as one of the chief coercions that drastically affect human health (Centers for Disease Control & Prevention, 2020; Teixeira-Lemos et al., 2011; Zimmet, 2000). There is an incredible increase in the number of diabetic patients over the past 10 years (Centers for Disease Control & Prevention, 2020; Teixeira-Lemos et al., 2011; Zimmet, 2000). In 2011, the International Diabetes Federation (IDF) estimated about 366 million diabetic patients globally and this number is expected to increase to 552 million in 2030 (Petrak et al., 2013). In 2015, IDF reported an increase in about 415 million diabetic patients (Ogurtsova et al., 2017). Furthermore, IDF reported that in 2015, diabetes was the cause of about 5 million deaths throughout the world at the rate of one person per every 6 s (IDF Diabetes Atlas, 2015). The Centers for Disease Control and Prevention (CDC) in the United States reported that 34.2 million diabetes cases in 2020 (Centers for Disease & Prevention, 2020; Teixeira-Lemos et al., 2011; Zimmet, 2000). WHO has predicted that by 2030, diabetes will be the seventh lethal disease among humans (Mathers & Loncar, 2006). In 2017, US National Diabetes Statistics Report indicated that 132,000 children and adolescents younger than 18 years (0.18% of total US population younger than 18 years of age) and 193,000 children and adolescents younger than 20 years (0.24% of total US population younger than 20 years of age) were affected by diabetes (Centers for Disease Control & Prevention, 2017). The case numbers will be similar or even worse if the total population of kids throughout the world is considered. This recent report indicates that diabetes is not a deficiency that affects only middle age group (30–50 years of age) but also spreading widely among youngsters and children below 20 years.

Diabetes directly distorts cellular metabolic functions and increases the chance for other complications such as obesity, hypertension, and cardiovascular diseases to set in. Several studies have shown that obesity is significantly associated with diabetes (Collins et al., 1994; Dunstan et al., 2002; Midthjell et al., 1999; Mokdad et al., 2001). According to WHO, obesity is defined as a condition in which the body mass index (BMI) of a person is ≥ 30 (De Lorenzo et al., 2003). Overweight (BMI between 25 and 30 kg/m^2) and obesity (BMI ≥ 30 kg/m^2) have become common among people, especially youngsters, which in most cases associated with diabetes (Spreghini et al., 2019). Furthermore, certain studies conducted among adults between 18 and 74 years revealed that an increase in waist circumference and BMI was independently associated with diabetes (German et al., 2020; Qiao & Nyamdorj, 2010). Similarly, hypertension is also a critical complication that is commonly associated with diabetic patients. Hypertension is defined as the condition of high blood pressure (above systolic pressure of 140 mmHg and diastolic pressure of 90 mmHg as reported by Joint National Committee on Prevention, Detection, Evaluation and Treatment of High Blood pressure) (Chobanian et al., 2003). In 2004, the study to help improve early evaluation and management of risk factors leading to diabetes (SHIELD) denoted in their survey (64% were adults) that hypertension is associated with higher BMI in diabetic patients (Bays et al., 2007). Recent studies in different regions and among several races show that diabetic patients are prone to hypertension and the cases are increasing with changes in life style and food habits (Castrejón-Pérez et al., 2017; De Boer et al., 2017; Zhang et al., 2020). Cardiovascular and kidney diseases are other major complications associated with diabetes, and recent studies reveal that most of the diabetic patients are exposed to these complications that eventually lead to death of patient (Cooper et al., 2020; Mende, 2017; Mostafa et al., 2018; Wanner et al., 2017).

These literatures and recent statistics alarm researchers to focus more on diagnosis and treatment of diabetes by using novel technologies. This chapter gives an overview of diabetes, its types, conventional, and latest treatments employed to control and cure diabetes. In addition, several statistical data and comparison between conventional and recent diabetes treatment methods were included to effectively deliver the current scenario in diabetes treatment.

Types of diabetes

Diabetes can be classified into several types, depending on the secretion, function, and availability of insulin as well as the function of pancreas in diabetic patients. Diabetic patients with damaged insulin-producing β-cells, common in autoimmune diabetes, are categorized under type 1 diabetes, and patients with β-cell dysfunctions, resulting in reduced cellular insulin response, are categorized under type 2 diabetes (Jeevanandam, 2017). Apart from types 1 and 2, gestational diabetes (Lavery et al., 2017), exocrine pancreas (Murtaugh & Keefe, 2015), endocrinopathies (Samis et al., 2016), drug-induced diabetes (Pai et al., 2017), genetic defects of β-cell function (Perelis et al., 2015; Roep et al., 2020), genetic defects of insulin action (Kase et al., 2015), Alzheimer's disease (AD; Lourenco et al., 2015), and rare forms of autoimmune-mediated diabetes (Morgan & Richardson, 2014) are also considered as other specific diabetic types (Kerner & Brückel, 2014). Among several diabetic types, types 1 and 2 are the most common and gestational diabetes is common among pregnant women. Fig. 1.1 describes the classification of various common and minor diabetic types.

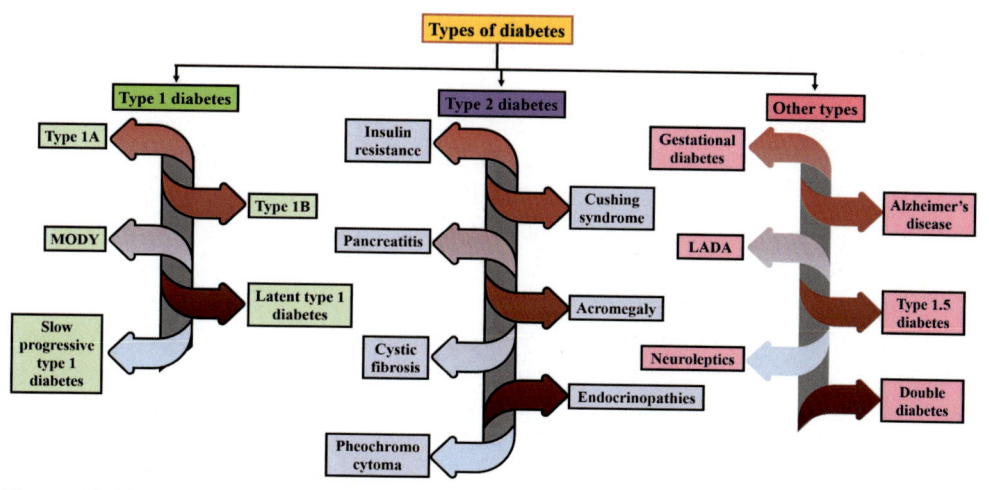

Figure 1.1 Various types of diabetes.

Type 1 diabetes

Type 1 diabetes is a diabetic condition in which the patient suffers from complete insulin deficiency due to the destruction of pancreatic β-cell by T-cell (American Diabetes Association, 2010). Generally, this form of diabetes is common among children and adolescent youngsters. However, this diabetic type is also predominant in adults with symptoms such as weight loss, polyuria, and polydipsia. The main pathophysiology is the declination in the capacity of insulin secretion, which eventually results in hyperglycemia with a proclivity to develop ketoacidosis (De Ferranti et al., 2014). In 2005, a statistical report described that 1 in 600 children across the world has type 1 diabetes and they are diagnosed at an early age (Stang & Story, 2017). This number is expected to increase based on previous reports (Dabelea, 2009). Type 1 diabetic patients are subclassified into type 1A, type 1B, and MODY. About 70%—90% of type 1 diabetes patients are grouped under type 1A, which is the classic type 1 (autoimmune) diabetes and the patients release self-reactive autoantibodies to mitigate the condition (Eisenbarth, 2007). Contradictorily, type 1B diabetic patients are categorized as patients with idiopathic diabetes history, and the specific reasons and pathogenesis for the prevalence are unclear (Gianani et al., 2010; Hollstein et al., 2020). Maturity onset diabetes of the young (MODY) is a subclass of type 1B diabetes in which the patients will have monogenic diabetic form, especially in children and youngsters (Christensen et al., 2020; Hattersley et al., 2009). However, these subclassifications are not typical in many countries due to poor characterizations, hence they are generally categorized as type 1 (Atkinson et al., 2014; Dabelea et al., 2007).

The genetic and familial studies from the past decades reported with ample elucidation that 80% of type 1 diabetes occurs due to hereditary, which confirms the crucial role of genetics in the prevalence of diabetes (Stankov et al., 2013). It can be noted that 40%—50% of type 1 diabetes cases are due to the impairment of human leukocyte antigen (HLA) class II genes such as HLA-DRB1, HLA-DQA1, and HLA-DQB1 loci (Noble & Erlich, 2012) along with HLA class I genes (Nejentsev et al., 2007). Apart from genetic and hereditary reasons, environment, microbiota, histone modification, and microRNA regulation are also found to be leading factors that contribute to the prevalence of type 1 diabetes (Wang, Leng, et al., 2017; Wang, Xie, et al., 2017). Environmental factors such as pollutants (Hathout et al., 2002; Langer et al., 2002), foods (Butalia et al., 2016), stress (Sepa & Ludvigsson, 2006), infectious agents (Lietzen et al., 2018; Nielsen et al., 2016), gut microflora (Paun et al., 2016; Wen et al., 2008), and even drugs (Hu et al., 2016, 2017) may also be the reason for the genetic modifications that lead to type 1 diabetes (Cerna, 2020; Wang, Leng, et al., 2017; Wang, Xie, et al., 2017). Similarly, histones play a major role in epigenetic modifications such as acetylation and methylation, and recent literatures suggest that histone modification leads to autoimmune diseases such as type 1 diabetes (Balcerczyk et al., 2017; Yin et al., 2017). Likewise, a noncoding small RNA class with 21—23 base pairs called microRNA and its regulation may also be the reason for the prevalence of type 1 diabetes in some patients (Sebastiani et al., 2017; Yang et al., 2015). However, it is difficult to identify the exact cause and pathogenesis of type 1 diabetes as either one or combinations of all these factors may lead to diabetic conditions.

Type 2 diabetes

A diabetic condition in which the patient suffers from physiological insulin resistance mechanism through the malfunction or blockage of insulin receptors or from inadequate insulin production is termed as type 2 diabetes (Rosenbloom et al., 1999). Compared to type 1, type 2 diabetes contributes to an impaired insulin activity and secretion mediated highly complex metabolic condition that leads to an increase in the level of blood glucose (Das & Elbein, 2006). It was estimated that over 400 million people across the world are suffering from type 2 diabetes (Zimmet et al., 2016). It was also reported by WHO that type 2 diabetes are common among diabetic patients with conditions such as obesity as well as people with unhealthy lifestyle and habits such as smoking (Surugue, 2016; Tangvarasittichai, 2015). The development of postprandial and fasting hyperglycemia is the prime contributor of type 2 diabetes, which induces many life-threatening comorbidities and complications in patients (Moon et al., 2020; Stratton et al., 2000). Insulin resistance is the main causative factor for the prevalence of type 2 diabetes and leads to increased glucose levels that results in diabetes mediated obesity (Riddy et al., 2018; Tang, 2011; Yaturu, 2011). Insulin resistance can be defined as the declined ability of insulin to stimulate the cellular insulin signaling cascade and subsequently activate insulin-mediated cellular processes (DeFronzo, 2009; Javeed & Matveyenko, 2018). Insulin resistance in diabetic patients is established by reduction in glucose transporter stimulation, reduction in adipocytes, suppression of hepatic glucose output, and reduction in skeletal muscle metabolism (Yaturu, 2011). Several recent research reports have identified the relationship between accumulation of fat, type 2 diabetes, and insulin resistance in intraabdomen (visceral obesity) (Shi et al., 2017). Thus it is clear that type 2 diabetes is more common among obese people and is not hereditary unlike type 1 diabetes.

Several factors have been identified to be responsible for the onset of type 2 diabetes in patients. Formation of reactive oxygen species (ROS) and peroxidation of lipids leads to oxidative stress in cells and these ROS-induced cells damage via oxidative stress are confirmed in many type 2 diabetic cases due to the reduced antioxidant defenses mechanism in patients (Seghrouchni et al., 2002; VanderJagt et al., 2001). Mechanisms such as polyol pathway (Ramana et al., 2003), $PKC_\beta 1/2$ kinase (Bhisitkul, 2006; Inoguchi et al., 2000), formation of advanced glycation end-products (Barrett et al., 2017; Nakamura et al., 1993), and glucose autoxidation (Bonnefont-Rousselot, 2002; Seebacher et al., 2015) were found to induce hyperglycemic condition via oxidative stress in type 2 diabetic patients (Brownlee, 2005). In addition, conditions such as increased free fatty acids, leptin, and other minor factors may also contribute for the overproduction of ROS in type 2 diabetic patients (Tangvarasittichai, 2015). Strong evidence is reported in literatures to show that the disruption of insulin signals and dysregulation of adipocytokines that leads to the pathogenesis of insulin resistance and eventually type 2 diabetes are mainly caused by oxidative stresses (Furukawa et al., 2017; Houstis et al., 2006). These oxidative stresses were reported to induce dysfunction of β-cells (Panigrahy et al., 2017; Robertson & Harmon, 2006), glucose-induced β-cells toxicity (Scullion et al., 2016; Tiedge et al., 1997), lipid-induced β-cells toxicity (Biden et al., 2014; Robertson et al., 2004), and dyslipidemia (excess postprandial lipemia) (Simons et al., 2016; Taskinen, 2003). The oxidative stress that leads to ROS production and lipid peroxidation were identified to be due to the lack of physical activity, modifications in life style, and history of smoking, apart from pharmacological interventions and other genetic reasons of metabolic disorder (Surugue, 2016; Tangvarasittichai, 2015). Thus it has been recommended by the American Diabetes Association and the American Heart Association that increased physical activity, life style modifications along with nutrition supplement therapy, weight loss, and reducing smoking habit can possibly reduce complications of type 2 diabetes (American Diabetes Association, 2002; Khavandi et al., 2017; Tangvarasittichai et al., 2009).

Other diabetic types

Several research articles revealed that types 1 and 2 are predominant among diabetic patients and they occupy around 90% of the diabetic population. However, there are other minor diabetic types that are reported in recent studies. Gestational diabetes is one such important diabetic type that is reported in literatures, other than type 1 and 2 diabetes. Gestational diabetes is a diabetic condition in pregnant women during pregnancy and was estimated that it affects about 8.3% of pregnancies across the world (Linnenkamp et al., 2014). Recent literatures reported that women affected by gestational diabetes during pregnancy have higher chances of developing postpregnancy type 2 diabetes and are exposed to risk of gestational hypertensive disorders and pre-eclampsia in subsequent pregnancies (Wang, Leng, et al., 2017; Wang, Xie, et al., 2017). Women with gestational diabetes have higher probability of delivering infants through cesarean (Luck et al., 2017) and infants born from such pregnancies are reported to be at high risk of type 2 diabetes and obesity in their later stage of life (Sobngwi et al., 2003; Xiong et al., 2001). Recently, it has been confirmed that gestational diabetes is associated with autism spectrum disorders (Lavery et al., 2017; Xu et al., 2014). Another diabetic type similar to gestational diabetes is latent autoimmune diabetes in adults (LADA) and it is estimated to be widespread among 10% of diabetic patients, close to type 1 diabetes. LADA is a term used to describe the subgroup of adult phenotypic patients of positive type 2 diabetes with to glutamic acid decarboxylase autoantibody and without weight loss and ketoacidosis (Centers for Disease Control & Prevention, 2020; Teixeira-Lemos et al., 2011; Zimmet, 2000). LADA is different from type 1 and type 2 as they are predominant and prevalent in adults rather than children and youngsters (Palmer & Hirsch, 2003). The LADA diabetic patients are further subcategorized into slowly progressive type 1 diabetes (Kobayashi, 1994; Tanaka et al., 2015), double diabetes (Cervin et al., 2008; Reinauer et al., 2017), latent type 1 diabetes (Palmer & Hirsch, 2003; Pilla et al., 2018), and type 1.5 diabetes (Gilliam et al., 2005; Naik & Palmer, 2003). Other conditions such as pancreatitis (Li et al., 2014), hemochromatosis (Raju & Venkataramappa, 2018), cystic fibrosis (Barrio, 2015), caused by exocrine pancreas disease (Meier & Giese, 2015), Cushing syndrome (Colao et al., 2014), pheochromocytoma (Pareek et al., 2017), acromegaly (Weiss et al., 2017), triggered by endocrinopathies (Burek et al., 2016), glucocorticoids (Rafacho et al., 2014), alpha-interferons (Spiegel, 1987), neuroleptics (Klockgether & Dichgans, 1990), pentamidine drug-induced hyperglycemic conditions (Fathallah et al., 2015) and maturity onset diabetes of the young (MODY) caused by genetic β-cell function defects (Pertusa et al., 2015) were also considered as minor diabetic types. Thus a condition where increased blood glucose level was identified is considered to be diabetes and it was classified into various types depending upon the reason behind the escalation of blood glucose level.

Pathophysiology of diabetes

All the different diabetes types are commonly diagnosed or caused by the deficiency of insulin secretion and excess glucagon secretion. Insulin deficiency leads to decreased glucose uptake in muscles and other tissues, increased protein catabolism, and increased lipolysis. The decrease in glucose uptake leads to hyperglycemia, glucosuria (excess sugar in urine), depletion of electrolyte and water (osmotic diuresis), which eventually leads to dehydration in cells, tissues, and organs, coma in patients, and death. Similarly, increased protein catabolism can lead to increased secretion of amino acids in plasma and can result in either increased gluconeogenesis or increased nitrogen loss in urine in the form of urea. Likewise, increased lipolysis can lead to increased oxidation of free fatty acid, increased ketogenesis and ketosis, which eventually leads to diabetic ketoacidosis (DKA), coma in patients, and death. However, each type of diabetes undergoes specific pathophysiological pathways that are important to classify and diagnose them. In type 1 diabetes, glucose is manufactured in the liver with limited glycogen storage. Gluconeogenesis (glucose formation by lipid and protein breakdown) becomes uncontrollable when insulin is deficient, leading to elevated blood glucose levels and eventually halting the uptake of blood glucose in muscle cells via glucose transporter 4 (GLUT4), which can cause other complications as mentioned earlier. It is noteworthy that the physiological conditions of older diabetic patients (age above 60) can lead to hyperosmolar hyperglycemic nonketotic syndrome, where the excess sugar in the patients' body is excreted out via urine. Additional and more specific information on

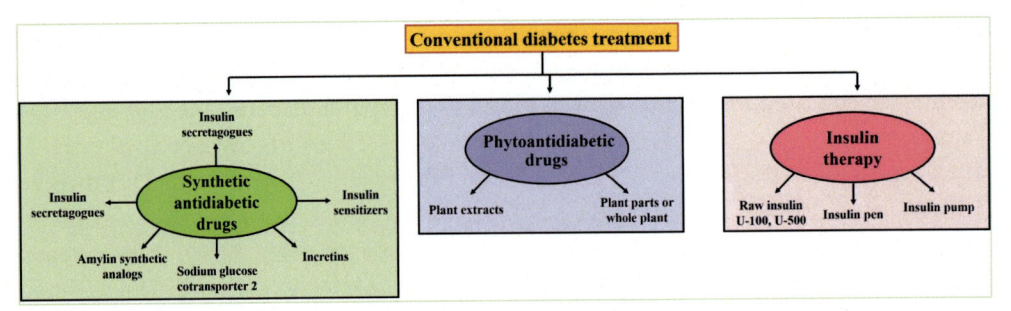

Figure 1.2 Types of conventional diabetes treatment methods.

the pathophysiology of each diabetic type is available in the work published by Moini in 2019 (Moini & Moini, 2019).

Conventional treatment for diabetes mellitus

The widespread of diabetes among humans, ranging from kids to adults, extends the scope of diabeticians and researchers to discover drugs and treatment procedures to reduce and control the severity of diabetes and to cure them. Synthetic antidiabetic drugs (Bösenberg & van Zyl, 2008; Inzucchi et al., 2015), phytoantidiabetic drugs (Gothai et al., 2016; Perez G et al., 1998), and insulin therapy (Lind et al., 2017; Pickup et al., 2017) are the conventional treatment procedures, as shown in Fig. 1.2, that are widely used, until now, among diabetic patients to control hyperglycemia and diabetes.

Synthetic antidiabetic drugs

Synthetic antidiabetic drugs that can inhibit sodium glucose cotransporter 2 (SGLT2), α-glucosidase, novel insulin secretagogues, insulin sensitizers, incretins, and amylin synthetic analogs are widely and conventionally used among diabetic patients to control hyperglycemic condition (Gupta et al., 2016). SGLT2 inhibitors are a novel oral antihyperglycemic agent that controls and regulates renal glucose level by hindering a glucose reabsorption transporter cell SGLT2 and by activating insulin independent excretion of plasma glucose through urine (Mosley et al., 2015). These inhibitors help in reabsorption of glucose in renal proximal tubular cells which enhance the excretion of glucose via urine (glycosuria) and improve glycemic control (Linden et al., 2017). Phlorizin (Bays et al., 2007), canagliflozin (Nomura et al., 2010), and dapagliflozin (Kaku et al., 2013) are some of the compounds that are used as SGLT2 inhibitors. However, apart from reducing increased blood glucose level, these inhibitors may also lead to side effects such as infection urinary tract, weight loss, and genital infections (Hedrington & Davis, 2015; Liu et al., 2015). Likewise, oral glucose lowering drugs such as α-glucosidase inhibitors were employed to control the level of intestinal glucose absorption for counterbalancing postprandial hyperglycemia which consequently improves insulin sensitivity and helps in β-cells stress release (Joshi et al., 2015; Turner et al., 2014). These pseudocarbohydrate inhibitors reduce rate of intestinal glucose absorption rate by reducing the activity of intestinal enzyme groups such as α-glucosidase that possess ability to digest polysaccharides into monosaccharides (Chan et al., 2018; Lebovitz, 1997). Commercial α-glucosidase inhibitor drugs such as acarbose (Van De Laar et al., 2005; Yew et al., 2020), miglitol (Hamada et al., 2020; Scott & Spencer, 2000), and vogliobose (Derosa & Maffioli, 2012; Ueno et al., 2015) are widely used by type 2 diabetic patients in the late 1990s. Recently, thiosugars and iminosugars showed promising results in controlling hyperglycemia and other related disorders (Brás et al., 2014). However, the major drawback of using α-glucosidase inhibitor is that they cause side effects such as abdominal discomfort, flatulence, nausea, edema, and diarrhea (Osonoi, 2015).

Other than inhibitors to control hyperglycemia, insulin secretagogues, insulin sensitizers, and incretins helps to control and regulate insulin secretion and function in diabetic patients. Sulfonylurea drugs such as tolbutamide; first-generation sulfonylureas namely chloropropamide, tolazamide, and acetonexamide; second-generation sulfonylureas namely gliclazide, glipizide, and glyburide; and third-generation sulfonylureas namely glimepiride (Aquilante, 2010; Sola et al., 2015) are widely used as antidiabetic drugs that can effectively reduce 1%—2% of glycated hemoglobin (HbA$_{1C}$) along with the combination of other antihyperglycemic drugs (Hashim, 2017). Similarly, nonsulfonylurea drugs such as meglitinides including glinide group of nateglinide and repaglinide are phenylalanine and benzoamide derivatives that are especially designed for type 2 diabetic patients to reduce the adverse effect of hypoglycemia (Charan Kumar & Murthy, 2016; Keilson et al., 2000). Inspite of advantages, sulfonylurea drugs also possess disadvantages such as weight gain, hypoglycemia, hemolytic anemia, cardiovascular agranulocytosis, thrombocytopenia, skin rashes, cholestatic jaundice, and mortality among diabetic hyperglycemic patients (Grunberger, 2017; Krentz & Bailey, 2005). Nonsulfonylurea drugs help to overcome the problems and possess low adverse effect risks than sulfonylurea drugs; however, they also lead to hypoglycemic condition similar to sulfonylurea drugs (Fowler, 2007). Insulin sensitizers such as biguanides including buformin, metformin, phenformin (Hampp et al., 2014), and thiazolidinediones including troglitazone, pioglitazone, and rosiglitazone (Soccio et al., 2014; Turner et al., 2014; Yokoi, 2010) are oral hypoglycemic drugs that used to enhance insulin sensitivity for reducing insulin resistance in type 2 diabetic patients. Both these types of insulin sensitizer drug may lead to side effects such as weight gain and hypoglycemia among patients (Kahn et al., 2000; Kohlroser et al., 2000; Miller et al., 2013). However, they are used as the first line of oral therapy drugs along with the combination of other hypoglycemic drugs (Bridges et al., 2014; Shaw et al., 2005).

Incretins are hormonal groups secreted by enteroendocrine cells that play a key role in maintaining the energy and glucose homeostasis in the body and alteration in these hormonal expressions lead to glucose intolerance and gastrointestinal tract dysfunction in type 2 diabetic patients (Sanyal, 2013). Thus endogenous incretins such as glucose-dependent insulinotropic peptide and glucagon like peptide-1 (GLP-1) are used to counter-alter the incretin expressions (Gasbjerg et al., 2020; Ross & Ekoé, 2010). It can be noted that the plasma half-life of endogenous incretins is short due to the presence of dipeptidyl peptidase-4 (DPP-4) in the blood (Holst, 2007). Thus GLP-1R agonists such as exenatide, albiglutide, dulaglutide, liraglutide, taspoglutide, and lizisenatide were used as an alternate to damages or altered incretins whereas sitagliptin, saxagliptin, alogliptin, vildagliptin, teneligliptin, and linagliptin were used as DPP-4 inhibitors to reduce blood glucose level in type 2 diabetic conditions (Kushwaha et al., 2014; Neumiller, 2009). Along with advantages in regulating hyperglycemic conditions, GLP-1R agonist and DPP-4 inhibitors lead to side effects such as hypertension, cardiovascular and neurological disorders, obesity, and dyslipidaemia (Seufert & Gallwitz, 2014). Glucose lowering ability of a neuroendocrine polypeptide hormone called amylin has led to the approval of synthetic amylin analogs that can be used in insulin therapy of both type 1 and 2 diabetes (Buse et al., 2002; Hoogwerf et al., 2008; Nussbaumer et al., 2020). Pramlintide is a synthetic amylin analog that is used in combination with insulin therapy among diabetic patients with problems in achieving glycemic control (Hoogwerf et al., 2008; Ratner et al., 2002). However, nausea, gastroparesis, and vomiting are some of the side effects that are associated with the intake of pramlintide among diabetic patients (Adeghate & Kalász, 2011). In addition, latest synthetic drugs such as AD and non-AD combinations such as sitagliptin/metformin, linagliptin/metformin, glipizide/metformin, sitagliptin/simvastatin, and dopamine receptor agonist such as bromocriptine are available for diabetic patients to keep blood glucose under control (Hampp et al., 2014). However, it can be noted that each synthetic antidiabetic drug has led to side effects in patients which is their major drawback in using them as a long-lasting glycemic control agent.

Phytoantidiabetic drugs

It can be noted that all the synthetic antidiabetic drugs are prepared by chemical methods, which involve the use of toxic chemicals that led to side effects in patients. Thus plant-based antidiabetic compounds are introduced to overcome the challenge of side effects and to enhance the blood glucose controllability (Chawla et al., 2013). Plants are the first medicine used to control diabetic complications. In 1550 BC, Papyrus Ebers recommended a high-fiber ocher and wheat grain diet for diabetic patients (Ebbell, 1937). Resveratrol, berberine, ginseng (panax), and cinnamon are some of the widely used phytoantidiabetic compounds (Gupta et al., 2016). 3,5,4'-Trihydroxystilbene, commonly known as resveratrol, is a polyphenol available in plant species such as peanuts, red wine grapes, berries, and *Polygonum cuspidatum* (Japanese knotweed) (Jang et al., 1997). This compound possesses properties such as antiobesity, antioxidative, antitumorigenic, cardio and neuro protection, antimuscle atrophy, and anti-inflammatory (Allen et al., 2018; Bastianetto et al., 2015; Frombaum et al., 2012; Zordoky et al., 2015). Moreover, resveratrol medication protects diabetic patients from the adverse effects of standard antidiabetic drugs via their antioxidative and anti-inflammatory properties (Burgess et al., 2011; Tan et al., 2012). Resveratrol controls ROS production in pancreatic cells and avoids the toxic effects that are caused by excess lipid and glucose (Szkudelska et al., 2009). In addition, resveratrol activates ATP-sensitive, voltage-gated K^+ channels, and Ca^{2+}-regulated K^+ channels to increase the secretion of insulin in diabetic patients (Chen et al., 2007). However, clinical trials revealed that resveratrol failed to alter insulin resistance in some cases, which is the major drawback of these compound in scale-up production and commercialization (Movahed et al., 2013; Szkudelski & Szkudelska, 2015).

Likewise, berberine compound is available in herbs such as *Berberis aquifolium* (Oregon grape), *Berberis aristata* (tree turmeric), *Coptis chinensis* (golden thread), and *Berberis vulgaris*, which possesses the ability to control lipid and glucose metabolism. Berberines are reported to be better than metformin, which activate AMP-activated protein kinase, regulate fat metabolism, and enhance insulin-independent glucose uptake in skeletal and adipose cells (Joshi et al., 2015; Turner et al., 2014). In addition, berberine downregulates hepatic gluconeogenic gene expression to regulate blood glucose level (Castrejón-Pérez et al., 2017; De Boer et al., 2017; Zhang et al., 2020), controls metabolism of lipids, and reduces triglyceride concentration in tissues and serum (Balcerczyk et al., 2017; Yin et al., 2017). Similarly, extract from ginseng obtained from plants such as *Panax ginseng* (Chinese/Korean red ginseng), *Panax quiquefolius* (American ginseng), *Panax japonicus* (Japanese ginseng), and *Eleutherococcus senticosus* (Siberian ginseng) are proved to have antihypertensive, antidiabetic, and antitumor properties via animal studies and clinical trials (Choi et al., 2013). Antidiabetic characteristics and plasma glucose reduction including HbA_1c in diabetic patients are the benefits of ginseng and ginsensoids extracted from ginseng that are described via several in vivo and in vitro studies (Wei et al., 2015; Yuan et al., 2012). Extracts of *Cinnamoni cassia*, commonly known as cinnamon, also proved to regulate type 2 diabetes—related hyperlipidemia or hyperglycemia (Kouzi et al., 2015; Shen et al., 2014). Other experiments with cinnamon extract products such as cinnamaldehyde and proanthocyanidins also possess antihyperglycemic and antidiabetic properties (Jiao et al., 2013; Sharma et al., 2020). Alkaloids such as vindolinine, catharanthine, vindoline, leurosine (Hernandez-Galicia et al., 2002), diphenylamine (Karawya et al., 1984; Swanston-Flatt et al., 1989), tecomine, tecostanine alkaloids from *Tecoma stans* leaves (Hammouda & Amer, 1966; Meela et al., 2017), trigonelline (Adams et al., 2014), insulin-like peptides (Paula et al., 2017), charantin (Ahlam et al., 2013), glycoside, bengalenoside (Baharvand-Ahmadi et al., 2016), neomyrtillin (Gaikwad et al., 2014), galactomannan gum (Naito et al., 2016), hypoglycins (Rynbrandt et al., 1972), peptidoglycans, and polysaccharides (Li et al., 2014) are some of the traditional plant extracts that show efficiency in diabetes treatment (Rasouli et al., 2020). However, clinical trials revealed that the cinnamon treatment does not show any significant difference in plasma glucose level, lipid profile, and insulin sensitivity (Allen et al., 2018; Bastianetto et al., 2015; Frombaum et al., 2012; Zordoky et al., 2015). The major drawback of all these plant extracts is their heterogenicity and uncertainty in clinical trials, which suggests extensive studies toward phytodrug-based diabetic treatments.

Other than plant extracts, whole plants such as *Momoridica charantia* (bitter gourd) (Rudá-Kučerová et al., 2015), *Abies balsamea* (L.) Mill (Canada balsam) (Leduc et al., 2006), *Lagerstroemia speciosa* (*Banaba* in Philippines) (Miura et al., 2012), *Trigonella foenum graecum* (fenugreek) (Saxena & Vikram, 2004), *Tinospora cordifolia* (Gulancha) (Patel & Mishra, 2011), and *Salvia officinalis* (mint family) (Eidi et al., 2005) are highly used in traditional medicine practices for diabetes treatment. Similarly, raw onion bulbs (*Allium cepa*), garlic cloves (Dureshahwar et al., 2017; Nasiri et al., 2017), leaf infusions of *Catharanthus roseus*, seeds of *Lupinus temis* (Mishkinsky et al., 1974), *Coccinia indica* (ivy gourd) (Kaushik et al., 2017), *Momordica foetida* (Tsabang et al., 2016), *Gymnema sylvestre* (Tiwari et al., 2017), *Vaccinium myrtillus* leaves (Sidorova et al., 2017), *Ficus benghalensis* (banyan tree) (Rathi et al., 2016), *Cyamopsis tetragonolobus* (Gandhi et al., 2014), *Amorphophallus konjac* tubers (Tester & Al-Ghazzewi, 2016), *Blighia sapida* (Oloyede et al., 2014), *Galega officinalis* (Nagalievska et al., 2018), *Ilex guayusa* (Pardau et al., 2017), *Sarcopoterium spinosum* (Elyasiyan et al., 2017), *Medicago sativa* (Martínez et al., 2016), and *Coleus forskohlii* (Badmaev et al., 2015) are some of the plant species used for efficient diabetic treatments. However, reports of renal and hepatic necrosis mediated fatal cases in South Africa, after consuming an uncharacterized herbal medicine to reduce diabetic complications (Stickel & Shouval, 2015) showed that phytodrugs also possess some toxic effects. Thus it is highly recommended to utilize combination of synthetic and phytodrugs to effectively reduce the complication of high blood glucose and diabetes in patients.

Insulin therapy

For the past two decades, insulin therapy has been the most commonly used treatment method among diabetic patients. It has been reported from several studies that glycemic control can be efficiently achieved in diabetic patients by intensive insulin therapy (Lane et al., 2014; Miele et al., 2015). Single dose of 80 U/d insulin is the insulin regimen that is commonly used for insulin therapy (Harris, 1996; Rady et al., 2005). U–500 (Lane et al., 2014; Miele et al., 2015) and U–100 insulin are the other nonpopular insulin therapy methods, which use different concentrations of insulin to reduce the severity of insulin resistance, reduce blood glucose level, and control diabetic complications (Hood et al., 2015; Kabul et al., 2016). Since, raw and high concentration of insulin causes complications such as overweight (Movva et al., 2015), insulin pump, and insulin pens are introduced to deliver required amount of insulin in diabetic patients. The precise and flexible insulin delivering ability of continuous subcutaneous insulin infusion pump among diabetic patients in their day-to-day life makes them a popular insulin therapy method for more than 25 years (Raskin et al., 2003). In insulin pump therapy, a rapid-acting, individualized, short insulin at a basal rate of 0.5−1.5 U/h dosage was delivered into diabetic patients using a programmed external pump (Bode et al., 2002; Isharwal et al., 2009). Patients with wide glucose level fluctuations and severe hypoglycemia utilize insulin pump to control and regulate blood glucose level (DeWitt & Hirsch, 2003). Commercial insulin pumps such as Animas Ping (Grunberger et al., 2014), Deltec Cozmo (Thrasher et al., 2014), Disetronic Spirit (Guilhem et al., 2009), Medtronic Paradigm 522/722 (Grunberger et al., 2010), and Insulet Omnipod (Zisser, 2010) are available in market, which served diabetic patients in effective insulin therapy (Soylar et al., 2020). Compared to regular insulin and multiple daily injection (MDI) therapy, rapid acting insulin analogs such as aspart, glulisine, and lispro through pump therapy help in decreasing late postprandial hypoglycemia risk, controlling postprandial hyperglycemia, and reducing severe hypoglycemia occurrence (Juvenile Diabetes Research Foundation Continuous Glucose Monitoring Study Group, 2008). In addition, new "smart" insulin pumps have preprogrammable bolus dose of insulin delivery facility according to expected carbohydrate intake which makes it easier to deliver smaller incremental insulin doses via pump therapy, compared to other insulin therapy methods (Weinzimer & Tamborlane, 2008). Thus the smart insulin pump therapy helps the diabetic patients to have normal lives without skipping or simplifying their meal and other unintended activities (Kirk, 2003). In spite of all these advantages, the risk of DKA development is the major drawback in using pump therapy (Sherr & Tamborlane, 2008). However, pump therapy shows higher advantages than MDI therapy, when patients are provided with appropriate training to use the pump according to the carbohydrate intake (Lebovitz, 1998).

Insulin pen is used as an alternative to insulin pump for insulin therapy among diabetic patients, especially for visually impaired patients (D'Eliseo et al., 2000). Insulin pens are more convenient with superior insulin dosing accuracy than conventional vial and syringe method along with less reported pain and greater social acceptability (Cobden et al., 2007). Statistics show that around three-quarters of diabetic patients in Japan and two-thirds in Europe are prescribed to use insulin pen therapy by doctors (Bhargava, 2007). Vial and syringe methods are the traditional routes to administer subcutaneous insulin which possess several disadvantages such as poor dose accuracy, fear of injections, lengthy training time, difficulty in transportation, and lack of social acceptance (Peyrot et al., 2005; Summers et al., 2004). Insulin pens with enhanced mealtime flexibility, delivery convenience, improved portability, and dosage accuracy are introduced to overcome the drawback of traditional vial and syringe method (Sommavilla et al., 2008). Less reported injection pain, improved cost of care, and improved self-management behavior in patients are the other advantages in using insulin pen for insulin therapy (Davis et al., 2008, 2009; McKay et al., 2009). Several prefilled disposable insulin pen devices namely Humalog, KwikPen with insulin lipro and insulin protamine suspension, original Humalog pen with insulin lipro, Humalog Mix75/25 and Humalog Mix50/50 pen with insulin lipro protamine suspension/insulin lispro injection, Humulin N pen with human insulin isophane suspension, Humulin 70/30 pen human insulin isophane suspension/human insulin injection, Lantus SoloSTAR with insulin glargine, Apidra SoloSTAR with insulin glulisine, Levemir FlexPen with insulin Levemir, aspart, Novolog FlexPen, and Novolog Mix 70/30 FlexPen with insulin aspart protamine suspension/insulin aspart are available in market. Similarly, refillable insulin pens such as Autopen classic with insulin lispro protamine suspension/insulin lispro injection, Autopen 24 with insulin glargine and insulin glulisine, HumaPen, LUXURA, and half-dose LUXURE with insulin lispro, protamine suspension/insulin lispro injection, HumaPen Memoir as HumaPen LUXURA, NovoPen 3 with insulin aspart, Levemir, aspart protamine suspension/insulin aspart, human insulin isophane suspension, NovoPen 4 and NovoPen Junior as NovoPen 3 and OptiClick with insulin glargine and glulisine are also commercially available in markets to help diabetic patients to control glycemic conditions (Pearson, 2010). However, both insulin pen and pump therapy have drawbacks such as increased hypoglycemic rate (Swinnen et al., 2009), risk of cardiovascular disease (Holman et al., 2008), risk of hyperinsulinemia, and weight gain in type 2 diabetes patients (Jeevanandam et al., 2015). Thus novel insulin pen and pump therapy along with conventional drugs were advised by doctors to enhance the efficiency of insulin therapy to fight and cure diabetes.

Latest insulin therapy routes and drugs for diabetes management

The previous section summarizes the advantages of each conventional and traditional diabetic treatment methods along with their disadvantages and side effects. Even though there are advantages in each method, the side effects associated with them creates opportunities for researchers to generate new and improved treatment methods. Such opportunities include promoting self-monitoring of diabetes using recent technological developments in smartphones, glucometers, and glucose monitoring systems. Lancing devices draw blood from the patient with a painful prick to check the blood glucose level (Saikley et al., 2015). Genteel is the latest FDA approved product, which reduces the painful pricking procedure by avoiding the deeper pain nerve via pricking and reaching only the blood capillaries (Crosby & Rattan, 2016). It is the first FDA approved Class II lancing device and a recent patient survey showed that it is satisfactory to over 82% of the diabetic patients (Wynn, 2017). Similarly, FDA has approved a new arm indicator for Guardian Sensor 3, which will help diabetic patients to wear the sensor on the upper arm to provide flexible and accurate insulin delivery (Aviad, 2018). Abbott FreeStyle Libre is a continuous glucose monitoring system, similar to Genteel, which is an FDA approved, nonpricking device that recently gained popularity among US diabetic patients (Ólafsdóttir et al., 2017). Recently, Roche Diabetes Care introduced MySugr diabetes app which is intended to help diabetic patients to monitor glucose via smartphones, along with other wireless glucose monitoring systems (Arnhold et al., 2014). In 2017, FDA-approved a smart insulin pen named InPen created by the Companion Medical company. It is a Bluetooth enabled, 0.5 unit of refillable insulin pen with a smartphone app. Its weightless and long battery life characteristics make it a promising diabetes management device among youngsters (Dubois, 2018). Other than electronic glucose monitoring and delivery systems, scientists are also working on pancreas replacement or transplantation. In 2018, scientists from University of Miami confirmed the presence of progenitor cells within human pancreas that can be stimulated to develop into glucose-responsive β cells (Qadir et al., n.d.). New diabetic drugs to control blood glucose level and other diabetic complications such as Victoza (Mathieu et al., 2014), insulin aspart Fiasp (Chaplin, 2017), and Xultophy by Novo Nordisk (Harris, 1996; Rady et al., 2005) and Soliqua from Sanofi, which is a GLP-1 receptor agonist hormone to normalize blood glucose level (Marathe et al., 2017), have been recently approved by FDA. Furthermore, orally administered Semaglutide (glucagon-like peptide receptor agonist) named Rybelsus (Anderson et al., 2020), nasal glucagon powder named Baqsimi (Kaur et al., 2020), and liraglutide named Victoza (Chen et al., 2007) were also approved by FDA in 2019 for the treatment of diabetes. These latest FDA-approved drugs and glucose monitoring systems show promise that efficient diabetes self-management is achievable. However, the side effects of each drug and delivery method, patient acceptability, and cost will determine the future of these new drugs and insulin therapy systems.

Conclusion

This chapter presents a summary of the different diabetes types, their classification criteria, pathophysiology, statistics to forecast the severity of each diabetes types, and emerging technologies for diabetes treatment. The statistics show that diabetes is spreading rapidly across the world and more children and youngsters are increasingly being affected by diabetes. This chapter discusses several conventional drugs, phyto-based drugs, and insulin therapy methods that can be used and/or are being used to control blood glucose levels, stimulate insulin secretion, and reduce diabetic complications. However, there are enormous side effects and drawbacks associated with the application of these methods, representing opportunities for researchers to develop new and improved technologies.

Some new and novel drugs that have been identified are undergoing clinical trials. Nanomedicine is emerging as a promising approach to address some of the challenges associated with conventional diabetes diagnostics, monitoring, and treatment methods.

References

Adams, G. G., Imran, S., Wang, S., Mohammad, A., Kok, M. S., Gray, D. A., Channell, G. A., & Harding, S. E. (2014). The hypoglycemic effect of pumpkin seeds, Trigonelline (TRG), Nicotinic acid (NA), and D-Chiro-inositol (DCI) in controlling glycemic levels in diabetes mellitus. *Critical Reviews in Food Science and Nutrition, 54*(10), 1322−1329.

Adeghate, E., & Kalász, H. (2011). Amylin analogues in the treatment of diabetes mellitus: Medicinal chemistry and structural basis of its function. *The Open Medicinal Chemistry Journal, 5*, 78.

Ahlam, M. S., Ganai, B., Zargar, M., & Seema, A. (2013). *In vivo study of anti-diabetic activity of* Eremurus himalaicus (M. Phil. thesis). University of Kashmir.

Allen, E. N., Potdar, S., Tapias, V., Parmar, M., Mizuno, C. S., Rimando, A., & Cavanaugh, J. E. (2018). Resveratrol and pinostilbene confer neuroprotection against aging-related deficits through an ERK1/2-dependent mechanism. *The Journal of Nutritional Biochemistry, 54*, 77−86.

Anderson, S. L., Beutel, T. R., & Trujillo, J. M. (2020). Oral semaglutide in type 2 diabetes. *Journal of Diabetes and Its Complications, 34*(4), 107520.

Aquilante, C. L. (2010). Sulfonylurea pharmacogenomics in Type 2 diabetes: The influence of drug target and diabetes risk polymorphisms. *Expert Review of Cardiovascular Therapy, 8*(3), 359−372.

Arnhold, M., Quade, M., & Kirch, W. (2014). Mobile applications for diabetics: A systematic review and expert-based usability evaluation considering the special requirements of diabetes patients age 50 years or older. *Journal of Medical Internet Research, 16*(4).

American Diabetes Association. (2002). Management of dyslipidemia in adults with diabetes. *Diabetes Care, 25*(Suppl. 1), s74−s77.

American Diabetes Association. (2010). Diagnosis and classification of diabetes mellitus. *Diabetes Care, 33* (Suppl. 1), S62.

Atkinson, M. A., Eisenbarth, G. S., & Michels, A. W. (2014). Type 1 diabetes. *The Lancet, 383*(9911), 69−82.

Aviad, M. (2018). *FDA approves medtronic's guardian sensor 3 for arm use* (Vol. 2018). Diabetes Media Foundation. <https://asweetlife.org/fda-approves-medtronics-guardian-sensor-3-for-arm-use/>.

Badmaev, V., Hatakeyama, Y., Yamazaki, N., Noro, A., Mohamed, F., Ho, C.-T., & Pan, M.-H. (2015). Preclinical and clinical effects of *Coleus forskohlii*, Salacia reticulata and Sesamum indicum modifying pancreatic lipase inhibition in vitro and reducing total body fat. *Journal of Functional Foods, 15*, 44−51.

Baharvand-Ahmadi, B., Bahmani, M., Tajeddini, P., Naghdi, N., & Rafieian-Kopaei, M. (2016). An ethno-medicinal study of medicinal plants used for the treatment of diabetes. *Journal of Nephropathology, 5*(1), 44.

Balcerczyk, A., Chriett, S., & Pirola, L. (2017). Insulin action, insulin resistance, and their link to histone acetylation. *Handbook of nutrition, diet, and epigenetics* (pp. 1−22). Springer.

Barrett, E. J., Liu, Z., Khamaisi, M., King, G. L., Klein, R., Klein, B. E., Hughes, T. M., Craft, S., Freedman, B. I., & Bowden, D. W. (2017). Diabetic microvascular disease: An Endocrine Society scientific statement. *The Journal of Clinical Endocrinology & Metabolism, 102*(12), 4343−4410.

Barrio, R. (2015). Management of endocrine disease: Cystic fibrosis-related diabetes: Novel pathogenic insights opening new therapeutic avenues. *European Journal of Endocrinology, 172*(4), R131−R141.

Bastianetto, S., Ménard, C., & Quirion, R. (2015). Neuroprotective action of resveratrol. *Biochimica et Biophysica Acta (BBA)-Molecular Basis of Disease, 1852*(6), 1195−1201.

Bays, H. E., Chapman, R., & Grandy, S. (2007). The relationship of body mass index to diabetes mellitus, hypertension and dyslipidaemia: Comparison of data from two national surveys. *International Journal of Clinical Practice, 61*(5), 737−747.

Bhargava, A. (2007). Insulin therapy: The question this issue. *Insulin, 2*(2), 92−94.

Bhisitkul, R. (2006). Vascular endothelial growth factor biology: Clinical implications for ocular treatments. *British Journal of Ophthalmology, 90*(12), 1542−1547.

Biden, T. J., Boslem, E., Chu, K. Y., & Sue, N. (2014). Lipotoxic endoplasmic reticulum stress, β cell failure, and type 2 diabetes mellitus. *Trends in Endocrinology & Metabolism, 25*(8), 389–398.

Bode, B. W., Sabbah, H. T., Gross, T. M., Fredrickson, L. P., & Davidson, P. C. (2002). Diabetes management in the new millennium using insulin pump therapy. *Diabetes/Metabolism Research and Reviews, 18*(S1).

Bonnefont-Rousselot, D. (2002). Glucose and reactive oxygen species. *Current Opinion in Clinical Nutrition & Metabolic Care, 5*(5), 561–568.

Bösenberg, L. H., & van Zyl, D. G. (2008). The mechanism of action of oral antidiabetic drugs: A review of recent literature. *Journal of Endocrinology, Metabolism and Diabetes of South Africa, 13*(3), 80–88.

Brás, N. F., Cerqueira, N. M., Ramos, M. J., & Fernandes, P. A. (2014). Glycosidase inhibitors: A patent review (2008–2013). *Expert Opinion on Therapeutic Patents, 24*(8), 857–874.

Bridges, H. R., Jones, A. J., Pollak, M. N., & Hirst, J. (2014). Effects of metformin and other biguanides on oxidative phosphorylation in mitochondria. *Biochemical Journal, 462*(3), 475–487.

Brownlee, M. (2005). The pathobiology of diabetic complications: A unifying mechanism. *Diabetes, 54*(6), 1615–1625.

Burek, C., Rose, N., Barbesino, G., Wang, J., Steck, A. K., Eisenbarth, G. S., Yu, L., De Vincentiis, L., Ricciuti, A., & De Remigis, A. (2016). Endocrinopathies: Chronic thyroiditis, addison disease, pernicious anemia, Graves' disease, diabetes, and hypophysitis. *Manual of molecular and clinical laboratory immunology* (8th ed., pp. 930–953). American Society of Microbiology.

Burgess, T. A., Robich, M. P., Chu, L. M., Bianchi, C., & Sellke, F. W. (2011). Improving glucose metabolism with resveratrol in a swine model of metabolic syndrome through alteration of signaling pathways in the liver and skeletal muscle. *Archives of Surgery, 146*(5), 556–564.

Buse, J. B., Weyer, C., & Maggs, D. G. (2002). Amylin replacement with pramlintide in type 1 and type 2 diabetes: A physiological approach to overcome barriers with insulin therapy. *Clinical Diabetes, 20*(3), 137–144.

Butalia, S., Kaplan, G. G., Khokhar, B., & Rabi, D. M. (2016). Environmental risk factors and type 1 diabetes: Past, present, and future. *Canadian Journal of Diabetes, 40*(6), 586–593.

Castrejón-Pérez, R. C., Gutiérrez-Robledo, L. M., Cesari, M., & Pérez-Zepeda, M. U. (2017). Diabetes mellitus, hypertension and frailty: A population-based, cross-sectional study of Mexican older adults. *Geriatrics & Gerontology International, 17*(6), 925–930.

Centers for Disease Control & Prevention. (2017). *National diabetes statistics report.* Centers for Disease Control and Prevention, US Department of Health and Human Services.

Centers for Disease Control & Prevention. (2020). *National diabetes statistics report.* Centers for Disease Control and Prevention, US Department of Health and Human Services.

Cerna, M. (2020). Epigenetic regulation in etiology of type 1 diabetes mellitus. *International Journal of Molecular Sciences, 21*(1), 36.

Cervin, C., Lyssenko, V., Bakhtadze, E., Lindholm, E., Nilsson, P., Tuomi, T., Cilio, C. M., & Groop, L. (2008). Genetic similarities between latent autoimmune diabetes in adults, type 1 diabetes, and type 2 diabetes. *Diabetes, 57*(5), 1433–1437.

Chan, C.-W., Yu, C.-L., Lin, J.-C., Hsieh, Y.-C., Lin, C.-C., Hung, C.-Y., Li, C.-H., Liao, Y.-C., Lo, C.-P., & Huang, J.-L. (2018). Glitazones and alpha-glucosidase inhibitors as the second-line oral anti-diabetic agents added to metformin reduce cardiovascular risk in Type 2 diabetes patients: A nationwide cohort observational study. *Cardiovascular Diabetology, 17*(1), 20.

Chaplin, S. (2017). Fiasp: A new faster-acting insulin aspart formulation for diabetes. *Prescriber, 28*(7), 37–38.

Charan Kumar, C., & Murthy, S. (2016). A review on management of blood glucose in type 2 diabetes mellitus. *International Journal of Plant Sciences, 6*, 114–120.

Chawla, R., Thakur, P., Chowdhry, A., Jaiswal, S., Sharma, A., Goel, R., Sharma, J., Priyadarshi, S. S., Kumar, V., & Sharma, R. K. (2013). Evidence based herbal drug standardization approach in coping with challenges of holistic management of diabetes: A dreadful lifestyle disorder of 21st century. *Journal of Diabetes & Metabolic Disorders, 12*(1), 35.

Chen, W.-P., Chi, T.-C., Chuang, L.-M., & Su, M.-J. (2007). Resveratrol enhances insulin secretion by blocking KATP and KV channels of beta cells. *European Journal of Pharmacology, 568*(1–3), 269–277.

Chobanian, A. V., Bakris, G. L., Black, H. R., Cushman, W. C., Green, L. A., Izzo, J. L., Jones, D. W., Materson, B. J., Oparil, S., & Wright, J. T. (2003). Seventh report of the joint national committee on prevention, detection, evaluation, and treatment of high blood pressure. *Hypertension, 42*(6), 1206–1252.

Choi, J., Kim, T.-H., Choi, T.-Y., & Lee, M. S. (2013). Ginseng for health care: A systematic review of randomized controlled trials in Korean literature. *PLoS One*, *8*(4), e59978.

Christensen, A. S., Hædersdal, S., Støy, J., Storgaard, H., Kampmann, U., Forman, J. L., Seghieri, M., Holst, J. J., Hansen, T., & Knop, F. K. (2020). Efficacy and safety of glimepiride with or without linagliptin treatment in patients with HNF1A diabetes (maturity-onset diabetes of the young type 3): A randomized, double-blinded, placebo-controlled, crossover trial (GLIMLINA). *Diabetes Care*, *43*(9), 2025–2033.

Cobden, D., Lee, W. C., Balu, S., Joshi, A. V., & Pashos, C. L. (2007). Health outcomes and economic impact of therapy conversion to a biphasic insulin analog pen among privately insured patients with type 2 diabetes mellitus. *Pharmacotherapy: The Journal of Human Pharmacology and Drug Therapy*, *27*(7), 948–962.

Colao, A., De Block, C., Gaztambide, M. S., Kumar, S., Seufert, J., & Casanueva, F. F. (2014). Managing hyperglycemia in patients with Cushing's disease treated with pasireotide: Medical expert recommendations. *Pituitary*, *17*(2), 180–186.

Collins, V. R., Dowse, G. K., Toelupe, P. M., Imo, T. T., Aloaina, F. L., Spark, R. A., & Zimmet, P. Z. (1994). Increasing prevalence of NIDDM in the Pacific island population of Western Samoa over a 13-year period. *Diabetes Care*, *17*(4), 288–296.

Cooper, M. E., Rosenstock, J., Kadowaki, T., Seino, Y., Wanner, C., Schnaidt, S., Clark, D., & Johansen, O. E., CARMELINA investigators. (2020). Cardiovascular and kidney outcomes of linagliptin treatment in older people with type 2 diabetes and established cardiovascular disease and/or kidney disease: A prespecified subgroup analysis of the randomized, placebo-controlled CARMELINA® trial. *Diabetes, Obesity and Metabolism*, *22*(7), 1062–1073.

Crosby, A. J., & Rattan, S. (2016). *Compliant syringe system, a method of making, and of using the same*. Google Patents.

D'Eliseo, P., Blaauw, J., Milicevic, J., Wyatt, J., Ignaut, D., & Malone, J. (2000). Patient acceptability of a new 3.0 mL pre-filled insulin pen. *Current Medical Research and Opinion*, *16*(2), 125–133.

Dabelea, D. (2009). The accelerating epidemic of childhood diabetes. *The Lancet*, *373*(9680), 1999–2000.

Dabelea, D., Bell, R. A., D'Agostino, J. R., Imperatore, G., Johansen, J. M., Linder, B., Liu, L. L., Loots, B., Marcovina, S., & Mayer-Davis, E. J. (2007). Incidence of diabetes in youth in the United States. *JAMA: The Journal of the American Medical Association*, *297*(24), 2716–2724.

Das, S. K., & Elbein, S. C. (2006). The genetic basis of type 2 diabetes. *Cellscience*, *2*(4), 100.

Davis, E. M., Bebee, A., Crawford, L., & Destache, C. (2009). Nurse satisfaction using insulin pens in hospitalized patients. *The Diabetes Educator*, *35*(5), 799–809.

Davis, E. M., Christensen, C. M., Nystrom, K. K., Foral, P. A., & Destache, C. (2008). Patient satisfaction and costs associated with insulin administered by pen device or syringe during hospitalization. *American Journal of Health-System Pharmacy*, *65*(14), 1347–1357.

De Boer, I. H., Bangalore, S., Benetos, A., Davis, A. M., Michos, E. D., Muntner, P., Rossing, P., Zoungas, S., & Bakris, G. (2017). Diabetes and hypertension: A position statement by the American Diabetes Association. *Diabetes Care*, *40*(9), 1273–1284.

De Ferranti, S. D., De Boer, I. H., Fonseca, V., Fox, C. S., Golden, S. H., Lavie, C. J., Magge, S. N., Marx, N., McGuire, D. K., & Orchard, T. J. (2014). Type 1 diabetes mellitus and cardiovascular disease: A scientific statement from the American Heart Association and American Diabetes Association. *Circulation*, *130*(13), 1110–1130.

De Lorenzo, A., Deurenberg, P., Pietrantuono, M., Di Daniele, N., Cervelli, V., & Andreoli, A. (2003). How fat is obese? *Acta Diabetologica*, *40*(1), s254–s257.

DeFronzo, R. A. (2009). From the triumvirate to the ominous octet: A new paradigm for the treatment of type 2 diabetes mellitus. *Diabetes*, *58*(4), 773–795.

Derosa, G., & Maffioli, P. (2012). α-Glucosidase inhibitors and their use in clinical practice. *Archives of Medical Science: AMS*, *8*(5), 899.

DeWitt, D. E., & Hirsch, I. B. (2003). Outpatient insulin therapy in type 1 and type 2 diabetes mellitus: Scientific review. *JAMA: The Journal of the American Medical Association*, *289*(17), 2254–2264.

Dubois, W. (2018). Field-testing companion medical's new "InPen" smart insulin pen. In *Diabetes mine* (Vol. 2018). Healthline. <https://www.healthline.com/diabetesmine/companion-medical-inpen-review#1>.

Dunstan, D. W., Zimmet, P. Z., Welborn, T. A., De Courten, M. P., Cameron, A. J., Sicree, R. A., Dwyer, T., Colagiuri, S., Jolley, D., & Knuiman, M. (2002). The rising prevalence of diabetes and impaired glucose tolerance: The Australian Diabetes, Obesity and Lifestyle Study. *Diabetes Care, 25* (5), 829–834.

Dureshahwar, K., Mubashir, M., & Une, H. D. (2017). Quantification of quercetin obtained from *Allium cepa* Lam. leaves and its effects on streptozotocin-induced diabetic neuropathy. *Pharmacognosy Research, 9*(3), 287.

Ebbell, B. (1937). *The Papyrus Ebers: The greatest Egyptian medical document.* Levin & Munksgaard.

Eidi, M., Eidi, A., & Zamanizadeh, H. (2005). Effect of *Salvia officinalis* L. leaves on serum glucose and insulin in healthy and streptozotocin-induced diabetic rats. *Journal of Ethnopharmacology, 100*(3), 310–313.

Eisenbarth, G. S. (2007). Update in type 1 diabetes. *The Journal of Clinical Endocrinology & Metabolism, 92* (7), 2403–2407.

Elyasiyan, U., Nudel, A., Skalka, N., Rozenberg, K., Drori, E., Oppenheimer, R., Kerem, Z., & Rosenzweig, T. (2017). Anti-diabetic activity of aerial parts of *Sarcopoterium spinosum. BMC Complementary and Alternative Medicine, 17*(1), 356.

Fathallah, N., Slim, R., Larif, S., Hmouda, H., & Salem, C. B. (2015). Drug-induced hyperglycaemia and diabetes. *Drug Safety, 38*(12), 1153–1168.

Fowler, M. J. (2007). Diabetes treatment, part 2: Oral agents for glycemic management. *Clinical Diabetes, 25*(4), 131–134.

Frombaum, M., Le Clanche, S., Bonnefont-Rousselot, D., & Borderie, D. (2012). Antioxidant effects of resveratrol and other stilbene derivatives on oxidative stress and NO bioavailability: Potential benefits to cardiovascular diseases. *Biochimie, 94*(2), 269–276.

Furukawa, S., Fujita, T., Shimabukuro, M., Iwaki, M., Yamada, Y., Nakajima, Y., Nakayama, O., Makishima, M., Matsuda, M., & Shimomura, I. (2017). Increased oxidative stress in obesity and its impact on metabolic syndrome. *The Journal of Clinical Investigation, 114*(12), 1752–1761.

Gaikwad, S. B., Krishna Mohan, G., & Sandhya Rani, M. (2014). Phytochemicals for diabetes management. *Pharmaceutical Crops, 5*(1).

Gandhi, G. R., Vanlalhruaia, P., Stalin, A., Irudayaraj, S. S., Ignacimuthu, S., & Paulraj, M. G. (2014). Polyphenols-rich *Cyamopsis tetragonoloba* (L.) Taub. beans show hypoglycemic and β-cells protective effects in type 2 diabetic rats. *Food and Chemical Toxicology, 66*, 358–365.

Gasbjerg, L. S., Bergmann, N. C., Stensen, S., Christensen, M. B., Rosenkilde, M. M., Holst, J. J., Nauck, M., & Knop, F. K. (2020). Evaluation of the incretin effect in humans using GIP and GLP-1 receptor antagonists. *Peptides, 125*, 170183.

German, C. A., Laughey, B., Bertoni, A. G., & Yeboah, J. (2020). Associations between BMI, waist circumference, central obesity and outcomes in type II diabetes mellitus: The ACCORD Trial. *Journal of Diabetes and Its Complications, 34*(3), 107499.

Gianani, R., Campbell-Thompson, M., Sarkar, S., Wasserfall, C., Pugliese, A., Solis, J., Kent, S., Hering, B., West, E., & Steck, A. (2010). Dimorphic histopathology of long-standing childhood-onset diabetes. *Diabetologia, 53*(4), 690–698.

Gilliam, L. K., Brooks-Worrell, B. M., Palmer, J. P., Greenbaum, C. J., & Pihoker, C. (2005). Autoimmunity and clinical course in children with type 1, type 2, and type 1.5 diabetes. *Journal of Autoimmunity, 25*(3), 244–250.

Gothai, S., Ganesan, P., Park, S.-Y., Fakurazi, S., Choi, D.-K., & Arulselvan, P. (2016). Natural phytobioactive compounds for the treatment of type 2 diabetes: Inflammation as a target. *Nutrients, 8*(8), 461.

Juvenile Diabetes Research Foundation Continuous Glucose Monitoring Study Group. (2008). Continuous glucose monitoring and intensive treatment of type 1 diabetes. *New England Journal of Medicine, 359*(14), 1464–1476.

Grunberger, G. (2017). Should side effects influence the selection of antidiabetic therapies in type 2 diabetes? *Current Diabetes Reports, 17*(4), 21.

Grunberger, G., Abelseth, J., Bailey, T., Bode, B., Handelsman, Y., Hellman, R., Jovanovič, L., Lane, W., Raskin, P., & Tamborlane, W. (2014). Consensus statement by the American Association of Clinical Endocrinologists/American College of Endocrinology insulin pump management task force. *Endocrine Practice, 20*(5), 463–489.

Grunberger, G., Bailey, T., Cohen, A., Flood, T., Handelsman, Y., Hellman, R., Jovanovič, L., Moghissi, E., & Orzeck, E. (2010). Statement by the American Association of Clinical Endocrinologists Consensus Panel on insulin pump management. *Endocrine Practice, 16*(5), 746–762.

Guilhem, I., Balkau, B., Lecordier, F., Malécot, J.-M., Elbadii, S., Leguerrier, A.-M., Poirier, J.-Y., Derrien, C., & Bonnet, F. (2009). Insulin pump failures are still frequent: A prospective study over 6 years from 2001 to 2007. *Diabetologia, 52*(12), 2662–2664.

Gupta, P., Bala, M., Gupta, S., Dua, A., Dabur, R., Injeti, E., & Mittal, A. (2016). Efficacy and risk profile of anti-diabetic therapies: Conventional vs traditional drugs—A mechanistic revisit to understand their mode of action. *Pharmacological Research, 113*, 636–674.

Hamada, Y., Goto, M., Nishimura, G., Nagasaki, H., Seino, Y., Kamiya, H., & Nakamura, J. (2020). The alpha-glucosidase inhibitor miglitol increases hepatic CYP7A1 activity in association with altered short-chain fatty acid production in the gut of obese diabetic mice. *Metabolism Open, 5*, 100024.

Hammouda, Y., & Amer, M. S. (1966). Antidiabetic effect of tecomine and tecostanine. *Journal of Pharmaceutical Sciences, 55*(12), 1452–1454.

Hampp, C., Borders-Hemphill, V., Moeny, D. G., & Wysowski, D. K. (2014). Use of antidiabetic drugs in the US, 2003–2012. *Diabetes Care, 37*(5), 1367–1374.

Harris, M. I. (1996). Medical care for patients with diabetes: Epidemiologic aspects. *Annals of Internal Medicine, 124*(1_Part_2), 117–122.

Hashim, R. H. (2017). Effects of metformin, glyburide and their combination on lipid profile in NIDDM patients. *Al-Qadisiyah Medical Journal, 10*(17), 67–77.

Hathout, E. H., Beeson, W. L., Nahab, F., Rabadi, A., Thomas, W., & Mace, J. W. (2002). Role of exposure to air pollutants in the development of type 1 diabetes before and after 5 yr of age. *Pediatric Diabetes, 3*(4), 184–188.

Hattersley, A., Bruining, J., Shield, J., Njolstad, P., & Donaghue, K. C. (2009). The diagnosis and management of monogenic diabetes in children and adolescents. *Pediatric Diabetes, 10*(s12), 33–42.

Hedrington, M. S., & Davis, S. N. (2015). The role of empagliflozin in the management of type 2 diabetes by patient profile. *Therapeutics and Clinical Risk Management, 11*, 739.

Hernandez-Galicia, E., Aguilar-Contreras, A., Aguilar-Santamaria, L., Roman-Ramos, R., Chavez-Miranda, A., Garcia-Vega, L., Flores-Saenz, J., & Alarcon-Aguilar, F. (2002). Studies on hypoglycemic activity of Mexican medicinal plants. *Proceedings of the Western Pharmacology Society* (Vol. 45, pp. 118–124). The Society.

Hollstein, T., Schulte, D. M., Schulz, J., Glück, A., Ziegler, A. G., Bonifacio, E., Wendorff, M., Franke, A., Schreiber, S., & Bornstein, S. R. (2020). Autoantibody-negative insulin-dependent diabetes mellitus after SARS-CoV-2 infection: A case report. *Nature Metabolism, 2*(10), 1021–1024.

Holman, R. R., Paul, S. K., Bethel, M. A., Matthews, D. R., & Neil, H. A. W. (2008). 10-year follow-up of intensive glucose control in type 2 diabetes. *New England Journal of Medicine, 359*(15), 1577–1589.

Holst, J. J. (2007). The physiology of glucagon-like peptide 1. *Physiological Reviews, 87*(4), 1409–1439.

Hood, R. C., Arakaki, R. F., Wysham, C., Li, Y. G., Settles, J. A., & Jackson, J. A. (2015). Two treatment approaches for human regular U-500 insulin in patients with type 2 diabetes not achieving adequate glycemic control on high-dose U-100 insulin therapy with or without oral agents: A randomized, titration-to-target clinical trial. *Endocrine Practice, 21*(7), 782–793.

Hoogwerf, B. J., Doshi, K. B., & Diab, D. (2008). Pramlintide, the synthetic analogue of amylin: Physiology, pathophysiology, and effects on glycemic control, body weight, and selected biomarkers of vascular risk. *Vascular Health and Risk Management, 4*(2), 355.

Houstis, N., Rosen, E. D., & Lander, E. S. (2006). Reactive oxygen species have a causal role in multiple forms of insulin resistance. *Nature, 440*(7086), 944.

Hu, Y., Jin, P., Peng, J., Zhang, X., Wong, F. S., & Wen, L. (2016). Different immunological responses to early-life antibiotic exposure affecting autoimmune diabetes development in NOD mice. *Journal of Autoimmunity, 72*, 47–56.

Hu, Y., Wong, F. S., & Wen, L. (2017). Antibiotics, gut microbiota, environment in early life and type 1 diabetes. *Pharmacological Research, 119*, 219–226.

IDF Diabetes Atlas (2015). *International diabetes federation, Brussels, 2015*. Accessed: 05.03.14.

Inoguchi, T., Li, P., Umeda, F., Yu, H. Y., Kakimoto, M., Imamura, M., Aoki, T., Etoh, T., Hashimoto, T., & Naruse, M. (2000). High glucose level and free fatty acid stimulate reactive oxygen species production through protein kinase C−dependent activation of NAD (P) H oxidase in cultured vascular cells. *Diabetes, 49*(11), 1939−1945.

Inzucchi, S. E., Bergenstal, R. M., Buse, J. B., Diamant, M., Ferrannini, E., Nauck, M., Peters, A. L., Tsapas, A., Wender, R., & Matthews, D. R. (2015). Management of hyperglycaemia in type 2 diabetes, 2015: A patient-centred approach. Update to a position statement of the American Diabetes Association and the European Association for the Study of Diabetes. *Diabetologia, 58*(3), 429−442.

Isharwal, S., Misra, A., Wasir, J., & Nigam, P. (2009). Diet & insulin resistance: A review & Asian Indian perspective. *The Indian Journal of Medical Research, 129*, 485−499.

Jang, M., Cai, L., Udeani, G. O., Slowing, K. V., Thomas, C. F., Beecher, C. W., Fong, H. H., Farnsworth, N. R., Kinghorn, A. D., & Mehta, R. G. (1997). Cancer chemopreventive activity of resveratrol, a natural product derived from grapes. *Science (New York, N.Y.), 275*(5297), 218−220.

Javeed, N., & Matveyenko, A. V. (2018). Circadian etiology of type 2 diabetes mellitus. *Physiology, 33*(2), 138−150.

Jeevanandam, J. (2017). *Enhanced synthesis and delivery of magnesium oxide nanoparticles for reverse insulin resistance in type 2 diabetes mellitus.* Curtin University.

Jeevanandam, J., Danquah, M. K., Debnath, S., Meka, V. S., & Chan, Y. S. (2015). Opportunities for nanoformulations in type 2 diabetes mellitus treatments. *Current Pharmaceutical Biotechnology, 16*(10), 853−870.

Jiao, L., Zhang, X., Huang, L., Gong, H., Cheng, B., Sun, Y., Li, Y., Liu, Q., Zheng, L., & Huang, K. (2013). Proanthocyanidins are the major anti-diabetic components of cinnamon water extract. *Food and Chemical Toxicology, 56*, 398−405.

Joshi, S. R., Standl, E., Tong, N., Shah, P., Kalra, S., & Rathod, R. (2015). Therapeutic potential of α-glucosidase inhibitors in type 2 diabetes mellitus: An evidence-based review. *Expert Opinion on Pharmacotherapy, 16*(13), 1959−1981.

Kabul, S., Hood, R. C., Duan, R., DeLozier, A. M., & Settles, J. (2016). Patient-reported outcomes in transition from high-dose U-100 insulin to human regular U-500 insulin in severely insulin-resistant patients with type 2 diabetes: Analysis of a randomized clinical trial. *Health and Quality of Life Outcomes, 14*(1), 139.

Kahn, C. R., Chen, L., & Cohen, S. E. (2000). Unraveling the mechanism of action of thiazolidinediones. *The Journal of Clinical Investigation, 106*(11), 1305−1307.

Kaku, K., Inoue, S., Matsuoka, O., Kiyosue, A., Azuma, H., Hayashi, N., Tokudome, T., Langkilde, A., & Parikh, S. (2013). Efficacy and safety of dapagliflozin as a monotherapy for type 2 diabetes mellitus in Japanese patients with inadequate glycaemic control: A phase II multicentre, randomized, double-blind, placebo-controlled trial. *Diabetes, Obesity and Metabolism, 15*(5), 432−440.

Karawya, M., Wahab, S. A., El-Olemy, M., & Farrag, N. (1984). Diphenylamine, an antihyperglycemic agent from onion and tea. *Journal of Natural Products, 47*(5), 775−780.

Kase, E. T., Feng, Y. Z., Badin, P.-M., Bakke, S. S., Laurens, C., Coue, M., Langin, D., Gaster, M., Thoresen, G. H., & Rustan, A. C. (2015). Primary defects in lipolysis and insulin action in skeletal muscle cells from type 2 diabetic individuals. *Biochimica et Biophysica Acta (BBA)-Molecular and Cell Biology of Lipids, 1851*(9), 1194−1201.

Kaur, J., Famta, P., Khurana, N., Vyas, M., & Khatik, G. L. (2020). Pharmacotherapy of type 2 diabetes. *Obesity and diabetes* (pp. 679−694). Springer.

Kaushik, U., Aeri, V., Showkat, R. M., & Ali, M. (2017). Cucurbitane-type triterpenoids from the blood glucose-lowering extracts of *Coccinia indica* and *Momordica balsamina* fruits. *Pharmacognosy Magazine, 13*(Suppl 1), S115.

Keilson, L., Mather, S., Walter, Y. H., Subramanian, S., & McLeod, J. F. (2000). Synergistic effects of nateglinide and meal administration on insulin secretion in patients with type 2 diabetes mellitus. *The Journal of Clinical Endocrinology & Metabolism, 85*(3), 1081−1086.

Kerner, W., & Brückel, J. (2014). Definition, classification and diagnosis of diabetes mellitus. *Experimental and Clinical Endocrinology & Diabetes, 122*(07), 384−386.

Khavandi, M., Duarte, F., Ginsberg, H. N., & Reyes-Soffer, G. (2017). Treatment of dyslipidemias to prevent cardiovascular disease in patients with type 2 diabetes. *Current Cardiology Reports, 19*(1), 7.

Kirk, S. E. (2003). Insulin pump therapy for type 2 diabetes. *Current Diabetes Reports, 3*(5), 373—377.

Klockgether, T., & Dichgans, J. (1990). *Do neuroleptic drugs still have a place in neurological therapy?* Springer.

Kobayashi, T. (1994). Subtype of insulin-dependent diabetes mellitus (IDDM) in Japan: Slowly progressive IDDM—The clinical characteristics and pathogenesis of the syndrome. *Diabetes Research and Clinical Practice, 24*, S95—S99.

Kohlroser, J., Mathai, J., Reichheld, J., Banner, B. F., & Bonkovsky, H. L. (2000). Hepatotoxicity due to troglitazone: Report of two cases and review of adverse events reported to the United States Food and Drug Administration. *The American Journal of Gastroenterology, 95*(1), 272.

Kouzi, S., Yang, S., Nuzum, D., & Dirks-Naylor, A. (2015). Natural supplements for improving insulin sensitivity and glucose uptake in skeletal muscle. *Frontiers in Bioscience (Elite Edition), 7*, 94—106.

Krentz, A. J., & Bailey, C. J. (2005). Oral antidiabetic agents. *Drugs, 65*(3), 385—411.

Kushwaha, R. N., Haq, W., & Katti, S. (2014). *Discovery of 17 gliptins in 17-years of research for the treatment of type 2 diabetes: A synthetic overview.*

Lane, W., Weinrib, S., Rappaport, J., & Hale, C. (2014). The effect of addition of liraglutide to high-dose intensive insulin therapy: A randomized prospective trial. *Diabetes, Obesity and Metabolism, 16*(9), 827—832.

Langer, P., Tajtáková, M., Guretzki, H.-J., Kočan, A., Petrík, J., Chovancová, J., Drobná, B., Jursa, S., Pavúk, M., & Trnovec, T. (2002). High prevalence of anti-glutamic acid decarboxylase (anti-GAD) antibodies in employees at a polychlorinated biphenyl production factory. *Archives of Environmental Health: An International Journal, 57*(5), 412—415.

Lavery, J., Friedman, A., Keyes, K., Wright, J., & Ananth, C. (2017). Gestational diabetes in the United States: Temporal changes in prevalence rates between 1979 and 2010. *BJOG: An International Journal of Obstetrics & Gynaecology, 124*(5), 804—813.

Lebovitz, H. E. (1997). Alpha-glucosidase inhibitors. *Endocrinology and Metabolism Clinics, 26*(3), 539—551.

Lebovitz, H. E. (1998). *Therapy for diabetes mellitus and related disorders.* American Diabetes Association.

Leduc, C., Coonishish, J., Haddad, P., & Cuerrier, A. (2006). Plants used by the Cree Nation of Eeyou Istchee (Quebec, Canada) for the treatment of diabetes: A novel approach in quantitative ethnobotany. *Journal of Ethnopharmacology, 105*(1—2), 55—63.

Li, L., Shen, J., Bala, M. M., Busse, J. W., Ebrahim, S., Vandvik, P. O., Rios, L. P., Malaga, G., Wong, E., & Sohani, Z. (2014). Incretin treatment and risk of pancreatitis in patients with type 2 diabetes mellitus: Systematic review and *meta*-analysis of randomised and non-randomised studies. *BMJ (Clinical Research ed.), 348*, g2366.

Lietzen, N., An, L.T., Jaakkola, M.K., Kallionpää, H., Oikarinen, S., Mykkänen, J., Knip, M., Veijola, R., Ilonen, J., & Toppari, J. (2018). Enterovirus-associated changes in blood transcriptomic profiles of children with genetic susceptibility to type 1 diabetes. *Diabetologia,* 1—8.

Lind, M., Polonsky, W., Hirsch, I. B., Heise, T., Bolinder, J., Dahlqvist, S., Schwarz, E., Ólafsdóttir, A. F., Frid, A., & Wedel, H. (2017). Continuous glucose monitoring vs conventional therapy for glycemic control in adults with type 1 diabetes treated with multiple daily insulin injections: The GOLD randomized clinical trial. *JAMA: The Journal of the American Medical Association, 317*(4), 379—387.

Linden, M. A., Ross, T. T., Hamilton, K. L., Miller, B. F., Braun, B., & Esler, W. P. (2017). Comparison of the Therapeutic Effects of the Combination of Exercise Training with Either a Sodium-glucose Cotransporter 2 Inhibitor or Metformin in a Murine Model of Type 2 Diabetes. *The FASEB Journal, 31*(1 Suppl.), 1020.10-1020.10.

Linnenkamp, U., Guariguata, L., Beagley, J., Whiting, D., & Cho, N. (2014). The IDF Diabetes Atlas methodology for estimating global prevalence of hyperglycaemia in pregnancy. *Diabetes Research and Clinical Practice, 103*(2), 186—196.

Liu, X.-Y., Zhang, N., Chen, R., Zhao, J.-G., & Yu, P. (2015). Efficacy and safety of sodium—glucose cotransporter 2 inhibitors in type 2 diabetes: A *meta*-analysis of randomized controlled trials for 1 to 2 years. *Journal of Diabetes and its Complications, 29*(8), 1295—1303.

Lourenco, M. V., Ferreira, S. T., & De Felice, F. G. (2015). Neuronal stress signaling and eIF2α phosphorylation as molecular links between Alzheimer's disease and diabetes. *Progress in Neurobiology, 129*, 37—57.

Luck, M. L., Savitsky, L. M., Speranza, R., Cheng, Y. W., & Caughey, A. B. (2017). 972: The rates of cesarean section in gestational diabetes stratified by race. *American Journal of Obstetrics & Gynecology, 216*(1), S548.

Marathe, P. H., Gao, H. X., & Close, K. L. (2017). American Diabetes Association standards of medical care in diabetes 2017. *Journal of Diabetes, 9*(4), 320−324.

Martínez, R., Kapravelou, G., Porres, J. M., Melesio, A. M., Heras, L., Cantarero, S., Gribble, F. M., Parker, H., Aranda, P., & López-Jurado, M. (2016). *Medicago sativa* L., a functional food to relieve hypertension and metabolic disorders in a spontaneously hypertensive rat model. *Journal of Functional Foods, 26*, 470−484.

Mathers, C. D., & Loncar, D. (2006). Projections of global mortality and burden of disease from 2002 to 2030. *PLoS Medicine, 3*(11), e442.

Mathieu, C., Rodbard, H., Cariou, B., Handelsman, Y., Philis-Tsimikas, A., Ocampo Francisco, A., Rana, A., & Zinman, B. (2014). A comparison of adding liraglutide vs a single daily dose of insulin aspart to insulin degludec in subjects with type 2 diabetes (BEGIN: VICTOZA ADD-ON). *Diabetes, Obesity and Metabolism, 16*(7), 636−644.

McKay, M., Compion, G., & Lytzen, L. (2009). A comparison of insulin injection needles on patients' perceptions of pain, handling, and acceptability: A randomized, open-label, crossover study in subjects with diabetes. *Diabetes Technology & Therapeutics, 11*(3), 195−201.

McPhee, S. J., Papadakis, M. A., & Rabow, M. W. (2010). *Current medical diagnosis & treatment*. McGraw-Hill Medical.

Meela, M. M., Mdee, L. K., & Eloff, J. N. (2017). *Tecoma stans* (Bignoniaceae), leaf extracts, fractions and isolated compound have promising activity against fungal phytopathogens. *Suid-Afrikaanse Tydskrif Vir Natuurwetenskap En Tegnologie, 36*(1), 7.

Meier, J. J., & Giese, A. (2015). Diabetes associated with pancreatic diseases. *Current Opinion in Gastroenterology, 31*(5), 400−406.

Mende, C. W. (2017). Diabetes and kidney disease: The role of sodium−glucose cotransporter-2 (SGLT-2) and SGLT-2 inhibitors in modifying disease outcomes. *Current Medical Research and Opinion, 33*(3), 541−551.

Midthjell, K., Krüger, O., Holmen, J., Tverdal, A., Claudi, T., Bjørndal, A., & Magnus, P. (1999). Rapid changes in the prevalence of obesity and known diabetes in an adult Norwegian population. The Nord-Trøndelag Health Surveys: 1984−1986 and 1995−1997. *Diabetes Care, 22*(11), 1813−1820.

Miele, L., Bosetti, C., Turati, F., Rapaccini, G., Gasbarrini, A., La Vecchia, C., Boccia, S., & Grieco, A. (2015). Diabetes and insulin therapy, but not metformin, are related to hepatocellular cancer risk. *Gastroenterology research and practice*.

Miller, R. A., Chu, Q., Xie, J., Foretz, M., Viollet, B., & Birnbaum, M. J. (2013). Biguanides suppress hepatic glucagon signalling by decreasing production of cyclic AMP. *Nature, 494*(7436), 256.

Mishkinsky, J., Goldschmied, A., Joseph, B., Ahronson, Z., & Sulman, F. (1974). Hypoglycaemic effect of *Trigonella foenum graecum* and *Lupinus termis* (leguminosae) seeds and their major alkaloids in alloxan-diabetic and normal rats. *Archives Internationales de Pharmacodynamie et de Thérapie, 210*(1), 27−37.

Miura, T., Takagi, S., & Ishida, T. (2012). Management of diabetes and its complications with banaba (*Lagerstroemia speciosa* L.) and corosolic acid. *Evidence-Based Complementary and Alternative Medicine, 2012*.

Moini, J., & Moini, J. (2019). Chapter 3: Pathophysiology of diabetes. *Epidemiology of diabetes* (pp. 25−43). Elsevier. Available from https://doi.org/10.1016/B978-0-12-816864-6.00003-1.

Mokdad, A. H., Bowman, B. A., Ford, E. S., Vinicor, F., Marks, J. S., & Koplan, J. P. (2001). The continuing epidemics of obesity and diabetes in the United States. *JAMA: The Journal of the American Medical Association, 286*(10), 1195−1200.

Moon, J., Kim, J. Y., Yoo, S., & Koh, G. (2020). Fasting and postprandial hyperglycemia: Their predictors and contributions to overall hyperglycemia in Korean patients with type 2 diabetes. *Endocrinology and Metabolism, 35*(2), 290−297.

Morgan, N. G., & Richardson, S. J. (2014). Enteroviruses as causative agents in type 1 diabetes: Loose ends or lost cause? *Trends in Endocrinology & Metabolism, 25*(12), 611−619.

Mosley, J. F., II, Smith, L., Everton, E., & Fellner, C. (2015). Sodium-glucose linked transporter 2 (SGLT2) inhibitors in the management of type-2 diabetes: A drug class overview. *Pharmacy and Therapeutics, 40*(7), 451.

Mostafa, S., Coleman, R., Agbaje, O., Gray, A., Holman, R., & Bethel, M. (2018). Modelling incremental benefits on complications rates when targeting lower HbA1c levels in people with Type 2 diabetes and cardiovascular disease. *Diabetic Medicine, 35*(1), 72−77.

Movahed, A., Nabipour, I., Lieben Louis, X., Thandapilly, S. J., Yu, L., Kalantarhormozi, M., Rekabpour, S. J., & Netticadan, T. (2013). Antihyperglycemic effects of short term resveratrol supplementation in type 2 diabetic patients. *Evidence-Based Complementary and Alternative Medicine, 2013*, 851267.

Movva, N., Theckedath, B. G., & Gilden, J. L. (2015). Difficulties in managing patients with insulin resistance: Alternatives to U-500 insulin. *Diabetes Case Studies: Real Problems, Practical Solutions, 109.*

Murtaugh, L. C., & Keefe, M. D. (2015). Regeneration and repair of the exocrine pancreas. *Annual Review of Physiology, 77,* 229−249.

Nagalievska, M., Sabadashka, M., Hachkova, H., & Sybirna, N. (2018). *Galega officinalis* extract regulate the diabetes mellitus related violations of proliferation, functions and apoptosis of leukocytes. *BMC Complementary and Alternative Medicine, 18*(1), 4.

Naik, R. G., & Palmer, J. P. (2003). Latent autoimmune diabetes in adults (LADA). *Reviews in Endocrine and Metabolic Disorders, 4*(3), 233−241.

Naito, Y., Ichikawa, H., Akagiri, S., Uchiyama, K., Takagi, T., Handa, O., Yasukawa, Z., Tokunaga, M., Ishihara, N., & Okubo, T. (2016). Identification of cysteinylated transthyretin, a predictive biomarker of treatment response to partially hydrolyzed guar gum in type 2 diabetes rats, by surface-enhanced laser desorption/ionization time-of-flight mass spectrometry. *Journal of Clinical Biochemistry and Nutrition, 58*(1), 23−33.

Nakamura, Y., Horii, Y., Nishino, T., Shiiki, H., Sakaguchi, Y., Kagoshima, T., Dohi, K., Makita, Z., Vlassara, H., & Bucala, R. (1993). Immunohistochemical localization of advanced glycosylation end products in coronary atheroma and cardiac tissue in diabetes mellitus. *The American Journal of Pathology, 143*(6), 1649.

Nasiri, A., Ziamajidi, N., Abbasalipourkabir, R., Goodarzi, M. T., Saidijam, M., Behrouj, H., & Asl, S. S. (2017). Beneficial effect of aqueous garlic extract on inflammation and oxidative stress status in the kidneys of type 1 diabetic rats. *Indian Journal of Clinical Biochemistry, 32*(3), 329−336.

Nejentsev, S., Howson, J. M., Walker, N. M., Szeszko, J., Field, S. F., Stevens, H. E., Reynolds, P., Hardy, M., King, E., & Masters, J. (2007). Localization of type 1 diabetes susceptibility to the MHC class I genes HLA-B and HLA-A. *Nature, 450*(7171), 887.

Neumiller, J. J. (2009). Differential chemistry (structure), mechanism of action, and pharmacology of GLP-1 receptor agonists and DPP-4 inhibitors. *Journal of the American Pharmacists Association: JAPhA, 49,* S16−S29.

Nielsen, P. R., Kragstrup, T. W., Deleuran, B. W., & Benros, M. E. (2016). Infections as risk factor for autoimmune diseases—a nationwide study. *Journal of Autoimmunity, 74,* 176−181.

Noble, J. A., & Erlich, H. A. (2012). Genetics of type 1 diabetes. *Cold Spring Harbor Perspectives in Medicine, 2*(1), a007732.

Nomura, S., Sakamaki, S., Hongu, M., Kawanishi, E., Koga, Y., Sakamoto, T., Yamamoto, Y., Ueta, K., Kimata, H., & Nakayama, K. (2010). Discovery of canagliflozin, a novel C-glucoside with thiophene ring, as sodium-dependent glucose cotransporter 2 inhibitor for the treatment of type 2 diabetes mellitus. *Journal of Medicinal Chemistry, 53*(17), 6355−6360.

Nussbaumer, R., Meyer-Gerspach, A. C., Peterli, R., Peters, T., Beglinger, C., Chiappetta, S., Drewe, J., & Wölnerhanssen, B. (2020). First-phase insulin and amylin after bariatric surgery: A prospective randomized trial on patients with insulin resistance or diabetes after gastric bypass or sleeve gastrectomy. *Obesity Facts,* 1−12.

Ogurtsova, K., da Rocha Fernandes, J. D., Huang, Y., Linnenkamp, U., Guariguata, L., Cho, N. H., Cavan, D., Shaw, J. E., & Makaroff, L. E. (2017). IDF Diabetes Atlas: Global estimates for the prevalence of diabetes for 2015 and 2040. *Diabetes Research and Clinical Practice, 128,* 40−50.

Ólafsdóttir, A. F., Attvall, S., Sandgren, U., Dahlqvist, S., Pivodic, A., Skrtic, S., Theodorsson, E., & Lind, M. (2017). A clinical trial of the accuracy and treatment experience of the flash glucose monitor FreeStyle Libre in adults with type 1 diabetes. *Diabetes Technology & Therapeutics, 19*(3), 164−172.

Oloyede, O., Ajiboye, T., Abdussalam, A., & Adeleye, A. (2014). *Blighia sapida* leaves halt elevated blood glucose, dyslipidemia and oxidative stress in alloxan-induced diabetic rats. *Journal of Ethnopharmacology, 157,* 309−319.

Osonoi, T. (2015). Alpha-glucosidase inhibitor. Nihon Rinsho. *Japanese Journal of Clinical Medicine, 73*(3), 390–394.

Pai, S., Munshi, R., & Nayak, C. (2017). Fatal dapsone hypersensitivity syndrome with hypothyroidism and steroid-induced diabetes mellitus. *Indian Journal of Pharmacology, 49*(5), 396.

Palmer, J. P., & Hirsch, I. B. (2003). What's in a name: Latent autoimmune diabetes of adults, type 1.5, adult-onset, and type 1 diabetes. *Diabetes Care, 26*, 536–538.

Panigrahy, S. K., Bhatt, R., & Kumar, A. (2017). Reactive oxygen species: Sources, consequences and targeted therapy in type 2 diabetes. *Journal of Drug Targeting, 25*(2), 93–101.

Pardau, M. D., Pereira, A. S., Apostolides, Z., Serem, J. C., & Bester, M. J. (2017). Antioxidant and anti-inflammatory properties of *Ilex guayusa* tea preparations: A comparison to *Camellia sinensis* teas. *Food & Function, 8*(12), 4601–4610.

Pareek, A. S., Garger, Y. B., Joshi, P. M., Romero, C. M., & Seth, A. K. (2017). Secondary causes of diabetes mellitus. *Principles of diabetes mellitus* (311).

Patel, M. B., & Mishra, S. (2011). Hypoglycemic activity of alkaloidal fraction of *Tinospora cordifolia*. *Phytomedicine: International Journal of Phytotherapy and Phytopharmacology, 18*(12), 1045–1052.

Paula, P., Oliveira, J., Sousa, D., Alves, B., Carvalho, A., Franco, O., & Vasconcelos, I. (2017). Insulin-like plant proteins as potential innovative drugs to treat diabetes—The Moringa oleifera case study. *New Biotechnology, 39*, 99–109.

Paun, A., Yau, C., & Danska, J. S. (2016). Immune recognition and response to the intestinal micro-biome in type 1 diabetes. *Journal of Autoimmunity, 71*, 10–18.

Pearson, T. L. (2010). *Practical aspects of insulin pen devices.* Sage.

Perelis, M., Marcheva, B., Ramsey, K. M., Schipma, M. J., Hutchison, A. L., Taguchi, A., Peek, C. B., Hong, H., Huang, W., & Omura, C. (2015). Pancreatic β cell enhancers regulate rhythmic transcription of genes controlling insulin secretion. *Science (New York, N.Y.), 350*(6261), aac4250.

Perez G, R., Zavala S, M., Perez G, S., & Perez G, C. (1998). Antidiabetic effect of compounds isolated from plants. *Phytomedicine: International Journal of Phytotherapy and Phytopharmacology, 5*(1), 55–75.

Pertusa, S., Montserrat, P., David, M., & Vanes, A. (2015). MODY 2 diabetes: An unusual presentation of diabetes. *Endocrinology & Metabolism International Journal, 2*(4), 00032.

Petrak, F., Herpertz, S., Albus, C., Hermanns, N., Hiemke, C., Hiller, W., Kronfeld, K., Kruse, J., Kulzer, B., & Ruckes, C. (2013). Study protocol of the Diabetes and Depression Study (DAD): A multi-center randomized controlled trial to compare the efficacy of a diabetes-specific cognitive behavioral group therapy vs sertraline in patients with major depression and poorly controlled diabetes mellitus. *BMC Psychiatry, 13*(1), 206.

Peyrot, M., Rubin, R. R., Lauritzen, T., Skovlund, S. E., Snoek, F. J., Matthews, D. R., Landgraf, R., & Kleinebreil, L. (2005). Resistance to insulin therapy among patients and providers: Results of the cross-national Diabetes Attitudes, Wishes, and Needs (DAWN) study. *Diabetes Care, 28*(11), 2673–2679.

Pickup, J. C., Reznik, Y., & Sutton, A. J. (2017). Glycemic control during continuous subcutaneous insulin infusion vs multiple daily insulin injections in type 2 diabetes: Individual patient data *meta*-analysis and *meta*-regression of randomized controlled trials. *Diabetes Care, 40*(5), 715–722.

Pilla, S. J., Maruthur, N. M., Schweitzer, M. A., Magnuson, T. H., Potter, J. J., Clark, J. M., & Lee, C. J. (2018). The role of laboratory testing in differentiating type 1 diabetes from type 2 diabetes in patients undergoing bariatric surgery. *Obesity Surgery, 28*(1), 25–30.

Qadir, M.M.F., Álvarez-Cubela, S., Klein, D., Lanzoni, G., García-Santana, C., Montalvo, A., Pláceres-Uray, F., Mazza, E.M.C., Ricordi, C., Inverardi, L.A., Pastori, R.L., & Domínguez-Bendala, J. (n.d.). P2RY1/ALK3-expressing cells within the adult human exocrine pancreas are BMP-7 expand-able and exhibit progenitor-like characteristics. *Cell Reports, 22*(9), 2408–2420. https://doi.org/10.1016/j.celrep.2018.02.006

Qiao, Q., & Nyamdorj, R. (2010). Is the association of type II diabetes with waist circumference or waist-to-hip ratio stronger than that with body mass index? *European Journal of Clinical Nutrition, 64*(1), 30–34.

Rady, M. Y., Johnson, D. J., Patel, B. M., Larson, J. S., & Helmers, R. A. (2005). Influence of individual characteristics on outcome of glycemic control in intensive care unit patients with or without diabetes mellitus. *Mayo Clinic Proceedings, 80*, 1558–1567.

Rafacho, A., Ortsäter, H., Nadal, A., & Quesada, I. (2014). Glucocorticoid treatment and endocrine pancreas function: Implications for glucose homeostasis, insulin resistance and diabetes. *Journal of Endocrinology, 223*(3), R49—R62.

Raju, K., & Venkataramappa, S. M. (2018). Primary hemochromatosis presenting as Type 2 diabetes mellitus: A case report with review of literature. *International Journal of Applied and Basic Medical Research, 8*(1), 57.

Ramana, K. V., Chandra, D., Srivastava, S., Bhatnagar, A., & Srivastava, S. K. (2003). Nitric oxide regulates the polyol pathway of glucose metabolism in vascular smooth muscle cells. *The FASEB Journal, 17*(3), 417—425.

Raskin, P., Bode, B. W., Marks, J. B., Hirsch, I. B., Weinstein, R. L., McGill, J. B., Peterson, G. E., Mudaliar, S. R., & Reinhardt, R. R. (2003). Continuous subcutaneous insulin infusion and multiple daily injection therapy are equally effective in type 2 diabetes: A randomized, parallel-group, 24-week study. *Diabetes Care, 26*(9), 2598—2603.

Rasouli, H., Yarani, R., Pociot, F., & Popović-Djordjević, J. (2020). Anti-diabetic potential of plant alkaloids: Revisiting current findings and future perspectives. *Pharmacological Research*, 104723.

Rathi, P., Pant, K., Dixit, R., Kumar, A., Pal, R., & Nath, R. (2016). An evaluation of the preventive role of ficus religiosa on high fat diet and low dose streptozotocin induced diabetes mellitus in rats. *Journal of Diabetes & Metabolic Disorders, 3*(1), 1—5.

Ratner, R. E., Want, L. L., Fineman, M. S., Velte, M. J., Ruggles, J. A., Gottlieb, A., Weyer, C., & Kolterman, O. G. (2002). Adjunctive therapy with the amylin analogue pramlintide leads to a combined improvement in glycemic and weight control in insulin-treated subjects with type 2 diabetes. *Diabetes Technology & Therapeutics, 4*(1), 51—61.

Reinauer, C., Bollow, E., Fröhlich-Reiterer, E., Laubner, K., Bergis, D., Schöfl, C., Kempe, H.-P., Hummel, M., Hennes, P., & Gollisch, K. (2017). Polycystic ovary syndrome (PCOS) in juvenile and adult type 1 diabetes in a German/Austrian cohort. *Experimental and Clinical Endocrinology & Diabetes, 125*(10), 661—668.

Riddy, D. M., Delerive, P., Summers, R. J., Sexton, P. M., & Langmead, C. J. (2018). G protein—coupled receptors targeting insulin resistance, obesity, and type 2 diabetes mellitus. *Pharmacological Reviews, 70*(1), 39—67.

Robertson, R. P., Harmon, J., Tran, P. O. T., & Poitout, V. (2004). β-cell glucose toxicity, lipotoxicity, and chronic oxidative stress in type 2 diabetes. *Diabetes, 53*(Suppl. 1), S119—S124.

Robertson, R. P., & Harmon, J. S. (2006). Diabetes, glucose toxicity, and oxidative stress: A case of double jeopardy for the pancreatic islet β cell. *Free Radical Biology and Medicine, 41*(2), 177—184.

Roep, B. O., Thomaidou, S., van Tienhoven, R., & Zaldumbide, A. (2020). Type 1 diabetes mellitus as a disease of the β-cell (do not blame the immune system?). *Nature Reviews Endocrinology*, 1—12.

Rosenbloom, A. L., Joe, J. R., Young, R. S., & Winter, W. E. (1999). Emerging epidemic of type 2 diabetes in youth. *Diabetes Care, 22*(2), 345—354.

Ross, S. A., & Ekoé, J.-M. (2010). Incretin agents in type 2 diabetes. *Canadian Family Physician, 56*(7), 639—648.

Rudá-Kučerová, J., Kotolova, H., & Koupý, D. (2015). Effectiveness of phytotherapy in supportive treatment of type 2 diabetes mellitus III. Momordica (*Momordica charantia*). *Ceska a Slovenska Farmacie: Casopis Ceske Farmaceuticke Spolecnosti a Slovenske Farmaceuticke Spolecnosti, 64*(4), 126—132.

Rynbrandt, R. H., Dutton, F. E., & Schmidt, F. L. (1972). exo-Bicyclo [3.1. 0] hexane-6-carboxylic acid and related compounds, oral hypoglycemic agents. *Journal of Medicinal Chemistry, 15*(4), 424—426.

Saikley, C. R., Hamerton-Kelly, P. R., Sia, S., Kennedy, D. F., Peterson, M. W., Tsang, S. C., Lortz, W. M., & Vivolo, J. A. (2015). *Automatic biological analyte testing meter with integrated lancing device and methods of use*. Google Patents.

Samis, J., Lee, P., Zimmerman, D., Arceci, R. J., Suttorp, M., & Hijiya, N. (2016). Recognizing endocrinopathies associated with tyrosine kinase inhibitor therapy in children with chronic myelogenous leukemia. *Pediatric Blood & Cancer, 63*(8), 1332—1338.

Sanyal, D. (2013). Diabetes is predominantly an intestinal disease. *Indian Journal of Endocrinology and Metabolism, 17*(Suppl. 1), S64.

Saxena, A., & Vikram, N. K. (2004). Role of selected Indian plants in management of type 2 diabetes: A review. *The Journal of Alternative & Complementary Medicine, 10*(2), 369—378.

Scott, L. J., & Spencer, C. M. (2000). Miglitol. *Drugs, 59*(3), 521−549.

Scullion, S. M., Hahn, C., Tyka, K., Flatt, P. R., McClenaghan, N. H., Lenzen, S., & Gurgul-Convey, E. (2016). Improved antioxidative defence protects insulin-producing cells against homocysteine toxicity. *Chemico-Biological Interactions, 256*, 37−46.

Sebastiani, G., Ventriglia, G., Stabilini, A., Socci, C., Morsiani, C., Laurenzi, A., Nigi, L., Formichi, C., Mfarrej, B., & Petrelli, A. (2017). Regulatory T-cells from pancreatic lymphnodes of patients with type-1 diabetes express increased levels of microRNA miR-125a-5p that limits CCR2 expression. *Scientific Reports, 7*(1), 6897.

Seebacher, N., Richardson, D., & Jansson, P. (2015). Glucose modulation induces reactive oxygen species and increases P-glycoprotein-mediated multidrug resistance to chemotherapeutics. *British Journal of Pharmacology, 172*(10), 2557−2572.

Seghrouchni, I., Drai, J., Bannier, E., Rivière, J., Calmard, P., Garcia, I., Orgiazzi, J., & Revol, A. (2002). Oxidative stress parameters in type I, type II and insulin-treated type 2 diabetes mellitus; insulin treatment efficiency. *Clinica Chimica Acta, 321*(1−2), 89−96.

Sepa, A., & Ludvigsson, J. (2006). Psychological stress and the risk of diabetes-related autoimmunity: A review article. *Neuroimmunomodulation, 13*(5−6), 301−308.

Seufert, J., & Gallwitz, B. (2014). The extra-pancreatic effects of GLP-1 receptor agonists: A focus on the cardiovascular, gastrointestinal and central nervous systems. *Diabetes, Obesity and Metabolism, 16*(8), 673−688.

Sharma, S., Mandal, A., Kant, R., Jachak, S., & Jagzape, M. (2020). Is cinnamon efficacious for glycaemic control in type-2 diabetes mellitus? *Diabetes, 30*, 32.

Shaw, R. J., Lamia, K. A., Vasquez, D., Koo, S.-H., Bardeesy, N., DePinho, R. A., Montminy, M., & Cantley, L. C. (2005). The kinase LKB1 mediates glucose homeostasis in liver and therapeutic effects of metformin. *Science (New York, N.Y.), 310*(5754), 1642−1646.

Shen, Y., Honma, N., Kobayashi, K., Jia, L. N., Hosono, T., Shindo, K., Ariga, T., & Seki, T. (2014). Cinnamon extract enhances glucose uptake in 3T3-L1 adipocytes and C2C12 myocytes by inducing LKB1-AMP-activated protein kinase signaling. *PLoS One, 9*(2), e87894.

Sherr, J., & Tamborlane, W. V. (2008). Past, present, and future of insulin pump therapy: Better shot at diabetes control. *Mount Sinai Journal of Medicine: A Journal of Translational and Personalized Medicine, 75* (4), 352−361.

Shi, L., Zhu, J., Yang, P., Tang, X., Yu, W., Pan, C., Shen, M., Zhu, D., Cheng, J., & Ye, X. (2017). Comparison of exenatide and acarbose on intra-abdominal fat content in patients with obesity and type-2 diabetes: A randomized controlled trial. *Obesity Research & Clinical Practice, 11*(5), 607−615.

Sidorova, Y., Shipelin, V., Mazo, V., Zorin, S., Petrov, N., & Kochetkova, A. (2017). Hypoglycemic and hypolipidemic effect of *Vaccinium myrtillus* L. leaf and *Phaseolus vulgaris* L. seed coat extracts in diabetic rats. *Nutrition (Burbank, Los Angeles County, Calif.), 41*, 107−112.

Simons, N., Dekker, J. M., van Greevenbroek, M. M., Nijpels, G., Leen, M., van der Kallen, C. J., Schalkwijk, C. G., Schaper, N. C., Stehouwer, C. D., & Brouwers, M. C. (2016). A common gene variant in glucokinase regulatory protein interacts with glucose metabolism on diabetic dyslipidemia: The combined CODAM and hoorn studies. *Diabetes Care, 39*(10), 1811−1817.

Sobngwi, E., Boudou, P., Mauvais-Jarvis, F., Leblanc, H., Velho, G., Vexiau, P., Porcher, R., Hadjadj, S., Pratley, R., & Tataranni, P. A. (2003). Effect of a diabetic environment in utero on predisposition to type 2 diabetes. *The Lancet, 361*(9372), 1861−1865.

Soccio, R. E., Chen, E. R., & Lazar, M. A. (2014). Thiazolidinediones and the promise of insulin sensitization in type 2 diabetes. *Cell Metabolism, 20*(4), 573−591.

Sola, D., Rossi, L., Schianca, G. P. C., Maffioli, P., Bigliocca, M., Mella, R., Corlianò, F., Fra, G. P., Bartoli, E., & Derosa, G. (2015). Sulfonylureas and their use in clinical practice. *Archives of Medical Science: AMS, 11*(4), 840.

Sommavilla, B., Jørgensen, C., & Jensen, K. (2008). Safety, simplicity and convenience of a modified pre-filled insulin pen. *Expert Opinion on Pharmacotherapy, 9*(13), 2223−2232.

Soylar, P., Kadioglu, B. U., & Kilic, K. (2020). Investigation of the barriers about insulin therapy in patients with type 2 diabetes. *Nigerian Journal of Clinical Practice, 23*(1), 98.

Spiegel, R. J. (1987). The alpha interferons: Clinical overview. *Seminars in Oncology* (14, pp. 1−12).

Spreghini, N., Cianfarani, S., Spreghini, M. R., Brufani, C., Morino, G. S., Inzaghi, E., Convertino, A., Fintini, D., & Manco, M. (2019). Oral glucose effectiveness and metabolic risk in obese children and adolescents. *Acta Diabetologica, 56*(8), 955−962.

Stang, J., & Story, M. (2017). *Guidelines for adolescent nutrition services (2005)*. University of Minnesota <http://www.Epi.Umn.Edu/Let/Pubs/Adol_book.Shtm>.

Stankov, K., Benc, D., & Draskovic, D. (2013). Genetic and epigenetic factors in etiology of diabetes mellitus type 1. *Pediatrics, 132*(6), 1112−1122.

Stickel, F., & Shouval, D. (2015). Hepatotoxicity of herbal and dietary supplements: An update. *Archives of Toxicology, 89*(6), 851−865.

Stratton, I. M., Adler, A. I., Neil, H. A. W., Matthews, D. R., Manley, S. E., Cull, C. A., Hadden, D., Turner, R. C., & Holman, R. R. (2000). Association of glycaemia with macrovascular and microvascular complications of type 2 diabetes (UKPDS 35): Prospective observational study. *BMJ (Clinical Research ed.), 321*(7258), 405−412.

Summers, K. H., Szeinbach, S. L., & Lenox, S. M. (2004). Preference for insulin delivery systems among current insulin users and nonusers. *Clinical Therapeutics, 26*(9), 1498−1505.

Surugue, L. (2016). World Health Day 2016: 422 million people live with diabetes worldwide. In *International Business Times*. IBTimes. <http://webcache.googleusercontent.com/search?q = cache: PGupkpsXdrcJ:http://www.ibtimes.co.uk/world-health-day-2016-422-million-people-live-diabetes-worldwide-1553465 + &cd = 18&hl = en&ct = clnk&gl = in>.

Swanston-Flatt, S. K., Day, C., Flatt, P. R., GoULD, B. J., & Bailey, C. (1989). Glycaemic effects of traditional European plant treatments for diabetes. Studies in normal and streptozotocin diabetic mice. *Diabetes Research (Edinburgh, Scotland), 10*(2), 69−73.

Swinnen, S. G., Hoekstra, J. B., & DeVries, J. H. (2009). Insulin therapy for type 2 diabetes. *Diabetes Care, 32*(Suppl. 2), S253−S259.

Szkudelska, K., Nogowski, L., & Szkudelski, T. (2009). Resveratrol, a naturally occurring diphenolic compound, affects lipogenesis, lipolysis and the antilipolytic action of insulin in isolated rat adipocytes. *The Journal of Steroid Biochemistry and Molecular Biology, 113*(1−2), 17−24.

Szkudelski, T., & Szkudelska, K. (2015). Resveratrol and diabetes: From animal to human studies. *Biochimica et Biophysica Acta (BBA)-Molecular Basis of Disease, 1852*(6), 1145−1154.

Tan, Z., Zhou, L.-J., Mu, P.-W., Liu, S.-P., Chen, S.-J., Fu, X.-D., & Wang, T.-H. (2012). Caveolin-3 is involved in the protection of resveratrol against high-fat-diet-induced insulin resistance by promoting GLUT4 translocation to the plasma membrane in skeletal muscle of ovariectomized rats. *The Journal of Nutritional Biochemistry, 23*(12), 1716−1724.

Tanaka, S., Ohmori, M., Awata, T., Shimada, A., Murao, S., Maruyama, T., Kamoi, K., Kawasaki, E., Nakanishi, K., & Nagata, M. (2015). Diagnostic criteria for slowly progressive insulin-dependent (type 1) diabetes mellitus (SPIDDM)(2012): Report by the Committee on Slowly Progressive Insulin-Dependent (Type 1) Diabetes Mellitus of the Japan Diabetes Society. *Diabetology International, 6*(1), 1−7.

Tang, C.-H. (2011). *Systemic POMC overexpression increases visceral fat accumulation in mice*.

Tangvarasittichai, S. (2015). Oxidative stress, insulin resistance, dyslipidemia and type 2 diabetes mellitus. *World Journal of Diabetes, 6*(3), 456−480. Available from https://doi.org/10.4239/wjd.v6.i3.456.

Tangvarasittichai, S., Lertsinthai, P., Taechasubamorn, P., Veerapun, O., & Tangvarasittichai, O. (2009). Effect of moderate-intensity exercise training on body weight, serum uric acid, serum hs-CRP, and insulin sensitivity in type 2 diabetic patients. *Siriraj Hospital Gazette, 61*, 310−313.

Taskinen, M.-R. (2003). Diabetic dyslipidaemia: From basic research to clinical practice. *Diabetologia, 46* (6), 733−749.

Teixeira-Lemos, E., Nunes, S., Teixeira, F., & Reis, F. (2011). Regular physical exercise training assists in preventing type 2 diabetes development: Focus on its antioxidant and anti-inflammatory properties. *Cardiovascular Diabetology, 10*(1), 12.

Tester, R. F., & Al-Ghazzewi, F. H. (2016). Beneficial health characteristics of native and hydrolysed konjac (*Amorphophallus konjac*) glucomannan. *Journal of the Science of Food and Agriculture, 96*(10), 3283−3291.

Thrasher, J., Bhargava, A., Rees, T., Wang, T., Guzman, C., & Glass, L. (2014). Insulin lispro with continuous subcutaneous insulin infusion is safe and effective in patients with type 2 diabetes: A randomized crossover trial of insulin lispro vs insulin aspart. *Endocrine Practice, 21*(3), 247–257.

Tiedge, M., Lortz, S., Drinkgern, J., & Lenzen, S. (1997). Relation between antioxidant enzyme gene expression and antioxidative defense status of insulin-producing cells. *Diabetes, 46*(11), 1733–1742.

Tiwari, P., Ahmad, K., & Hassan Baig, M. (2017). *Gymnema sylvestre* for diabetes: From traditional herb to future's therapeutic. *Current Pharmaceutical Design, 23*(11), 1667–1676.

Tsabang, N., Ngah, N., Estella, F., & Agbor, G. (2016). Herbal medicine and treatment of diabetes in Africa: Case study in Cameroon. *Diabetes Case Reports, 1*(112), 2.

Turner, L. W., Nartey, D., Stafford, R. S., Singh, S., & Alexander, G. C. (2014). Ambulatory treatment of type 2 diabetes in the US, 1997–2012. *Diabetes Care, 37*(4), 985–992.

Ueno, T., Takeuchi, H., Kawasaki, K., & Kubo, T. (2015). Changes in the gene expression profiles of the hypopharyngeal gland of worker honeybees in association with worker behavior and hormonal factors. *PLoS One, 10*(6), e0130206.

Van De Laar, F. A., Lucassen, P. L., Akkermans, R. P., van de Lisdonk, E. H., Rutten, G. E., & van Weel, C. (2005). α-Glucosidase inhibitors for patients with type 2 diabetes: Results from a Cochrane systematic review and *meta*-analysis. *Diabetes Care, 28*(1), 154–163.

VanderJagt, D. J., Harrison, J. M., Ratliff, D. M., Hunsaker, L. A., & Vander Jagt, D. L. (2001). Oxidative stress indices in IDDM subjects with and without long-term diabetic complications1. *Clinical Biochemistry, 34*(4), 265–270.

Wang, L., Leng, J., Liu, H., Zhang, S., Wang, J., Li, W., Li, W., Li, N., Zhang, T., & Hu, G. (2017). Abstract P274: Pregnancy-induced-hypertension and the risk of postpartum hypertension among gestational diabetes women. *Circulation, 135*, AP274.

Wang, Z., Xie, Z., Lu, Q., Chang, C., & Zhou, Z. (2017). Beyond genetics: What causes type 1 diabetes. *Clinical Reviews in Allergy & Immunology, 52*(2), 273–286.

Wanner, C., Lachin, J. M., Inzucchi, S. E., Fitchett, D., Mattheus, M., George, J. T., Woerle, H.-J., Broedl, U. C., von Eynatten, M., & Zinman, B. (2017). Empagliflozin and clinical outcomes in patients with type 2 diabetes, established cardiovascular disease and chronic kidney disease. *Circulation, 137*, 119–129.

Wei, S., Li, W., Yu, Y., Yao, F., Lixiang, A., Lan, X., Guan, F., Zhang, M., & Chen, L. (2015). Ginsenoside Compound K suppresses the hepatic gluconeogenesis via activating adenosine-5′ monophosphate kinase: A study in vitro and in vivo. *Life Sciences, 139*, 8–15.

Weinzimer, S. A., & Tamborlane, W. V. (2008). Sensor-augmented pump therapy in type 1 diabetes. *Current Opinion in Endocrinology, Diabetes and Obesity, 15*(2), 118–122.

Weiss, J., Wood, A. J., Zajac, J. D., Grossmann, M., Andrikopoulos, S., & Ekinci, E. I. (2017). Diabetic ketoacidosis in acromegaly: A rare complication precipitated by corticosteroid use. *Diabetes Research and Clinical Practice, 134*, 29–37.

Wen, L., Ley, R. E., Volchkov, P. Y., Stranges, P. B., Avanesyan, L., Stonebraker, A. C., Hu, C., Wong, F. S., Szot, G. L., & Bluestone, J. A. (2008). Innate immunity and intestinal microbiota in the development of Type 1 diabetes. *Nature, 455*(7216), 1109.

World Health Organization. (1999). *Definition, diagnosis and classification of diabetes mellitus and its complications: Report of a WHO consultation. Part 1: Diagnosis and classification of diabetes mellitus.*

Wynn, P. (2017). *New diabetes products for 2017: Lancing devices and diabetes drugs* (Vol. 2018). Diabetes Self-Management. <https://www.diabetesselfmanagement.com/diabetes-resources/tools-tech/new-diabetes-products-2017-lancing-devices-diabetes-drugs/>.

Xiong, X., Saunders, L., Wang, F., & Demianczuk, N. (2001). Gestational diabetes mellitus: Prevalence, risk factors, maternal and infant outcomes. *International Journal of Gynecology & Obstetrics, 75*(3), 221–228.

Xu, G., Jing, J., Bowers, K., Liu, B., & Bao, W. (2014). Maternal diabetes and the risk of autism spectrum disorders in the offspring: A systematic review and *meta*-analysis. *Journal of Autism and Developmental Disorders, 44*(4), 766–775.

Yang, M., Ye, L., Wang, B., Gao, J., Liu, R., Hong, J., Wang, W., Gu, W., & Ning, G. (2015). Decreased miR-146 expression in peripheral blood mononuclear cells is correlated with ongoing islet autoimmunity in type 1 diabetes patients 1 型糖尿病患者外周血单个核细胞 miR-146 表达下调与胰岛持续免疫失衡相关. *Journal of Diabetes, 7*(2), 158–165.

Yaturu, S. (2011). Obesity and type 2 diabetes. *Journal of Diabetes Mellitus, 1*(04), 79.

Yew, G. Y., Cuong, N. T., Yen, H. T., Thao, P. T. H., Thao, N. T., le Thanh, N. S., Trang, N. T. H., Trung, N. T., Afridi, R., & Anh, D. T. M. (2020). Selection, purification, and evaluation of acarbose − An α–glucosidase inhibitor from *Actinoplanes* sp. *Chemosphere*, 129167.

Yin, H., Wu, H., Zhao, M., Zhang, Q., Long, H., Fu, S., & Lu, Q. (2017). Histone demethylase JMJD3 regulates CD11a expression through changes in histone H3K27 tri-methylation levels in CD4 + T cells of patients with systemic lupus erythematosus. *Oncotarget, 8*(30), 48938.

Yokoi, T. (2010). Troglitazone. *Adverse drug reactions* (pp. 419−435). Springer.

Yuan, H.-D., Kim, J. T., Kim, S. H., & Chung, S. H. (2012). Ginseng and diabetes: The evidences from in vitro, animal and human studies. *Journal of Ginseng Research, 36*(1), 27.

Zhang, W., Liu, C., Ji, L., & Wang, J., ATTEND investigators. (2020). Blood pressure and glucose control and the prevalence of albuminuria and left ventricular hypertrophy in patients with hypertension and diabetes. *The Journal of Clinical Hypertension, 22*(2), 212−220.

Zimmet, P. (2000). Globalization, coca-colonization and the chronic disease epidemic: Can the Doomsday scenario be averted? *Journal of Internal Medicine, 247*(3), 301−310.

Zimmet, P., Alberti, K. G., Magliano, D. J., & Bennett, P. H. (2016). Diabetes mellitus statistics on prevalence and mortality: Facts and fallacies. *Nature Reviews Endocrinology, 12*(10), 616.

Zisser, H. C. (2010). The OmniPod Insulin Management System: The latest innovation in insulin pump therapy. *Diabetes Therapy, 1*(1), 10−24.

Zordoky, B. N., Robertson, I. M., & Dyck, J. R. (2015). Preclinical and clinical evidence for the role of resveratrol in the treatment of cardiovascular diseases. *Biochimica et Biophysica Acta (BBA)-Molecular Basis of Disease, 1852*(6), 1155−1177.

Further reading

Allen, R. W., Schwartzman, E., Baker, W. L., Coleman, C. I., & Phung, O. J. (2013). Cinnamon use in type 2 diabetes: An updated systematic review and *meta*-analysis. *The Annals of Family Medicine, 11*(5), 452−459.

Amos, A. F., McCarty, D. J., & Zimmet, P. (1997). The rising global burden of diabetes and its complications: Estimates and projections to the year 2010. *Diabetic Medicine, 14*(S5).

American Diabetes Association. (2017). 2. Classification and diagnosis of diabetes. *Diabetes Care, 40*(Suppl. 1), S11−S24.

Bays, H. (2013). Sodium glucose co-transporter type 2 (SGLT2) inhibitors: Targeting the kidney to improve glycemic control in diabetes mellitus. *Diabetes Therapy, 4*(2), 195−220.

Buzzetti, R., Tuomi, T., Mauricio, D., Pietropaolo, M., Zhou, Z., Pozzilli, P., & Leslie, R. D. (2020). Management of latent autoimmune diabetes in adults: A consensus statement from an international expert panel. *Diabetes, 69*(10), 2037−2047.

Chen, C. H., & Lu, T. K. (2020). Development and challenges of antimicrobial peptides for therapeutic applications. *Antibiotics, 9*(1), 24.

Gong, J., Hu, M., Huang, Z., Fang, K., Wang, D., Chen, Q., Li, J., Yang, D., Zou, X., & Xu, L. (2017). Berberine attenuates intestinal mucosal barrier dysfunction in type 2 diabetic rats. *Frontiers in Pharmacology, 8*, 42.

Harris, K., & Nealy, K. L. (2018). The clinical use of a fixed-dose combination of insulin degludec and liraglutide (Xultophy 100/3.6) for the treatment of type 2 diabetes. *Annals of Pharmacotherapy, 52*(1), 69−77.

Hikino, H., Takahashi, M., Otake, K., & Konno, C. (1986). Antidiabetes drugs. 20. Isolation and hypoglycemic activity of Eleutheran-A, Eleutheran-B, Eleutheran-C, Eleutheran-D, Eleutheran-E, Eleutheran-F, and Eleutheran-G-Glycans of Eleutherococcus-senticosus roots. *Journal of Natural Products, 49*(2), 293−297.

Kim, C. (2010). *Gestational diabetes during and after pregnancy*. Springer.

King, H., Aubert, R. E., & Herman, W. H. (1998). Global burden of diabetes, 1995−2025: Prevalence, numerical estimates, and projections. *Diabetes Care, 21*(9), 1414−1431.

Lane, W., Weinrib, S., & Rappaport, J. (2011). The effect of liraglutide added to U-500 insulin in patients with type 2 diabetes and high insulin requirements. *Diabetes Technology & Therapeutics, 13*(5), 592−595.

Li, X.-Z., & Zhang, S. (2016). Effervescent granules prepared using *Eucommia ulmoides* Oliv. and moso bamboo leaves: Hypoglycemic activity in HepG2 cells. *Evidence-Based Complementary and Alternative Medicine, 2016.*

Mehta, N., Stenholm, S., Männistö, S., Jousilahti, P., & Elo, I. (2020). Excess body weight, cigarette smoking, and type II diabetes incidence in the national FINRISK studies. *Annals of Epidemiology, 42,* 12−18.

Mirfeizi, M., Mirfeizi, S., MehdizadehTourzani, Z., & AsghariJafarabadi, M. (2015). Controlling diabetes mellitus type 2 with herbal medicines: A triple blind, randomized clinical trial of efficacy and safety. *Journal of Diabetes* (17, pp. A49−A50).

Nathan, D. M., Axelrod, L., Flier, J. S., & Carr, D. B. (1981). U-500 insulin in the treatment of antibody-mediated insulin resistance. *Annals of Internal Medicine, 94*(5), 653−656.

Prabhakar, P. K., & Doble, M. (2011). Effect of natural products on commercial oral antidiabetic drugs in enhancing 2-deoxyglucose uptake by 3T3-L1 adipocytes. *Therapeutic Advances in Endocrinology and Metabolism, 2*(3), 103−114.

Sirisidthi, K., Kosai, P., & Jiraungkoorskul, W. (2016). Antidiabetic activity of the lingzhi or reishi medicinal mushroom Ganoderma lucidum. *SA Pharmaceutical Journal, 83*(8), 45−47.

Tabeshpour, J., Imenshahidi, M., & Hosseinzadeh, H. (2017). A review of the effects of *Berberis vulgaris* and its major component, berberine, in metabolic syndrome. *Iranian Journal of Basic Medical Sciences, 20*(5), 557.

Turner, N., Li, J.-Y., Gosby, A., To, S. W., Cheng, Z., Miyoshi, H., Taketo, M. M., Cooney, G. J., Kraegen, E. W., & James, D. E. (2008). Berberine and its more biologically available derivative, dihydroberberine, inhibit mitochondrial respiratory complex I: A mechanism for the action of berberine to activate AMP-activated protein kinase and improve insulin action. *Diabetes, 57*(5), 1414−1418.

Yan, Z., Leng, S., Lu, F., Lu, X., Dong, H., & Gao, Z. (2008). Effects of berberine on expression of hepatocyte nuclear factor 4alpha and glucokinase activity in mouse primary hepatocytes. *Zhongguo Zhong Yao Za Zhi = Zhongguo Zhongyao Zazhi = China Journal of Chinese Materia Medica, 33*(18), 2105−2109.

Yin, J., Zhang, H., & Ye, J. (2008). Traditional Chinese medicine in treatment of metabolic syndrome. *Endocrine, Metabolic & Immune Disorders-Drug Targets (Formerly Current Drug Targets-Immune, Endocrine & Metabolic Disorders), 8*(2), 99−111.

Zhang, H., Wei, J., Xue, R., Wu, J.-D., Zhao, W., Wang, Z.-Z., Wang, S.-K., Zhou, Z.-X., Song, D.-Q., & Wang, Y.-M. (2010). Berberine lowers blood glucose in type 2 diabetes mellitus patients through increasing insulin receptor expression. *Metabolism: Clinical and Experimental, 59*(2), 285−292.

Zimmet, P. Z. (1995). The pathogenesis and prevention of diabetes in adults: Genes, autoimmunity, and demography. *Diabetes Care, 18*(7), 1050−1064.

CHAPTER 2

Nanoparticles and nanocomposites for controlled delivery of insulin

Introduction

Nanotechnology has turned out to be a significant entity in advancing several applications ranging from space science to health care (Parhizkar et al., 2018). Recent advancements in electron microscopic capabilities to view, analyze, and characterize materials has facilitated more understanding and application of nanotechnology. Nanomaterials are defined as materials with internal or external structures in nanoscale dimensions, whereas nanoparticles are three-dimensional nanoscale objects with external nanoscale dimensions (Jeevanandam et al., 2018). In other words, nanomaterials include both nanoparticles and nanoformulated particles. Nanosized materials and particles are usually below 100 nm size with unique dimensional properties. They possess extraordinarily large surface-to-volume ratios, which enable them to possess unique properties for desired applications (Kreuter, 2014). Currently, nanotechnologies are utilized in applications such as renewable energies (Hussein, 2015), oil and gas industries (Cheraghian & Hendraningrat, 2016; Peng et al., 2018), solar collectors (Hussein, 2016), wastewater treatment (Mahadik, 2017), agriculture (Singh et al., 2015), food industry (Dasgupta & Ranjan, 2018), packaging (Ramachandraiah et al., 2015), and textiles (Asif & Hasan, 2018). Since nanotechnology has shown diversity in application and enhanced bioavailability, it has been utilized in biomedical applications and health care products.

Nanoparticles and nanomaterials have boosted bioavailability (Tafazoli et al., 2017) and biocompatibility (Pisani et al., 2017), and green synthesized nanoparticles possess less toxicity compared to chemically synthesized nanoparticles (Jeevanandam et al., 2016). Further, they also possess antimicrobial properties, which help to reduce microbial infections and diseases (Kale et al., 2017; Rongione et al., 2017). They have been included as diagnostic (Bagheri et al., 2018), curative (Wang, McGuirk, et al., 2017), and therapeutic (Zhang et al., 2015) agents against several diseases. Nanotechnology has been successfully used in imaging (Bhandare & Narayana, 2014), treatment (Roeth et al., 2017), and diagnosis (Hou et al., 2017) during cancer therapy (Jeevanandam et al., 2020; Sarkar, 2018). Biomedical applications of nanotechnology has now been expanded to include other diseases. It has been used in the treatment and detection of diseases such as diabetes (Ashrafizadeh et al., 2020; Gupta, 2017), Alzheimer's, Parkinson (Soursou et al., 2015), neurodegenerative diseases (Pezzini et al., 2017), endodontics (Kishen, 2016),

Emerging Nanomedicines for Diabetes Mellitus Theranostics
DOI: https://doi.org/10.1016/B978-0-323-85396-5.00009-9

prosthodontics (Gopinadh et al., 2015), dermatology, cosmetics (Arif et al., 2015), plastic and reconstructive surgery (Petersen et al., 2014), infectious, inflammatory diseases (Ikoba et al., 2015), lung infections (Kuzmov & Minko, 2015), and periodontal diseases (Zupancic et al., 2015). This chapter seeks to discuss advances in the types of nanoparticles and nanocomposites used for controlled delivery of insulin to treat diabetes. Additionally, the mechanism of action of nanosized encapsulations in delivering insulin towards targeted diabetic cells is discussed.

Classification of nanoparticles

Generally, nanosized particles and materials are classified based on the materials utilized for synthesis and also dimensions. Other classifications are based on the unique properties of the nanoparticles.

Dimension-based classification

Nanosized particles and materials are classified into four categories; zero, one, two, and three-dimensional materials, based on dimensions (Saleh, 2020). In three-dimensional nanoparticles such as nanocrystals (Wu et al., 2017), electrons can travel in three dimensions, i.e. X, Y, and Z, whereas electrons travel in X and Y directions in two dimensional nanoparticles such as graphene (Politano et al., 2017) and nanosheets (Kong et al., 2017). Nanomaterials in which electrons travel only in X direction are called one dimensional nanoparticles (Ji et al., 2018), while zero dimensional nanoparticles are electronically confined in all the directions (Li et al., 2018). Examples of one dimensional nanoparticles are carbon nanotubes (Yamamoto & Fukuyama, 2018) and nanofibers (Ji et al., 2018), and zero dimensional nanoparticles are mainly quantum dots (Zeng et al., 2018).

Material-based classification

Nanosized materials and particles are also classified based on precursors used for synthesis and the resultant material formed. They are further subclassified into primary, secondary, and tertiary materials.

Primary nanomaterials: Primary nanomaterials are synthesized using one precursor and the resultant nanomaterial formed mostly has a unit composition. Examples of primary nanomaterials are metal nanoparticles (Iravani et al., 2018) and carbon nanoparticles (Martin-Gallego et al., 2018).

Secondary nanomaterials: Secondary nanomaterials are categorized under materials fabricated using two precursors and the resultant nanomaterial formed mostly has two compositions. Metal oxide nanoparticles are common examples of secondary nanomaterials (Almansoori et al., 2018). Other materials such as functionalized nanomaterials (Wang, Li, et al., 2017) and nanoceramics (Yue et al., 2018) also fall under this category.

Tertiary nanomaterials: Nanomaterials that are fabricated using three or more precursors and the resultant material has a multiple composition of materials are described as tertiary nanomaterials. $LiNi_{0.5}Mn_1.5O_4$ nanoparticles are an examples of tertiary nanomaterials (Jo et al., 2010). Nanocomposites are also grouped under tertiary material category as it involves multiple precursors (Baig et al., 2018).

Nano-doped materials: Certain nanomaterials are grouped under nano-doped materials where a nanomaterial is doped with, for example, an ion or particle to increase its electrical conductivity. Nano-doped materials can be categorized under both secondary and tertiary nanomaterials (Acquah et al., 2017).

Nanocomposites

Nanocomposites are fabricated when two or more materials with distinct properties are combined together to produce a new nanomaterial with characteristics that are unique and different from the individual materials (Zhao et al., 2015). Nanocomposites are considered to be the future of nanomaterial applications as it can be used to alter and enhance the property of nanomaterials (Kim et al., 2014). Also, it can be used to fabricate novel nanomaterials by incorporating the unique properties of individual materials, leading to the formation of nanocomposite material with superior properties (Hai et al., 2016). Generally, nanocomposites are classified as magnetic and nonmagnetic.

Magnetic nanocomposites

Nanocomposites that are fabricated using magnetic or superparamagnetic iron oxide nanoparticles are magnetic nanocomposites (Ghanbari et al., 2014). These magnetic nanocomposites are further subclassified into metals (Atar et al., 2015), metal oxides (Ojha et al., 2017), carbon (Qiu et al., 2015), and polymer (Park et al., 2016) based magnetic nanocomposites depending on the type of material combined with the iron oxides to form a composite material. Silver-iron oxide (Pant et al., 2017), titanium dioxide-iron oxide (Yusoff et al., 2016), carbon nanotube-iron oxide (Engel et al., 2018), and cellulose-iron oxide nanocomposites (Sadasivuni et al., 2016) are few important examples of magnetic nanocomposites.

Nonmagnetic nanocomposites

Nonmagnetic nanocomposites are synthesized without the incorporation of iron oxides. These nonmagnetic nanocomposites are also subclassified into metal-metal (Yang et al., 2014), metal oxide—metal (Kokate et al., 2016), metal oxide-metal oxide (Ahmad et al., 2016), carbon-metal/metal oxide (Raghavan et al., 2015) and polymer-metal/metal oxide (Al-Naamani et al., 2017) nanocomposites. Gold-copper, silver-zinc oxide, zinc oxide-silver-manganese oxide (Saravanan et al., 2015), graphene oxide-gold (Zheng et al., 2015), carbon nanotube-copper oxide (Chen et al., 2016), cellulose-silver (X. Li et al., 2018), and zinc oxide-starch (Vigneshwaran et al., 2006) nanocomposites are some examples of nonmagnetic nanocomposites.

Nanoparticles and nanomaterials for controlled delivery of insulin

Carbon nanotubes, quantum dots, microspheres, and nanoencapsulations have been used for controlled and targeted delivery of insulin (see Table 2.1). These nanosized particles and materials have gained importance as insulin controlled delivery platforms to address some of the challenges of conventional insulin therapy (Dyer et al., 2002; Sabu & Pramod, 2020).

Carbon nanotubes

Graphene layers that are rolled into one or more layers of seamless cylinders with open or closed ends are called as carbon nanotubes (CNT) (Odom et al., 2002; Sun et al., 2015). The hollow portions of CNTs have been exploited for drug loading and delivery applications (Tekade et al., 2017). CNTs have also been used as sensors for glucose (Barone et al., 2005) and insulin (Bisker et al., 2018) monitoring in blood to regulate their levels in diabetic patients. Insulin molecules have been conjugated onto the sides of single walled CNT via functionalization and activated amidation of diimides to create noninvasive insulin delivery systems (Ng et al., 2016). Also, CNTs incorporated with hydrogel needle-free devices have been developed for electro-stimulated skin delivery of larger or hydrophilic molecules such as insulin (Guillet et al., 2017). Several single walled and multi walled CNTs have been employed and are extensively researched as efficient drug delivery systems to diabetes and other disease. CNTs have demonstrated promise for developing efficient glucose sensors and insulin delivery systems if they are made to possess less cytotoxicity and high biocompatibility towards human cells (Chauhan et al., 2020).

Quantum dots

Quantum dots are typically 2−20 nm sized crystals that are zero dimensional and are electronically confined in all directions (Hassan et al., 2018). These materials are widely used in biomedical applications, especially in fabricating drug delivery systems (Chen et al., 2016). The void space in quantum dots has the potential to confine drug molecules, giving it a high drug loading capability (Dong et al., 2018). Quantum dots have also been used as glucose sensors to regulate blood glucose levels in diabetes patients (Cao et al., 2008; Cash & Clark, 2010; Tang et al., 2008). Nickel oxide quantum dots modified with zinc oxide nanorods have been used to fabricate a nonenzymatic flexible field-effect transistor based glucose sensor (Ahmad et al., 2016). Giudice et al. (2013) investigated that biotinylated ligands with streptavidin-conjugated quantum dots can

Table 2.1 Nanoparticles and nanocomposites for controlled insulin delivery.

Carbon nanotubes

Nanomaterials	Diabetes application	Reference
Single wall carbon nanotube	Glucose detection	Barone et al. (2005)
Single wall carbon nanotube	Insulin detection	Bisker et al. (2018)
Insulin conjugated single wall carbon nanotube	Noninvasive insulin delivery systems	Ng et al. (2016)
Needle-free hydrogels with carbon nanotube	Electro-stimulated skin delivery of insulin	Guillet et al. (2017)

Quantum dots

Quantum dots	Glucose sensors	Cao et al. (2008), Tang et al. (2008), Cash and Clark (2010)
Nickel oxide quantum dot modified with zinc oxide nanorods	Nonenzymatic flexible field-effect transistor-based glucose sensor	Ahmad and Hahn (2018)
Biotinylated ligands with streptavidin-conjugated quantum dots	Dissociation of insulin-insulin receptor complex and efficient glucose uptake	Giudice et al. (2013)

Microspheres

Eudragit L100 and S100	Enteral insulin delivery	Morishita et al. (1993)
Polyphosphazene	Enhanced insulin delivery	Caliceti et al. (2000)
PLGA/PEG	Enhanced insulin delivery	Yuksel et al. (2000)
Eudragit	Enhanced insulin delivery	Mundargi et al. (2011)
PLGA	Enhanced insulin delivery	Aguiar et al. (2004)
Chitosan	Enhanced insulin delivery	Wang, Gu, et al. (2006)
Gelatin	Enhanced insulin delivery	Wang, Tabata, et al. (2006)
Biodegradable salicylate-based poly (anhydride-ester)	Mucoadhesive insulin delivery	Delgado-Rivera et al. (2014)
Chitosan and polyvinyl alcohol	Mucoadhesive insulin delivery	Zadeh et al. (2017)
Glucose-responsive core-shell with photoactive shell	Bolus release of insulin	Bai et al. (2018)
Biodegradable salicylic acid-based poly (anhydride ester) (SAPAE) polymer injectable microsphere	Extended insulin delivery	Yu et al. (2015)
Insulin-loaded, porous poly-L-lactide microspheres	Inhalable pulmonary insulin delivery	Chen et al. (2015)

Nanoencapsulation

Nanolayer encapsulation of insulin-chitosan complexes	Oral insulin delivery	Song et al. (2014)
Glucose-responsive microgels	Targeted insulin delivery	Anderson et al. (2016)
Dendrimer encapsulated insulin formulations	Controlled and targeted insulin deliveryi	Dong et al. (2011), NOSE-TO-BRAIN (2018)

(Continued)

Table 2.1 (Continued)

Carbon nanotubes

Nanomaterials	Diabetes application	Reference
Liposomes	Controlled insulin delivery	Chono et al. (2009), Al-Remawi et al. (2017)
Micelles	Controlled insulin delivery	Alai et al. (2015), Wen et al. (2017)

Other nanoparticles

Gold nanoparticles	Transmucosal insulin delivery	Joshi et al. (2006)
Alginate/chitosan nanoparticles loaded with insulin	Insulin delivery system	Sarmento, Ribeiro, et al. (2007)
Mucoadhesive and permeable thiolated trimethyl chitosan nanoparticles	Oral insulin delivery	Yin et al. (2009)
Nanoparticles composed of chitosan and poly (γ-glutamic acid)	Oral insulin delivery	Lin et al. (2007)
Polymeric nanoparticles	Enhanced insulin delivery	Chen et al. (2011), Fonte et al. (2015)
Solid lipid nanoparticles	Oral insulin delivery	Sarmento, Martins, et al. (2007)
Polyurethane—alginate nanoparticles	Oral insulin delivery	Bhattacharyya et al. (2016)
Calcium phosphate nanoparticles	Oral insulin delivery	Ramachandran et al. (2009)
Mesoporous silica nanoparticles	Controlled insulin delivery	Zhao et al. (2009)
Dextran nanoparticles	Controlled insulin delivery	Gu et al. (2013)

Iron oxide nanocomposites

Insulin-loaded iron oxide-chitosan nanocomposite	Oral insulin delivery	Kebede et al. (2013)
Hybrid nanogels of chitosan-based luminescent/magnetic nanocomposites	Higher insulin loading capacity and insulin sensitivity	Shen et al. (2012)
Insulin loaded iron magnetic nanoparticle—graphene oxide composites	Protection of insulin molecules in biological fluids and enhanced insulin delivery	Turcheniuk et al. (2014)
Iron oxide—polymer nanocomposites	Controlled insulin delivery	Luo et al. (2016)
Iron oxide nanocomposites	Glucose and insulin sensors	Ahmad et al. (2017), Baghayeri et al. (2018), Nandwana et al. (2018)

(Continued)

Table 2.1 (Continued)

Carbon nanotubes

Nanomaterials	Diabetes application	Reference
Carbon-based nanocomposites		
Insulin-impregnated reduced graphene oxide modified electrode	Electrochemical release of human insulin	Teodorescu et al. (2015)
Polyolefin/graphene nanocomposites	Oral delivery of insulin	Zhang et al. (2012)
Insulin impregnated reduced graphene oxide—nickel hydroxide thin films nanocomposites	Glucose sensor and electrochemical release of insulin	Belkhalfa et al. (2016)
Reduced graphene oxide—iron pentacarbonyl nanocomposites-based hydrogels	Improved release of on-demand insulin	Teodorescu et al. (2017)
Pencil graphite electrode modified ruthenium oxide—graphene oxide	Micro-extraction of insulin	Ensafi et al. (2015)
Graphene/NiO	Glucose sensor	Li et al. (2014)
Silver nanoflower-reduced graphene oxide composite	Glucose sensor	Yagati, Choi, et al. (2016)
CO_3O_4-reduced graphene oxide nanocomposite	Glucose sensor	Xie et al. (2013)
Nickel hydroxide-graphene	Insulin sensor	Lin et al. (2014)
Graphene oxide—copper oxide	Insulin sensor	Song et al. (2013)
Reduced graphene oxide	Insulin sensor	Yagati, Park, et al. (2016)
Noniron oxide based nanocomposites		
Mesoporous silica/gold nanocomposites	Insulin detection	Xiong et al. (2015)
ZnO nanorods-gold	Glucose sensor	Wei et al. (2010)
Zinc oxide-chitosan-graft-poly (vinyl alcohol) core shell	Glucose sensor	Shukla et al. (2012)
ZnO-CuO hierarchical structure	Glucose sensor	Zhou et al. (2014)
Platinum-copper oxide	Glucose sensor	Dhara et al. (2014)
Copper oxide—titanium dioxide	Glucose sensor	Chen et al. (2012)
Titanium dioxide—cellulose	Glucose sensor	Maniruzzaman et al. (2012)
Platinum gold—manganese dioxide	Glucose sensor	Xiao et al. (2013)
Chitosan—zinc oxide	Insulin delivery	El-Mekawy and Jassas (2017)
Polygalacturonic acid (PGLA) hypoglycemic organic polymer	Dual-functional oral insulin carrier and delivery system	Zhang et al. (2017)

help promote endocytosis and intracellular dissociation of human insulin-insulin receptor complex in live cells to regulate the degradation of insulin and cell surface receptors, leading to improved glucose uptake and reduce blood glucose level (Giudice et al., 2013). Carbon (Pardo et al., 2018), graphene (Dong et al., 2018), proton-resistant (Mohs et al., 2009), cadmium selenide (Selvan et al., 2005), and zinc selenide (Liu et al., 2007) quantum dots have also been explored for drug delivery applications, and thus can be used as insulin delivery platforms (Wang, Wang, et al., 2020).

Microspheres and nanoencapsulations

Spherical shaped particles with diameter in the range of 1 to 1000 μm are called microspheres. They can be solid, hollow, differ widely in density, and are used in various biomedical applications (Guiot, 2018). Microsphere formulation of synthetic insulin is currently gaining enormous research attention due to their versatility and ability for controlled delivery of insulin (I. Y. Kim et al., 2014). Morishta et al. 1993 investigated three different microsphere formulations of eudragit L100 and S100 for delivery of insulin (Morishita et al., 1993). Later, various microsphere formulations of insulin such as polyphosphazene (Caliceti et al., 2000), PLGA/PEG (Yuksel et al., 2000), eudragit (Mundargi et al., 2011), PLGA (Aguiar et al., 2004), chitosan (L.-Y. Wang, Gu, et al., 2006) and gelatin microspheres (Wang, Zhou, et al., 2020) were developed for enhanced delivery of insulin to diabetic cells. More recently, biodegradable microspheres using salicylate-based poly (anhydride-ester) (Delgado-Rivera et al., 2014), chitosan and polyvinyl alcohol as mucoadhesive microsphere (Zadeh et al., 2017), glucose-responsive core-shell microspheres with photo-active shell for regulated basal and bolus release of insulin (Bai et al., 2018), biodegradable salicylic acid-based poly (anhydride ester) (SAPAE) polymer injectable microspheres for the extended delivery of insulin (Yu et al., 2015) and insulin-loaded, porous poly-L-lactide microspheres for inhalable pulmonary insulin delivery (Chen et al., 2015) have been designed and fabricated for enhanced delivery of insulin. Also, insulin encapsulated into nano-sized polymers are gaining research attention in recent times. These nanoencapsulations possess enhanced ability to cross the blood brain barrier and deliver drug ingredients to targeted cells or organs (Inchaurraga et al., 2020; Kumari et al., 2014). Nanolayer encapsulation of insulin-chitosan complexes have been demonstrated to improve the efficiency of oral insulin delivery (Song et al., 2014). Also, glucose-responsive microgels (Anderson et al., 2016) and dendrimer encapsulated insulin formulations (J. Dong et al., 2018) have been developed and demonstrated to increase drug loading ability for controlled and targeted delivery of insulin. Liposomes (Al-Remawi et al., 2017; Chono et al., 2009) and micelles (Alai et al., 2015; Wen et al., 2017) have also been explored for use in controlled delivery of insulin. Thus, microspheres and nanoencapsulations hold a great promise for encapsulating raw, synthetic, and high concentrations of insulin for use in insulin pumps and pen therapy.

Other nanoparticles for insulin delivery

Apart from nanoformulated insulin, some unique nanoparticles have been used for controlled delivery of insulin. Gold nanoparticles have been used as an efficient carrier for transmucosal delivery of insulin (Joshi et al., 2006). Alginate/chitosan nanoparticles have been loaded with insulin and investigated as an improved insulin delivery system (Sarmento, Martins, et al., 2007). The results in diabetic rats showed that the nanoparticles were effective for oral insulin delivery applications. Mucoadhesive and permeable thiolated trimethyl chitosan nanoparticles (Yin et al., 2009) and nanoparticles composed of chitosan and poly (γ-glutamic acid) (Lin et al., 2007) have also been used for oral insulin delivery. Several polymeric nanoparticles have also been studied for enhanced delivery of insulin (Chen et al., 2011; Fonte et al., 2015). Solid lipid nanoparticles (Sarmento, Martins, et al., 2007), polyurethane—alginate nanoparticles (Bhattacharyya et al., 2016), and calcium phosphate nanoparticles (Ramachandran et al., 2009), mesoporous silica nanoparticles (Z. Zhao et al., 2015) and dextran nanoparticles (Gu et al., 2013) are examples of unique nanoparticles that have been used for controlled and targeted insulin delivery. Nanoparticles, nanomaterials, nanoencapsulations, and nanoformulated insulin have shown to be useful in type 1 diabetes treatment. They also demonstrate the ability to reduce insulin resistance complications in certain cases of type 2 diabetes (Li et al., 2019; Mohseni et al., 2019).

Nanocomposites in controlled insulin delivery

There are two types of nanocomposites (iron oxide-based and noniron oxide based) that have been explored for diabetes detection and treatment. Iron oxide and carbon-based nanocomposites are widely used as glucose and insulin sensors whereas noniron oxide nanocomposites support insulin delivery to diabetic cells.

Iron oxide and carbon-based nanocomposites

Kebede et al. (2013) synthesized shape and size-controlled iron oxide nanoparticles via a laser ablation technique and fabricated them into an insulin-loaded iron oxide-chitosan nanocomposite. The nanocomposite was investigated for oral delivery of insulin in type 2 diabetes management. The results showed that spherical nanocomposite loaded with insulin showed 51% blood glucose reduction in diabetic rats compared to the control group (Kebede et al., 2013). Also, a direct gelation method has been used to fabricate hybrid nanogels of chitosan-based luminescent/magnetic nanocomposites by using cadmium telluride, chitosan, and superparamagnetic iron oxides. The result showed that the nanocomposite possessed a high insulin loading capacity and improved insulin sensitivity in human normal hepatocyte L02 cell lines (Shen et al., 2012). Insulin loaded iron magnetic nanoparticle—graphene oxide composites were fabricated by Turcheniuk et al. (2014) and showed a high drug loading capacity. The researcher demonstrated that the nanocomposites supported the protection of insulin in biological fluids under in vivo conditions (Turcheniuk et al., 2014). In addition to the application of iron oxide-polymer nanocomposites for controlled delivery of insulin, different types of iron oxide nanocomposites have been investigated and employed as glucose and insulin sensors to control blood glucose levels (Ahmad et al., 2017; Baghayeri et al., 2018; Nandwana et al., 2018).

Carbon-based nanocomposites have also been investigated for enhanced delivery of insulin. An insulin-impregnated reduced graphene oxide modified electrode has been fabricated by Teodorescu et al. (2015) for electrochemical release of human insulin, and this supported the release of 70 $\pm$ 4% of human insulin into physiological medium (Teodorescu et al., 2015). Also, polyolefin/graphene nanocomposite has been used as a biodegradable thermo-sensitive, pH-dependent hydrogels for the oral delivery of insulin and the results showed that the composite possesses better insulin release profile (F. Zhang et al., 2015). More recently, insulin impregnated reduced graphene oxide−nickel hydroxide thin film nanocomposite has been developed and demonstrated to possess glucose sensing property and promote electrochemical release of insulin (Belkhalfa et al., 2016). Hydrogels synthesized using reduced graphene oxide−iron pentacarbonyl nanocomposites have been found to show an efficient on-demand insulin release ability via a photothermal triggering approach (Teodorescu et al., 2017). Similar to iron oxide nanocomposites, carbon based nanocomposites have been used as biosensors to detect blood glucose (Li et al., 2014; Xie et al., 2013; Yagati, Park, et al., 2016) and insulin (Song et al., 2014) (Y levels). With the enormous potential of iron oxide and carbon-based nanocomposites for diabetes theranostics, the unique cytotoxicity features of the nanocomposites need to be addressed to facilitate wide-scale diabetes therapy.

Noniron oxide-based nanocomposites

Noniron oxide nanocomposites are nano-sized composites materials synthesized using other metal or metal oxides. These nonmagnetic nanocomposites have also been used as biosensors for glucose and insulin monitoring applications. In 2015, Xiong et al. reported the synthesis of an aptamer-immobilized magnetic mesoporous silica/gold nanocomposites for effective detection of insulin (Xiong et al., 2015). Also, ZnO nanorods-gold (Wei et al., 2010), zinc oxide-chitosan-graft-poly (vinyl alcohol) core shell (Shukla et al., 2012), ZnO-CuO hierarchical structure (Zhou et al., 2014), platinum-copper oxide (Dhara et al., 2014), copper oxide−titanium dioxide (J. Chen et al., 2012), titanium dioxide−cellulose (Maniruzzaman et al., 2012) and platinum gold−manganese dioxide (Xiao et al., 2013) nanocomposites have also been explored for glucose sensing applications. These nanocomposites, in one form or others, have also been explored for insulin delivery applications. Chitosan–zinc oxide nanocomposite hydrogel has been fabricated as a flexible and smart cross-linked three-dimensional polymer for insulin delivery (El-Mekawy & Jassas, 2017). Also, polygalacturonic acid (PGLA) hypoglycemic organic polymer nanocomposite has been used as an insulin carrier to fabricate a dual-functional oral insulin delivery system and showed efficient insulin loading rate, release mode, thermostability and good in vitro and in vivo absorption in type 1 diabetic rats (L. Zhang et al., 2017). It is important to mention that only a few reported work is available on the application of noniron oxide nanocomposites for insulin delivery, Thus, this represents a huge research gap to promote the application of less toxic and noniron oxide nanocomposites for insulin delivery.

Route of administration for nanomaterial-based insulin delivery

Insulin can be delivered conventionally via subcutaneous, intravenous, and personal insulin pumps as well as mucosal and oral routes (Shah et al., 2016). In general, nanomaterials are used to deliver insulin via oral, transdermal, and mucosal routes. Hu and Luo (2018) outlined several polysaccharide-based nanoparticles, such as alginate, glucan, chitosan, and glucan nanoparticles for oral insulin delivery to overcome multiple barriers in the gastrointestinal tract (Hu & Luo, 2018). Sgorla et al. (2018) synthesized lipid-polymeric nanoparticles via a modified solvent emulsification-evaporation approach using hydroxypropylmethylcellulose acetate succinate and ethyl palmitate as a matrix. The resultant nanoparticles were 300 nm in size with a negative surface charge of -20 mV and 80% insulin encapsulation efficiency for oral insulin delivery. In vitro analysis revealed that the lipid nanoparticles were able to release 9 and 14% of insulin at pH 1.2 in 2 h and at pH 6.8 in 6 h respectively under stimulated gastrointestinal conditions without any cytotoxic effects towards intestinal epithelial cells (Sgorla et al., 2018). Zhang et al. (2017) incorporated insulin molecules into the 58–67 nm sized nanoparticulate backbone of polygalacturonic acid (PGLA) polymer with genipin and chitosan as crosslinker for oral insulin delivery applications. The insulin encapsulation ability of the nanoparticle was higher, compared to other insulin delivery systems, to reduce hyperglycemic effects, with 4.4% bioavailability and less toxicity towards intestinal epithelial cells (L. Zhang et al., 2017). Liu et al. (2016) reported the formation of self-assembled lecithin-chitosan nanoparticles as a potential oral delivery system for insulin. The insulin loaded with phospholipid nanoparticles was transformed into complexes with 180 nm size, 94% of insulin encapsulation efficiency, 4.5% insulin loading efficiency, and 6.01% bioavailability in diabetic rats to reduce hyperglycemic condition via oral administration (Y.-S. Liu et al., 2007).

In recent times, some nanoparticles have been identified to have the potential to deliver insulin via transdermal route. Jiang et al. (2019) fabricated polymer microneedles and integrated them with mesoporous bioactive glass nanoparticles that are responsive to glucose for effective transdermal insulin delivery. In this study, the insulin was loaded on the surface of the mesoporous nanoparticles via a pre-modification approach using a composite enzyme layer, which consists of glucose oxidase, polyethyleneimine, and catalase. The results showed that the pH-triggered transdermal delivery of insulin via mesoporous nanoparticles integrated with the microneedles lowered hypoglycemia risk in diabetic rats, compared to subcutaneous or intravenous insulin delivery (Jiang et al., 2019). Zhang et al. (2020) recently prepared a polymeric nanocarrier made up of carboxymethyl chitosan combined with a microneedle therapy system for improved transdermal insulin delivery. The study showed that the resultant nanocarriers were spherical in shape, 200 nm in size with insulin encapsulation ability of 83.78 $\pm$ 3.73% and penetrated the stratum corneum for transdermal insulin delivery at a passive diffusion rate of 2.77 $\pm$ 0.64 µg (cm^{-2} h^{-1}). The microneedle with the insulin-loaded nanoparticles showed 4.twofold transdermal rate in the in vitro permeation experiment, and the in vivo hypoglycemic experiment demonstrated the potential to be a painless transdermal insulin delivery (P. Zhang et al., 2020). Xu et al. (2017) synthesized mesoporous silica nanoparticles that are sensitive to hydrogen peroxide by integrating them with microneedle patches for glucose-monitored transdermal insulin delivery. In this work, the nanoparticles were fabricated by modifying 4- (imidazoyl carbamate) phenylboronic acid pinacol ester with a host-guest complexation between alpha-cyclodextrin. The study revealed that the transdermal injection of the insulin-integrated nanoparticles via microneedles into diabetic rats provided an increased hypoglycemic effect over time compared to the subcutaneous injection (Xu et al., 2017). He et al. (2018) utilized a sequential flash nano-complexation approach for scalable production of core-shell nanoparticles to enhance the mucosal transport of oral insulin delivery. In this study, L-pentratin, which is a cell-penetrating peptide that can act as a trans-epithelial transport enhancer, is complexed with insulin to form nanoparticles as the core and hyaluronic acid as the shell. The resultant core-shell nanostructures were identified to possess 97% high insulin encapsulation efficiency and 67% payload capacity. Further, the study showed that the insulin nanoparticle complex enhanced permeation via intestinal mucosal layer and elevated trans-epithelial insulin absorption. The study also emphasized that the nanocomplex insulin is effective in reducing the glucose level in type 1 diabetic rat models upon a single oral administration with 11% bioavailability compared to subcutaneous delivery of free-insulin forms (He et al., 2018). Wei et al. (2020) demonstrated the internalization mechanism of a phenylboronic acid-decorated nanoplatform to enhance the nasal delivery of insulin. In this study, phenylboronic acid-functionalized with dextran nanoplatforms were used to improve the permeability of insulin and boost its penetration. The study also identified that the endocytosis mechanism of the nanoplatforms involved clathrin- and lipid raft or caveolae-dependent endocytic pathways (Wei et al., 2010).

Mechanism of nanomaterial-based insulin delivery

There are two main approaches-involved in the controlled delivery of insulin into targeted cells using nanomaterials. The first approach covers drug loading of insulin into a hollow nanomaterial (Chopra et al., 2017; Zhang et al., 2001). This approach is made up of the synthesis of the porous nanomaterial, insulin loading into the pore chambers (Fonte et al., 2012; Paul & Sharma, 2001), and insulin encapsulation using a nanomaterial (Al-Qadi et al., 2012; Sajeesh & Sharma, 2006) to protect the insulin molecule. Upon reaching the targeted diabetic cell, the insulin encapsulated nanomaterial either enters into the cell or bind to the surface and disintegrate to release insulin. The second approach involves the coating of insulin over solid, nonporous nanomaterials (Morales et al., 2014; Shilo et al., 2015). The nanomaterial binds onto the surface or enter the target cell and disintegrate to release the insulin molecule. A schematic drawing showing both approaches is presented in Fig. 2.1.

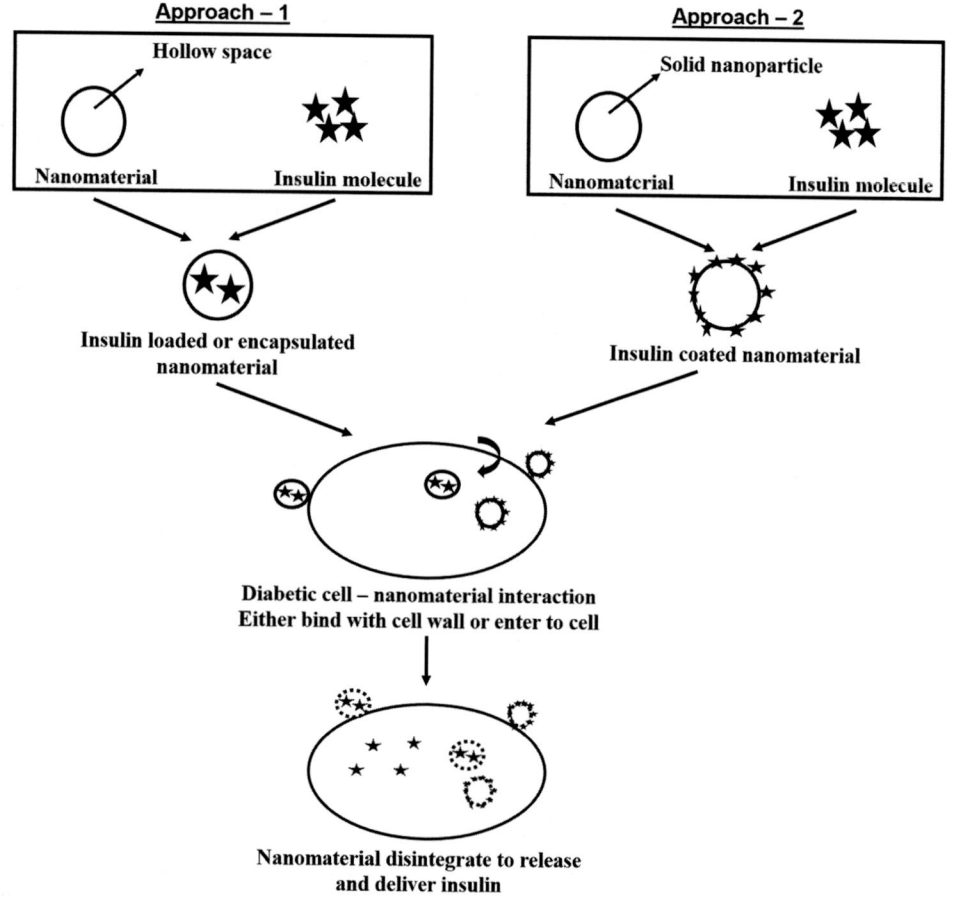

Figure 2.1 Schematics of insulin delivery approaches using nanomaterials.

Advantages and limitations of nanomaterials and nanoparticles

Nanomaterials and nanoparticles possess unique properties that make them highly beneficial for enhanced insulin delivery compared to conventional insulin delivery systems. For example, the high surface-to-volume ratio of nanomaterials makes them highly reactive in the environment and elevates their physical, chemical, and biological properties compared to their bulk counterparts or microparticles (Zhu & Xu, 2016). The highly reactive nature of nanoparticles makes it possible to apply them less quantities to achieve significant outcomes (Ealias & Saravanakumar, n.d.). Further, the surface charge of nanomaterials plays a crucial role in improving their ability to interact with other materials for tailored applications (Sadat et al., 2016). In biological applications, the smaller size of nanoparticles helps with effective internalization into the cells and provide enhanced bioactivity and bioavailability (Choi & Park, 2016; Wong & Wright, 2016). Nanomaterial-mediated delivery of insulin is highly beneficial as an effective delivery system as the encapsulation rate is high and their payload release is significantly higher compared to conventional drug delivery systems (J. Li et al., 2020), delivering insulin in a controlled manner over a prolonged time period.

Even though, nanomaterials possess several positive attributes for biological applications, especially as a drug delivery system, there are some limitations. The toxicity of nanomaterials to humans and the environment presents a major challenge (Huang et al., 2017). Physicochemical synthesis approaches to generate nanomaterials usually utilize toxic chemicals as precursors, reducing or stabilizing agents, which is then exposed to the environment and/or humans (Jeevanandam, Kulabhusan, et al., 2020; Sarkar, 2018). To partly address this challenge, green and biosynthesis approaches have been introduced through the use of enzymes and phytochemicals to synthesize nanoparticles (Sorbiun et al., 2018; Vidhya et al., 2020). However, the stability of these nanoparticles is lower as compared to chemically or physically synthesized nanoparticles (Suresh et al., 2020). Further, the fate of nanomaterials in the human body is still unknown and are under extensive research (Q. Liu et al., 2020). Whilst a significant number of research reports discuss the biomedical applications of nanomaterials in drug delivery, only a few reports discuss the circulation time of nanomaterials in the blood stream, their bioavailability and half-life, and whether or not they are excreted as nanoparticles or rejected by the immune system (Wang, Zhou, et al., 2020). In addition, nanomaterials may also cause chronic mutation in living organisms as it can internalize and interact with genetic materials (Anand et al., 2017). In the case of insulin delivery, the prolonged release of insulin by nanomaterials can reduce hyperglycemic conditions and result in hypoglycemia (El-Borady et al., 2020). The hypoglycemic condition can eventually lead to uncontrolled diabetes with symptoms such as fatigue, blurry vision, dizziness, and headache as well as other complications including cardiovascular ailments and neurodegenerative disorders (Bahman et al., 2019). Thus, extensive research is required to improve nanoparticle-based insulin delivery for widespread application in diabetes treatment.

Conclusion

Advancements in the field of nanotechnology have opened opportunities for applications in biomedicine and pharmaceutics, resulting in the fabrication of nanomaterials for controlled and targeted delivery of insulin. The variety of nanomaterials, their uniqueness in sizes, shapes, and biofunctional properties makes them promising candidates in creating new and improved insulin delivery systems for diabetes treatment. Currently, most of these nanomaterials have been investigated in cell lines and animals to characterize their insulin delivery efficacy and drug loading ability. Full-scale preclinical and clinical applications of nanomaterials in diabetes treatment is yet to be realized. Thus, there is a huge research opportunity to focus on nanomaterial-based insulin delivery systems using in vivo diabetic models to improve diabetes treatment in the future.

References

Acquah, C., Obeng, E. M., Agyei, D., Ongkudon, C. M., Moy, C. K., & Danquah, M. K. (2017). Nano-doped monolithic materials for molecular separation. *Separations, 4*(1), 2.

Aguiar, M., Rodrigues, J., Jr, & Silva Cunha, A. (2004). Encapsulation of insulin–cyclodextrin complex in PLGA microspheres: a new approach for prolonged pulmonary insulin delivery. *Journal of Microencapsulation, 21*(5), 553–564.

Ahmad, R., Ahn, M.-S., & Hahn, Y.-B. (2017). Fabrication of a non-enzymatic glucose sensor field-effect transistor based on vertically-oriented ZnO nanorods modified with Fe2O3. *Electrochemistry Communications, 77*, 107–111.

Ahmad, R., & Hahn, Y.-B. (2018). Nonenzymatic flexible field-effect transistor based glucose sensor fabricated using NiO quantum dots modified ZnO nanorods. *Journal of Colloid and Interface Science, 512*, 21–28.

Ahmad, W., Mehmood, U., Al-Ahmed, A., Al-Sulaiman, F. A., Aslam, M. Z., Kamal, M. S., & Shawabkeh, R. (2016). Synthesis of zinc oxide/titanium dioxide (ZnO/TiO2) nanocomposites by wet incipient wetness impregnation method and preparation of ZnO/TiO2 paste using poly (vinylpyrrolidone) for efficient dye-sensitized solar cells. *Electrochimica Acta, 222*, 473–480.

Alai, M. S., Lin, W. J., & Pingale, S. S. (2015). Application of polymeric nanoparticles and micelles in insulin oral delivery. *Journal of Food and Drug Analysis, 23*(3), 351–358.

Almansoori, Z., Khorshidi, B., Sadri, B., & Sadrzadeh, M. (2018). Parametric study on the stabilization of metal oxide nanoparticles in organic solvents: A case study with indium tin oxide (ITO) and heptane. *Ultrasonics Sonochemistry, 40*, 1003–1013.

Al-Naamani, L., Dobretsov, S., Dutta, J., & Burgess, J. G. (2017). Chitosan-zinc oxide nanocomposite coatings for the prevention of marine biofouling. *Chemosphere, 168*, 408–417.

Al-Qadi, S., Grenha, A., Carrión-Recio, D., Seijo, B., & Remuñán-López, C. (2012). Microencapsulated chitosan nanoparticles for pulmonary protein delivery: in vivo evaluation of insulin-loaded formulations. *Journal of Controlled Release, 157*(3), 383–390.

Al-Remawi, M., Elsayed, A., Maghrabi, I., Hamaidi, M., & Jaber, N. (2017). Chitosan/lecithin liposomal nanovesicles as an oral insulin delivery system. *Pharmaceutical Development and Technology, 22*(3), 390–398.

Anand, A. S., Prasad, D. N., Singh, S. B., & Kohli, E. (2017). Chronic exposure of zinc oxide nanoparticles causes deviant phenotype in Drosophila melanogaster. *Journal of Hazardous Materials, 327*, 180–186.

Anderson, D. G., Gu, Z., & Langer, R. S. (2016). Glucose-responsive microgels for closed loop insulin delivery. Google Patents.

Arif, T., Nisa, N., Amin, S. S., Shoib, S., Mushtaq, R., & Shawl, M. R. (2015). Therapeutic and diagnostic applications of nanotechnology in dermatology and cosmetics. *Journal of Nanomedicine & Biotherapeutic Discovery, 5*(3), 1.

Ashrafizadeh, H., Abtahi, S. R., & Oroojan, A. A. (2020). Trace element nanoparticles improved diabetes mellitus; a brief report. *Diabetes & Metabolic Syndrome: Clinical Research & Reviews, 14*, 443−445.

Asif, A. A. H., & Hasan, M. Z. (2018). Application of Nanotechnology in Modern Textiles: A Review.

Atar, N., Eren, T., Yola, M. L., Karimi-Maleh, H., & Demirdögen, B. (2015). Magnetic iron oxide and iron oxide@ gold nanoparticle anchored nitrogen and sulfur-functionalized reduced graphene oxide electrocatalyst for methanol oxidation. *RSC Advances, 5*(33), 26402−26409.

Baghayeri, M., Amiri, A., Alizadeh, Z., Veisi, H., & Hasheminejad, E. (2018). Non-enzymatic voltammetric glucose sensor made of ternary NiO/Fe3O4-SH/para-amino hippuric acid nanocomposite. *Journal of Electroanalytical Chemistry, 810*, 69−77.

Bagheri, S., Yasemi, M., Safaie-Qamsari, E., Rashidiani, J., Abkar, M., Hassani, M., Mirhosseini, S. A., & Kooshki, H. (2018). Using gold nanoparticles in diagnosis and treatment of melanoma cancer. *Artificial Cells, Nanomedicine, and Biotechnology*, 1−10.

Bahman, F., Greish, K., & Taurin, S. (2019). Nanotechnology in insulin delivery for management of diabetes. *Pharmaceutical Nanotechnology, 7*(2), 113−128.

Bai, M., He, J., Kang, L., Nie, J., & Yin, R. (2018). Regulated basal and bolus insulin release from glucose-responsive core-shell microspheres based on concanavalin A-sugar affinity. *International Journal of Biological Macromolecules, 113*, 889−899. Available from https://doi.org/10.1016/j.ijbiomac.2018.03.030.

Baig, Z., Mamat, O., & Mustapha, M. (2018). Recent progress on the dispersion and the strengthening effect of carbon nanotubes and graphene-reinforced metal nanocomposites: a review. *Critical Reviews in Solid State and Materials Sciences, 43*(1), 1−46.

Barone, P. W., Parker, R. S., & Strano, M. S. (2005). In vivo fluorescence detection of glucose using a single-walled carbon nanotube optical sensor: design, fluorophore properties, advantages, and disadvantages. *Analytical Chemistry, 77*(23), 7556−7562.

Belkhalfa, H., Teodorescu, F., Quéniat, G., Coffinier, Y., Dokhan, N., Sam, S., Abderrahmani, A., Boukherroub, R., & Szunerits, S. (2016). Insulin impregnated reduced graphene oxide/Ni (OH) 2 thin films for electrochemical insulin release and glucose sensing. *Sensors and Actuators B: Chemical, 237*, 693−701.

Bhandare, N., & Narayana, A. (2014). Applications of nanotechnology in cancer: a literature review of imaging and treatment. *Journal of Nuclear Medicine & Radiation Therapy, 5*(4), 1−9.

Bhattacharyya, A., Mukherjee, D., Mishra, R., & Kundu, P. P. (2016). Development of pH sensitive polyurethane−alginate nanoparticles for safe and efficient oral insulin delivery in animal models. *RSC Advances, 6*(48), 41835−41846.

Bisker, G., Bakh, N. A., Lee, M. A., Ahn, J., Park, M., O'Connell, E. B., Iverson, N. M., & Strano, M. S. (2018). Insulin detection using a corona phase molecular recognition site on single-walled carbon nanotubes. *ACS Sensors, 3*(2), 367−377.

Caliceti, P., Veronese, F. M., & Lora, S. (2000). Polyphosphazene microspheres for insulin delivery. *International Journal of Pharmaceutics, 211*(1−2), 57−65.

Cao, L., Ye, J., Tong, L., & Tang, B. (2008). A new route to the considerable enhancement of glucose oxidase (GOx) activity: the simple assembly of a complex from CdTe quantum dots and GOx, and its glucose sensing. *Chemistry-a European Journal, 14*(31), 9633−9640.

Cash, K. J., & Clark, H. A. (2010). Nanosensors and nanomaterials for monitoring glucose in diabetes. *Trends in Molecular Medicine, 16*(12), 584−593.

Chauhan, P. S., Yadav, D., Tayal, S., & Jin, J.-O. (2020). Therapeutic Advancements in the Management of Diabetes Mellitus with Special Reference to Nanotechnology. *Current Pharmaceutical Design, 26*(38), 4909−4916.

Chen, A.-Z., Tang, N., Wang, S.-B., Kang, Y.-Q., & Song, H.-F. (2015). Insulin-loaded poly-l-lactide porous microspheres prepared in supercritical CO2 for pulmonary drug delivery. *The Journal of Supercritical Fluids, 101*, 117−123. Available from https://doi.org/10.1016/j.supflu.2015.03.010.

Chen, J., Xu, L., Xing, R., Song, J., Song, H., Liu, D., & Zhou, J. (2012). Electrospun three-dimensional porous CuO/TiO2 hierarchical nanocomposites electrode for nonenzymatic glucose biosensing. *Electrochemistry Communications, 20*, 75−78.

Chen, M., Hou, C., Huo, D., Yang, M., & Fa, H. (2016). An ultrasensitive electrochemical DNA biosensor based on a copper oxide nanowires/single-walled carbon nanotubes nanocomposite. *Applied Surface Science, 364*, 703−709.

Chen, M.-C., Sonaje, K., Chen, K.-J., & Sung, H.-W. (2011). A review of the prospects for polymeric nanoparticle platforms in oral insulin delivery. *Biomaterials, 32*(36), 9826−9838.

Cheraghian, G., & Hendraningrat, L. (2016). A review on applications of nanotechnology in the enhanced oil recovery part A: effects of nanoparticles on interfacial tension. *International Nano Letters, 6*(2), 129−138.

Choi, J.-S., & Park, J.-S. (2016). Effects of paclitaxel nanocrystals surface charge on cell internalization. *European Journal of Pharmaceutical Sciences, 93*, 90−96.

Chono, S., Fukuchi, R., Seki, T., & Morimoto, K. (2009). Aerosolized liposomes with dipalmitoyl phosphatidylcholine enhance pulmonary insulin delivery. *Journal of Controlled Release, 137*(2), 104−109.

Chopra, S., Bertrand, N., Lim, J.-M., Wang, A., Farokhzad, O. C., & Karnik, R. (2017). Design of insulin-loaded nanoparticles enabled by multistep control of nanoprecipitation and zinc chelation. *ACS Applied Materials & Interfaces, 9*(13), 11440−11450.

Dasgupta, N., & Ranjan, S. (2018). *Nanotechnology in food sector. An Introduction to Food Grade Nanoemulsions* (pp. 1−18). Springer.

Delgado-Rivera, R., Rosario-Meléndez, R., Yu, W., & Uhrich, K. E. (2014). Biodegradable salicylate-based poly (anhydride-ester) microspheres for controlled insulin delivery. *Journal of Biomedical Materials Research. Part A, 102*(8), 2736−2742.

Dhara, K., Stanley, J., Ramachandran, T., & Nair, B. G. (2014). Pt-CuO nanoparticles decorated reduced graphene oxide for the fabrication of highly sensitive non-enzymatic disposable glucose sensor. *Sensors and Actuators B: Chemical, 195*, 197−205.

Dong, J., Wang, K., Sun, L., Sun, B., Yang, M., Chen, H., Wang, Y., Sun, J., & Dong, L. (2018). Application of graphene quantum dots for simultaneous fluorescence imaging and tumor-targeted drug delivery. *Sensors and Actuators B: Chemical, 256*, 616−623.

Dong, Z., Hamid, K. A., Gao, Y., Lin, Y., Katsumi, H., Sakane, T., & Yamamoto, A. (2011). Polyamidoamine dendrimers can improve the pulmonary absorption of insulin and calcitonin in rats. *Journal of Pharmaceutical Sciences, 100*(5), 1866−1878.

Dyer, A., Hinchcliffe, M., Watts, P., Castile, J., Jabbal-Gill, I., Nankervis, R., Smith, A., & Illum, L. (2002). Nasal delivery of insulin using novel chitosan based formulations: a comparative study in two animal models between simple chitosan formulations and chitosan nanoparticles. *Pharmaceutical Research, 19*(7), 998−1008.

Ealias, A. M., & Saravanakumar, M. P. (n.d.). A review on the classification, characterisation, synthesis of nanoparticles and their application (Vol. 263, p. 032019).

El-Borady, O. M., Othman, M. S., Atallah, H. H., & Moneim, A. E. A. (2020). Hypoglycemic potential of selenium nanoparticles capped with polyvinyl-pyrrolidone in streptozotocin-induced experimental diabetes in rats. *Heliyon, 6*(5), e04045.

El-Mekawy, R. E., & Jassas, R. S. (2017). Recent trends in smart and flexible three-dimensional cross-linked polymers: synthesis of chitosan−ZnO nanocomposite hydrogels for insulin drug delivery. *MedChemComm, 8*(5), 897−906.

Engel, M., Hadar, Y., Belkin, S., Lu, X., Elimelech, M., & Chefetz, B. (2018). Bacterial inactivation by a carbon nanotube−iron oxide nanocomposite: a mechanistic study using E. coli mutants. *Environmental Science: Nano.*

Ensafi, A. A., Khoddami, E., Rezaei, B., & Jafari-Asl, M. (2015). A supported liquid membrane for microextraction of insulin, and its determination with a pencil graphite electrode modified with RuO 2-graphene oxide. *Microchimica Acta, 182*(9), 1599−1607.

Fonte, P., Araújo, F., Silva, C., Pereira, C., Reis, S., Santos, H. A., & Sarmento, B. (2015). Polymer-based nanoparticles for oral insulin delivery: Revisited approaches. *Biotechnology Advances, 33*(6), 1342−1354.

Fonte, P., Soares, S., Costa, A., Andrade, J. C., Seabra, V., Reis, S., & Sarmento, B. (2012). Effect of cryoprotectants on the porosity and stability of insulin-loaded PLGA nanoparticles after freeze-drying. *Biomatter, 2*(4), 329−339.

Ghanbari, D., Salavati-Niasari, M., & Ghasemi-Kooch, M. (2014). A sonochemical method for synthesis of Fe3O4 nanoparticles and thermal stable PVA-based magnetic nanocomposite. *Journal of Industrial and Engineering Chemistry, 20*(6), 3970–3974.

Giudice, J., Jares-Erijman, E. A., & Leskow, F. C. (2013). Endocytosis and intracellular dissociation rates of human insulin–insulin receptor complexes by quantum dots in living cells. *Bioconjugate Chemistry, 24*(3), 431–442.

Gopinadh, A., Prakash, M., Lohitha, K., Kishore, K. K., Chowdary, A. S., & Dev, J. R. R. (2015). The changing phase of prosthodontics: Nanotechnology. *Journal of Dental and Allied Sciences, 4*(2), 78.

Gu, Z., Aimetti, A. A., Wang, Q., Dang, T. T., Zhang, Y., Veiseh, O., Cheng, H., Langer, R. S., & Anderson, D. G. (2013). Injectable nano-network for glucose-mediated insulin delivery. *ACS Nano, 7*(5), 4194–4201.

Guillet, J.-F., Flahaut, E., & Golzio, M. (2017). Hydrogel/Carbon nanotubes Needle-free device for Electrostimulated skin Delivery. *ChemPhysChem, 18*(19), 2715–2723.

Guiot, P. (2018). *Polymeric nanoparticles and microspheres*. CRC press.

Gupta, R. (2017). Diabetes treatment by nanotechnology. *J Biotechnol Biomater, 7*(268), 2.

Hai, Z., Gao, L., Zhang, Q., Xu, H., Cui, D., Zhang, Z., Tsoukalas, D., Tang, J., Yan, S., & Xue, C. (2016). Facile synthesis of core–shell structured PANI-Co3O4 nanocomposites with superior electrochemical performance in supercapacitors. *Applied Surface Science, 361*, 57–62.

Hassan, M., Gomes, V. G., Dehghani, A., & Ardekani, S. M. (2018). Engineering carbon quantum dots for photomediated theranostics. *Nano Research, 11*(1), 1–41.

He, Z., Liu, Z., Tian, H., Hu, Y., Liu, L., Leong, K. W., Mao, H.-Q., & Chen, Y. (2018). Scalable production of core–shell nanoparticles by flash nanocomplexation to enhance mucosal transport for oral delivery of insulin. *Nanoscale, 10*(7), 3307–3319. Available from https://doi.org/10.1039/C7NR08047F.

Hou, W., Xia, F., Alfranca, G., Yan, H., Zhi, X., Liu, Y., Peng, C., Zhang, C., de la Fuente, J. M., & Cui, D. (2017). Nanoparticles for multi-modality cancer diagnosis: simple protocol for self-assembly of gold nanoclusters mediated by gadolinium ions. *Biomaterials, 120*, 103–114.

Hu, Q., & Luo, Y. (2018). Recent advances of polysaccharide-based nanoparticles for oral insulin delivery. *International Journal of Biological Macromolecules, 120*, 775–782. Available from https://doi.org/10.1016/j.ijbiomac.2018.08.152.

Huang, Y.-W., Cambre, M., & Lee, H.-J. (2017). The toxicity of nanoparticles depends on multiple molecular and physicochemical mechanisms. *International Journal of Molecular Sciences, 18*(12), 2702.

Hussein, A. K. (2015). Applications of nanotechnology in renewable energies—A comprehensive overview and understanding. *Renewable and Sustainable Energy Reviews, 42*, 460–476.

Hussein, A. K. (2016). Applications of nanotechnology to improve the performance of solar collectors—recent advances and overview. *Renewable and Sustainable Energy Reviews, 62*, 767–792.

Ikoba, U., Peng, H., Li, H., Miller, C., Yu, C., & Wang, Q. (2015). Nanocarriers in therapy of infectious and inflammatory diseases. *Nanoscale, 7*(10), 4291–4305.

Inchaurraga, L., Martínez-López, A. L., Martin-Arbella, N., & Irache, J. M. (2020). Zein-based nanoparticles for the oral delivery of insulin. *Drug Delivery and Translational Research, 10*(6), 1601–1611.

Iravani, S., Thota, S., & Crans, D. C. (2018). Methods for preparation of metal nanoparticles. *Metal Nanoparticles: Synthesis and Applications in Pharmaceutical Sciences*, 15–31.

Jeevanandam, J., Barhoum, A., Chan, Y. S., Dufresne, A., & Danquah, M. K. (2018). Review on nanoparticles and nanostructured materials: history, sources, toxicity and regulations. *Beilstein Journal of Nanotechnology, 9*, 1050.

Jeevanandam, J., Chan, Y. S., & Danquah, M. K. (2016). Biosynthesis of metal and metal oxide nanoparticles. *ChemBioEng Reviews, 3*(2), 55–67.

Jeevanandam, J., Kulabhusan, P. K., Sabbih, G., Akram, M., & Danquah, M. K. (2020). Phytosynthesized nanoparticles as a potential cancer therapeutic agent. *3 Biotech, 10*(12), 1–26.

Jeevanandam, J., Sundaramurthy, A., Sharma, V., Murugan, C., Pal, K., Kodous, M. H. A., & Danquah, M. K. (2020). *Sustainability of one-dimensional nanostructures: fabrication and industrial applications. In Sustainable nanoscale engineering* (pp. 83–113). Elsevier.

Ji, X., Li, D., Lu, Q., Guo, E., Yao, L., & Liu, H. (2018). Electrospinning preparation of one-dimensional Co2 + -doped Li4Ti5O12 nanofibers for high-performance lithium ion battery. *Ionics*, 1–8.

Jiang, G., Xu, B., Zhu, J., Zhang, Y., Liu, T., & Song, G. (2019). Polymer microneedles integrated with glucose-responsive mesoporous bioactive glass nanoparticles for transdermal delivery of insulin. *Biomedical Physics & Engineering Express, 5*(4), 045038. Available from https://doi.org/10.1088/2057-1976/ab3202.

Jo, M., Lee, Y.-K., Kim, K. M., & Cho, J. (2010). Nanoparticle–nanorod core–shell LiNi0. 5Mn1. 5O4 spinel cathodes with high energy density for Li-ion batteries. *Journal of the Electrochemical Society, 157*(7), A841–A845.

Joshi, H. M., Bhumkar, D. R., Joshi, K., Pokharkar, V., & Sastry, M. (2006). Gold nanoparticles as carriers for efficient transmucosal insulin delivery. *Langmuir: the ACS Journal of Surfaces and Colloids, 22* (1), 300–305.

Kale, S. N., Kitture, R., Ghosh, S., Chopade, B. A., & Yakhmi, J. V. (2017). *Nanomaterials as enhanced antimicrobial agent/activity-enhancer for transdermal applications: A review. In Antimicrobial Nanoarchitectonics* (pp. 279–321). Elsevier.

Kebede, A., Singh, A. K., Rai, P. K., Giri, N. K., Rai, A. K., Watal, G., & Gholap, A. (2013). Controlled synthesis, characterization, and application of iron oxide nanoparticles for oral delivery of insulin. *Lasers in Medical Science, 28*(2), 579–587.

Kim, I. Y., Jo, Y. K., Lee, J. M., Wang, L., & Hwang, S.-J. (2014). Unique advantages of exfoliated 2D nanosheets for tailoring the functionalities of nanocomposites. *The Journal of Physical Chemistry Letters, 5*(23), 4149–4161.

Kishen, A. (2016). *Nanotechnology in Endodontics.* Springer.

Kokate, M., Garadkar, K., & Gole, A. (2016). Zinc-oxide-silica-silver nanocomposite: Unique one-pot synthesis and enhanced catalytic and anti-bacterial performance. *Journal of Colloid and Interface Science, 483*, 249–260.

Kong, X., Liu, Q., Zhang, C., Peng, Z., & Chen, Q. (2017). Elemental two-dimensional nanosheets beyond graphene. *Chemical Society Reviews, 46*(8), 2127–2157.

Kreuter, J. (2014). *Nanoparticles. In Colloidal drug delivery systems* (pp. 231–253). CRC Press.

Kumari, A., Singla, R., Guliani, A., & Yadav, S. K. (2014). Nanoencapsulation for drug delivery. *EXCLI Journal, 13*, 265.

Kuzmov, A., & Minko, T. (2015). Nanotechnology approaches for inhalation treatment of lung diseases. *Journal of Controlled Release, 219*, 500–518.

Li, H., Jin, K., Luo, M., Wang, X., Zhu, X., Liu, X., Jiang, T., Zhang, Q., Wang, S., & Pang, Z. (2019). Size dependency of circulation and biodistribution of biomimetic nanoparticles: red blood cell membrane-coated nanoparticles. *Cells, 8*(8), 881.

Li, J., Wu, H., Jiang, K., Liu, Y., Yang, L., & Park, H. J. (2020). Alginate calcium microbeads containing chitosan nanoparticles for controlled insulin release. *Applied Biochemistry and Biotechnology*, 1–16.

Li, S.-J., Xia, N., Lv, X.-L., Zhao, M.-M., Yuan, B.-Q., & Pang, H. (2014). A facile one-step electrochemical synthesis of graphene/NiO nanocomposites as efficient electrocatalyst for glucose and methanol. *Sensors and Actuators B: Chemical, 190*, 809–817.

Li, X., Li, G., Hu, Y., Kong, X., Su, X., Zhou, P., & Wang, Z. (2018). High-performance nanocomposites synergistically reinforced by two-dimensional montmorillonite and zero-dimensional nanoTiO 2. *Polymer Bulletin, 75*(4), 1457–1472.

Lin, Y., Hu, L., Li, L., & Wang, K. (2014). Facile synthesis of nickel hydroxide–graphene nanocomposites for insulin detection with enhanced electro-oxidation properties. *RSC Advances, 4*(86), 46208–46213.

Lin, Y.-H., Mi, F.-L., Chen, C.-T., Chang, W.-C., Peng, S.-F., Liang, H.-F., & Sung, H.-W. (2007). Preparation and characterization of nanoparticles shelled with chitosan for oral insulin delivery. *Biomacromolecules, 8*(1), 146–152.

Liu, L., Zhou, C., Xia, X., & Liu, Y. (2016). Self-assembled lecithin/chitosan nanoparticles for oral insulin delivery: preparation and functional evaluation. *International Journal of Nanomedicine, 11*, 761–769. Available from https://doi.org/10.2147/IJN.S96146.

Liu, Q., Guan, J., Qin, L., Zhang, X., & Mao, S. (2020). Physicochemical properties affecting the fate of nanoparticles in pulmonary drug delivery. *Drug Discovery Today, 25*(1), 150–159.

Liu, Y.-S., Sun, Y., Vernier, P. T., Liang, C.-H., Chong, S. Y. C., & Gundersen, M. A. (2007). pH-sensitive photoluminescence of CdSe/ZnSe/ZnS quantum dots in human ovarian cancer cells. *The Journal of Physical Chemistry C, 111*(7), 2872–2878.

Luo, Y. Y., Xiong, X. Y., Tian, Y., Li, Z. L., Gong, Y. C., & Li, Y. P. (2016). A review of biodegradable polymeric systems for oral insulin delivery. *Drug Delivery, 23*(6), 1882−1891.

Mahadik, S. (2017). Applications of nanotechnology in water and waste water treatment. *AADYA-Journal of Management and Technology (JMT), 7*, 187−191.

Maniruzzaman, M., Jang, S.-D., & Kim, J. (2012). Titanium dioxide−cellulose hybrid nanocomposite and its glucose biosensor application. *Materials Science and Engineering: B, 177*(11), 844−848.

Martin-Gallego, M., Yuste-Sanchez, V., Sanchez-Hidalgo, R., Verdejo, R., & Lopez-Manchado, M.A. (2018). Epoxy Nanocomposites filled with Carbon Nanoparticles. The Chemical Record.

Mohs, A. M., Duan, H., Kairdolf, B. A., Smith, A. M., & Nie, S. (2009). Proton-resistant quantum dots: stability in gastrointestinal fluids and implications for oral delivery of nanoparticle agents. *Nano Research, 2*(6), 500−508.

Mohseni, R., ArabSadeghabadi, Z., Ziamajidi, N., Abbasalipourkabir, R., & RezaeiFarimani, A. (2019). Oral administration of resveratrol-loaded solid lipid nanoparticle improves insulin resistance through targeting expression of SNARE proteins in adipose and muscle tissue in rats with type 2 diabetes. *Nanoscale Research Letters, 14*(1), 227.

Morales, J. O., Huang, S., Williams, R. O., III, & McConville, J. T. (2014). Films loaded with insulin-coated nanoparticles (ICNP) as potential platforms for peptide buccal delivery. *Colloids and Surfaces B: Biointerfaces, 122*, 38−45.

Morishita, I., Morishita, M., Takayama, K., Machida, Y., & Nagai, T. (1993). Enteral insulin delivery by microspheres in 3 different formulations using Eudragit L100 and S100. *International Journal of Pharmaceutics, 91*(1), 29−37.

Mundargi, R. C., Rangaswamy, V., & Aminabhavi, T. M. (2011). pH-Sensitive oral insulin delivery systems using Eudragit microspheres. *Drug Development and Industrial Pharmacy, 37*(8), 977−985.

Nandwana, V., Huang, W., Li, Y., & Dravid, V. P. (2018). One-pot green synthesis of Fe3O4/MoS2 0D/2D nanocomposites and their application in noninvasive point-of-care glucose diagnostics. *ACS Applied Nano Materials, 1*(4), 1949−1958.

Ng, C. M., Loh, H.-S., Muthoosamy, K., Sridewi, N., & Manickam, S. (2016). Conjugation of insulin onto the sidewalls of single-walled carbon nanotubes through functionalization and diimide-activated amidation. *International Journal of Nanomedicine, 11*, 1607.

NOSE-TO-BRAIN, D. T. (2018). Dendrimers: Nanosized multifunctional platform for drug delivery. *Drug Delivery, 8*(1), 20.

Odom, T. W., Huang, J., & Lieber, C. M. (2002). Single-walled carbon nanotubes. *Annals of the New York Academy of Sciences, 960*(1), 203−215.

Ojha, D. P., Joshi, M. K., & Kim, H. J. (2017). Photo-Fenton degradation of organic pollutants using a zinc oxide decorated iron oxide/reduced graphene oxide nanocomposite. *Ceramics International, 43*(1), 1290−1297.

Pant, B., Park, M., Lee, J. H., Kim, H.-Y., & Park, S.-J. (2017). Novel magnetically separable silver-iron oxide nanoparticles decorated graphitic carbon nitride nano-sheets: a multifunctional photocatalyst via one-step hydrothermal process. *Journal of Colloid and Interface Science, 496*, 343−352.

Pardo, J., Peng, Z., & Leblanc, R. M. (2018). Cancer targeting and drug delivery using carbon-based quantum dots and nanotubes. *Molecules (Basel, Switzerland), 23*(2), 378.

Parhizkar, M., Mahalingam, S., Homer-Vanniasinkam, S., & Edirisinghe, M. (2018). Latest developments in innovative manufacturing to combine nanotechnology with healthcare. Future Medicine.

Park, J., Kadasala, N. R., Abouelmagd, S. A., Castanares, M. A., Collins, D. S., Wei, A., & Yeo, Y. (2016). Polymer−iron oxide composite nanoparticles for EPR-independent drug delivery. *Biomaterials, 101*, 285−295.

Paul, W., & Sharma, C. P. (2001). Porous hydroxyapatite nanoparticles for intestinal delivery of insulin. *Trends Biomater Artif Organs, 14*, 37−38.

Peng, B., Tang, J., Luo, J., Wang, P., Ding, B., & Tam, K. C. (2018). Applications of nanotechnology in oil and gas industry: Progress and perspective. *The Canadian Journal of Chemical Engineering, 96*(1), 91−100.

Petersen, D. K., Naylor, T. M., & Ver Halen, J. P. (2014). Current and future applications of nanotechnology in plastic and reconstructive surgery. *Plast Aesthet Res, 1*, 43−50.

Pezzini, I., Mattoli, V., & Ciofani, G. (2017). Mitochondria and neurodegenerative diseases: The promising role of nanotechnology in targeted drug delivery. *Expert Opinion on Drug Delivery, 14*(4), 513–523.

Pisani, C., Rascol, E., Dorandeu, C., Charnay, C., Guari, Y., Chopineau, J., Devoisselle, J.-M., & Prat, O. (2017). Biocompatibility assessment of functionalized magnetic mesoporous silica nanoparticles in human HepaRG cells. *Nanotoxicology, 11*(7), 871–890.

Politano, A., Vitiello, M., Viti, L., Boukhvalov, D., & Chiarello, G. (2017). The role of surface chemical reactivity in the stability of electronic nanodevices based on two-dimensional materials "beyond graphene" and topological insulators. *FlatChem, 1*, 60–64.

Qiu, B., Wang, Y., Sun, D., Wang, Q., Zhang, X., Weeks, B. L., O'Connor, R., Huang, X., Wei, S., & Guo, Z. (2015). Cr (VI) removal by magnetic carbon nanocomposites derived from cellulose at different carbonization temperatures. *Journal of Materials Chemistry A, 3*(18), 9817–9825.

Raghavan, N., Thangavel, S., & Venugopal, G. (2015). Enhanced photocatalytic degradation of methylene blue by reduced graphene-oxide/titanium dioxide/zinc oxide ternary nanocomposites. *Materials Science in Semiconductor Processing, 30*, 321–329.

Ramachandraiah, K., Han, S. G., & Chin, K. B. (2015). Nanotechnology in meat processing and packaging: potential applications—a review. *Asian-Australasian Journal of Animal Sciences, 28*(2), 290.

Ramachandran, R., Paul, W., & Sharma, C. P. (2009). Synthesis and characterization of PEGylated calcium phosphate nanoparticles for oral insulin delivery. *Journal of Biomedical Materials Research, Part B: Applied Biomaterials, 88*(1), 41–48.

Roeth, A. A., Slabu, I., Baumann, M., Alizai, P. H., Schmeding, M., Guentherodt, G., Schmitz-Rode, T., & Neumann, U. P. (2017). Establishment of a biophysical model to optimize endoscopic targeting of magnetic nanoparticles for cancer treatment. *International Journal of Nanomedicine, 12*, 5933.

Rongione, N. A., Floerke, S. A., & Celik, E. (2017). *Developments in antibacterial disinfection techniques: applications of nanotechnology. Applying Nanotechnology for Environmental Sustainability* (pp. 185–203). IGI Global.

Sabu, C., & Pramod, K. (2020). *Advanced nanostructures for oral insulin delivery. In Nanoscience in Medicine* (Vol. 1, pp. 187–212). Springer.

Sadasivuni, K. K., Ponnamma, D., Ko, H.-U., Kim, H. C., Zhai, L., & Kim, J. (2016). Flexible NO2 sensors from renewable cellulose nanocrystals/iron oxide composites. *Sensors and Actuators B: Chemical, 233*, 633–638.

Sadat, S. M. A., Jahan, S. T., & Haddadi, A. (2016). Effects of size and surface charge of polymeric nanoparticles on in vitro and in vivo applications. *Journal of Biomaterials and Nanobiotechnology, 7*(2), 91.

Sajeesh, S., & Sharma, C. P. (2006). Cyclodextrin–insulin complex encapsulated polymethacrylic acid based nanoparticles for oral insulin delivery. *International Journal of Pharmaceutics, 325*(1–2), 147–154.

Saleh, T. A. (2020). Nanomaterials: Classification, properties, and environmental toxicities. *Environmental Technology & Innovation*, 101067.

Saravanan, R., Khan, M. M., Gupta, V. K., Mosquera, E., Gracia, F., Narayanan, V., & Stephen, A. (2015). ZnO/Ag/Mn2O3 nanocomposite for visible light-induced industrial textile effluent degradation, uric acid and ascorbic acid sensing and antimicrobial activity. *RSC Advances, 5*(44), 34645–34651.

Sarkar, A. (2018). Novel platinum compounds and nanoparticles as anticancer agents. *Pharmaceutical Patent Analyst, 7*(1), 33–46.

Sarmento, B., Martins, S., Ferreira, D., & Souto, E. B. (2007). Oral insulin delivery by means of solid lipid nanoparticles. *International Journal of Nanomedicine, 2*(4), 743.

Sarmento, B., Ribeiro, A., Veiga, F., Sampaio, P., Neufeld, R., & Ferreira, D. (2007). Alginate/chitosan nanoparticles are effective for oral insulin delivery. *Pharmaceutical Research, 24*(12), 2198–2206.

Selvan, S. T., Tan, T. T., & Ying, J. Y. (2005). Robust, non-cytotoxic, silica-coated cdse quantum dots with efficient photoluminescence. *Advanced Materials, 17*(13), 1620–1625.

Sgorla, D., Lechanteur, A., Almeida, A., Sousa, F., Melo, E., Bunhak, É., Mainardes, R., Khalil, N., Cavalcanti, O., & Sarmento, B. (2018). Development and characterization of lipid-polymeric nanoparticles for oral insulin delivery. *Expert Opinion on Drug Delivery, 15*(3), 213–222. Available from https://doi.org/10.1080/17425247.2018.1420050.

Shah, R. B., Patel, M., Maahs, D. M., & Shah, V. N. (2016). Insulin delivery methods: Past, present and future. *International Journal of Pharmaceutical Investigation, 6*(1), 1.

Shen, J.-M., Xu, L., Lu, Y., Cao, H.-M., Xu, Z.-G., Chen, T., & Zhang, H.-X. (2012). Chitosan-based luminescent/magnetic hybrid nanogels for insulin delivery, cell imaging, and antidiabetic research of dietary supplements. *International Journal of Pharmaceutics, 427*(2), 400–409. Available from https://doi.org/10.1016/j.ijpharm.2012.01.059.

Shilo, M., Berenstein, P., Dreifuss, T., Nash, Y., Goldsmith, G., Kazimirsky, G., Motiei, M., Frenkel, D., Brodie, C., & Popovtzer, R. (2015). Insulin-coated gold nanoparticles as a new concept for personalized and adjustable glucose regulation. *Nanoscale, 7*(48), 20489–20496.

Shukla, S., Deshpande, S. R., Shukla, S. K., & Tiwari, A. (2012). Fabrication of a tunable glucose biosensor based on zinc oxide/chitosan-graft-poly (vinyl alcohol) core-shell nanocomposite. *Talanta, 99*, 283–287.

Singh, S., Singh, B. K., Yadav, S., & Gupta, A. (2015). Applications of nanotechnology in agricultural and their role in disease management. *Research Journal of Nanoscience and Nanotechnology, 5*, 1–5.

Song, J., Xu, L., Zhou, C., Xing, R., Dai, Q., Liu, D., & Song, H. (2013). Synthesis of graphene oxide based CuO nanoparticles composite electrode for highly enhanced nonenzymatic glucose detection. *ACS Applied Materials & Interfaces, 5*(24), 12928–12934.

Song, L., Zhi, Z., & Pickup, J. C. (2014). Nanolayer encapsulation of insulin-chitosan complexes improves efficiency of oral insulin delivery. *International Journal of Nanomedicine, 9*, 2127.

Sorbiun, M., Shayegan Mehr, E., Ramazani, A., & Mashhadi Malekzadeh, A. (2018). Biosynthesis of metallic nanoparticles using plant extracts and evaluation of their antibacterial properties. *Nanochemistry Research, 3*(1), 1–16.

Soursou, G., Alexiou, A., Md Ashraf, G., Ali Siyal, A., Mushtaq, G., & Kamal, M. A. (2015). Applications of nanotechnology in diagnostics and therapeutics of alzheimer's and parkinson's disease. *Current Drug Metabolism, 16*(8), 705–712.

Sun, G., Zhang, X., Lin, R., Yang, J., Zhang, H., & Chen, P. (2015). Hybrid fibers made of molybdenum disulfide, reduced graphene oxide, and multi-walled carbon nanotubes for solid-state, flexible, asymmetric supercapacitors. *Angewandte Chemie International Edition, 54*(15), 4651–4656.

Suresh, M., Jeevanandam, J., Chan, Y. S., Danquah, M. K., & Kalaiarasi, J. M. V. (2020). Opportunities for metal oxide nanoparticles as a potential mosquitocide. *BioNanoScience, 10*(1), 292–310.

Tafazoli, M., Hojjati, S. M., Biparva, P., Kooch, Y., & Lamersdorf, N. (2017). Reduction of soil heavy metal bioavailability by nanoparticles and cellulosic wastes improved the biomass of tree seedlings. *Journal of Plant Nutrition and Soil Science, 180*(6), 683–693.

Tang, B., Cao, L., Xu, K., Zhuo, L., Ge, J., Li, Q., & Yu, L. (2008). A new nanobiosensor for glucose with high sensitivity and selectivity in serum based on fluorescence resonance energy transfer (FRET) between CdTe quantum dots and Au nanoparticles. *Chemistry-a European Journal, 14*(12), 3637–3644.

Tekade, R. K., Maheshwari, R., Soni, N., & Tekade, M. (2017). *Carbon nanotubes in targeting and delivery of drugs. In Nanotechnology-Based Approaches for Targeting and Delivery of Drugs and Genes* (pp. 389–426). Elsevier.

Teodorescu, F., Oz, Y., Quéniat, G., Abderrahmani, A., Foulon, C., Lecoeur, M., Sanyal, R., Sanyal, A., Boukherroub, R., & Szunerits, S. (2017). Photothermally triggered on-demand insulin release from reduced graphene oxide modified hydrogels. *Journal of Controlled Release, 246*, 164–173.

Teodorescu, F., Rolland, L., Ramarao, V., Abderrahmani, A., Mandler, D., Boukherroub, R., & Szunerits, S. (2015). Electrochemically triggered release of human insulin from an insulin-impregnated reduced graphene oxide modified electrode. *Chemical Communications, 51*(75), 14167–14170.

Turcheniuk, K., Khanal, M., Motorina, A., Subramanian, P., Barras, A., Zaitsev, V., Kuncser, V., Leca, A., Martoriati, A., & Cailliau, K. (2014). Insulin loaded iron magnetic nanoparticle–graphene oxide composites: synthesis, characterization and application for in vivo delivery of insulin. *RSC Advances, 4*(2), 865–875.

Vidhya, E., Vijayakumar, S., Prathipkumar, S., & Praseetha, P. K. (2020). Green way biosynthesis: Characterization, antimicrobial and anticancer activity of ZnO nanoparticles. *Gene Reports*, 100688.

Vigneshwaran, N., Kumar, S., Kathe, A., Varadarajan, P., & Prasad, V. (2006). Functional finishing of cotton fabrics using zinc oxide–soluble starch nanocomposites. *Nanotechnology, 17*(20), 5087.

Wang, J., Tabata, Y., & Morimoto, K. (2006). Aminated gelatin microspheres as a nasal delivery system for peptide drugs: evaluation of in vitro release and in vivo insulin absorption in rats. *Journal of Controlled Release, 113*(1), 31–37.

Wang, J., Wang, Z., Yu, J., Kahkoska, A. R., Buse, J. B., & Gu, Z. (2020). Glucose-responsive insulin and delivery systems: Innovation and translation. *Advanced Materials, 32*(13), 1902004.

Wang, L., Li, J., Zhou, X., Zheng, Q., & Cheng, X. (2017). Clinical application of carbon nanoparticles in curative resection for colorectal carcinoma. *OncoTargets and Therapy, 10*, 5585.

Wang, L.-Y., Gu, Y.-H., Su, Z.-G., & Ma, G.-H. (2006). Preparation and improvement of release behavior of chitosan microspheres containing insulin. *International Journal of Pharmaceutics, 311*(1−2), 187−195.

Wang, S., McGuirk, C. M., Ross, M. B., Wang, S., Chen, P., Xing, H., Liu, Y., & Mirkin, C. A. (2017). General and direct method for preparing oligonucleotide-functionalized metal−organic framework nanoparticles. *Journal of the American Chemical Society, 139*(29), 9827−9830.

Wang, Y., Zhou, C., Ding, Y., Liu, M., Tai, Z., Jin, Q., Yang, Y., Li, Z., Yang, M., & Gong, W. (2020). Red blood cell-hitchhiking chitosan nanoparticles for prolonged blood circulation time of vitamin K1. *International Journal of Pharmaceutics, 592*, 120084.

Wei, X., Duan, X., Zhang, Y., Ma, Z., Li, C., & Zhang, X. (2020). Internalization mechanism of phenylboronic-acid-decorated nanoplatform for enhanced nasal insulin delivery. *ACS Applied Bio Materials, 3*(4), 2132−2139. Available from https://doi.org/10.1021/acsabm.0c00002.

Wei, Y., Li, Y., Liu, X., Xian, Y., Shi, G., & Jin, L. (2010). ZnO nanorods/Au hybrid nanocomposites for glucose biosensor. *Biosensors and Bioelectronics, 26*(1), 275−278.

Wen, N., Gao, C., Lü, S., Xu, X., Bai, X., Wu, C., Ning, P., Zhang, S., & Liu, M. (2017). Novel amphiphilic glucose-responsive modified starch micelles for insulin delivery. *RSC Advances, 7*(73), 45978−45986.

Wong, A. C., & Wright, D. W. (2016). Size-dependent cellular uptake of DNA functionalized gold nanoparticles. *Small (Weinheim an der Bergstrasse, Germany), 12*(40), 5592−5600.

Wu, L., Willis, J. J., McKay, I. S., Diroll, B. T., Qin, J., Cargnello, M., & Tassone, C. J. (2017). High-temperature crystallization of nanocrystals into three-dimensional superlattices. *Nature, 548*(7666), 197.

Xiao, F., Li, Y., Gao, H., Ge, S., & Duan, H. (2013). Growth of coral-like PtAu−MnO2 binary nanocomposites on free-standing graphene paper for flexible nonenzymatic glucose sensors. *Biosensors and Bioelectronics, 41*, 417−423.

Xie, J., Cao, H., Jiang, H., Chen, Y., Shi, W., Zheng, H., & Huang, Y. (2013). Co3O4-reduced graphene oxide nanocomposite as an effective peroxidase mimetic and its application in visual biosensing of glucose. *Analytica Chimica Acta, 796*, 92−100.

Xiong, Y., Deng, C., Zhang, X., & Yang, P. (2015). Designed synthesis of aptamer-immobilized magnetic mesoporous silica/Au nanocomposites for highly selective enrichment and detection of insulin. *ACS Applied Materials & Interfaces, 7*(16), 8451−8456.

Xu, B., Jiang, G., Yu, W., Liu, D., Zhang, Y., Zhou, J., Sun, S., & Liu, Y. (2017). H2O2-Responsive mesoporous silica nanoparticles integrated with microneedle patches for the glucose-monitored transdermal delivery of insulin. *Journal of Materials Chemistry B, 5*(41), 8200−8208. Available from https://doi.org/10.1039/C7TB02082A.

Yagati, A. K., Choi, Y., Park, J., Choi, J.-W., Jun, H.-S., & Cho, S. (2016). Silver nanoflower−reduced graphene oxide composite based micro-disk electrode for insulin detection in serum. *Biosensors and Bioelectronics, 80*, 307−314.

Yagati, A. K., Park, J., & Cho, S. (2016). Reduced graphene oxide modified the interdigitated chain electrode for an insulin sensor. *Sensors, 16*(1), 109.

Yamamoto, T., & Fukuyama, H. (2018). Possible high thermoelectric power in semiconducting carbon nanotubes—A case study of doped one-dimensional semiconductors. *Journal of the Physical Society of Japan, 87*(2), 024707.

Yang, H., Liu, W., Ma, C., Zhang, Y., Wang, X., Yu, J., & Song, X. (2014). Gold−silver nanocomposite-functionalized graphene based electrochemiluminescence immunosensor using graphene quantum dots coated porous PtPd nanochains as labels. *Electrochimica Acta, 123*, 470−476.

Yin, L., Ding, J., He, C., Cui, L., Tang, C., & Yin, C. (2009). Drug permeability and mucoadhesion properties of thiolated trimethyl chitosan nanoparticles in oral insulin delivery. *Biomaterials, 30*(29), 5691−5700.

Yu, W., Bien-Aime, S., Li, J., Zhang, L., McCormack, E. S., Goldberg, I. D., Narayan, P., & Uhrich, K. E. (2015). Injectable microspheres for extended delivery of bioactive insulin and salicylic acid. *Journal of Bioactive and Compatible Polymers, 30*(3), 340−346.

Yue, Y., Hou, Y., Zheng, M., Fu, J., Yan, X., & Zhu, M. (2018). Energy harvesting characteristic in pie-zoelectric nanoceramics with high mechanical property. *Materials Letters, 227*, 21−24.

Yuksel, E., Weinfeld, A. B., Cleek, R., Wamsley, S., Jensen, J., Boutros, S., Waugh, J. M., Shenaq, S. M., & Spira, M. (2000). Increased free fat-graft survival with the long-term, local delivery of insu-lin, insulin-like growth factor-I, and basic fibroblast growth factor by PLGA/PEG microspheres. *Plastic and Reconstructive Surgery, 105*(5), 1712−1720.

Yusoff, N., Kumar, S. V., Rameshkumar, P., Pandikumar, A., Shahid, M. M., Ab Rahman, M., & Huang, N. M. (2016). A facile preparation of titanium dioxide-iron oxide@ silicon dioxide incorpo-rated reduced graphene oxide nanohybrid for electrooxidation of methanol in alkaline medium. *Electrochimica Acta, 192*, 167−176.

Zadeh, S., Rajabnezhad, S., Zandkarimi, M., Dahmardeh, S., & Mir, L. (2017). Mucoadhesive micro-spheres of chitosan and polyvinyl alcohol as a carrier for intranasal delivery of insulin: in vitro and in vivo studies. *MOJ Bioequiv Availab, 3*(2), 00030.

Zeng, Z., Xiao, F.-X., Phan, H., Chen, S., Yu, Z., Wang, R., Nguyen, T.-Q., & Tan, T. T. Y. (2018). Unraveling the cooperative synergy of zero-dimensional graphene quantum dots and metal nanocrys-tals enabled by layer-by-layer assembly. *Journal of Materials Chemistry A*.

Zhang, F., Smolen, J. A., Zhang, S., Li, R., Shah, P. N., Cho, S., Wang, H., Raymond, J. E., Cannon, C. L., & Wooley, K. L. (2015). Degradable polyphosphoester-based silver-loaded nanoparticles as therapeutics for bacterial lung infections. *Nanoscale, 7*(6), 2265−2270.

Zhang, L., Zhang, Y.-X., Qiu, J.-N., Li, J., Chen, W., & Guan, Y.-Q. (2017). Preparation and character-ization of hypoglycemic nanoparticles for oral insulin delivery. *Biomacromolecules, 18*(12), 4281−4291. Available from https://doi.org/10.1021/acs.biomac.7b01322.

Zhang, P., Zhang, Y., & Liu, C.-G. (2020). Polymeric nanoparticles based on carboxymethyl chitosan in combination with painless microneedle therapy systems for enhancing transdermal insulin delivery. *RSC Advances, 10*(41), 24319−24329. Available from https://doi.org/10.1039/D0RA04460A.

Zhang, Q., Shen, Z., & Nagai, T. (2001). Prolonged hypoglycemic effect of insulin-loaded polybutylcya-noacrylate nanoparticles after pulmonary administration to normal rats. *International Journal of Pharmaceutics, 218*(1−2), 75−80.

Zhang, Z., Gao, X., Zhang, A., Wu, X., Chen, L., He, C., Zhuang, X., & Chen, X. (2012). Biodegradable pH-dependent thermo-sensitive hydrogels for oral insulin delivery. *Macromolecular Chemistry and Physics, 213*(7), 713−719.

Zhao, Y., Trewyn, B. G., Slowing, I. I., & Lin, V. S.-Y. (2009). Mesoporous silica nanoparticle-based double drug delivery system for glucose-responsive controlled release of insulin and cyclic AMP. *Journal of the American Chemical Society, 131*(24), 8398−8400.

Zhao, Z., Sun, Y., & Dong, F. (2015). Graphitic carbon nitride based nanocomposites: a review. *Nanoscale, 7*(1), 15−37.

Zheng, Y., Wang, A., Lin, H., Fu, L., & Cai, W. (2015). A sensitive electrochemical sensor for direct phoxim detection based on an electrodeposited reduced graphene oxide−gold nanocomposite. *RSC Advances, 5*(20), 15425−15430.

Zhou, C., Xu, L., Song, J., Xing, R., Xu, S., Liu, D., & Song, H. (2014). Ultrasensitive non-enzymatic glucose sensor based on three-dimensional network of ZnO-CuO hierarchical nanocomposites by electrospinning. *Scientific Reports, 4*, 7382.

Zhu, Q.-L., & Xu, Q. (2016). Immobilization of ultrafine metal nanoparticles to high-surface-area mate-rials and their catalytic applications. *Chem, 1*(2), 220−245.

Zupancic, S., Kocbek, P., Baumgartner, S., & Kristl, J. (2015). Contribution of nanotechnology to improved treatment of periodontal disease. *Current Pharmaceutical Design, 21*(22), 3257−3271.

Further reading

Chen, F., Gao, W., Qiu, X., Zhang, H., Liu, L., Liao, P., Fu, W., & Luo, Y. (2018). Graphene quantum dots in biomedical applications: Recent advances and future challenges. *Frontiers in Laboratory Medicine, 1*(4), 192−199.

Hossain, K. M. Z., Patel, U., & Ahmed, I. (2015). Development of microspheres for biomedical applications: A review. *Progress in Biomaterials*, *4*(1), 1—19.

Kim, J. U., Shahbaz, H. M., Lee, H., Kim, T., Yang, K., Roh, Y. H., & Park, J. (2020). Optimization of phytic acid–crosslinked chitosan microspheres for oral insulin delivery using response surface methodology. *International Journal of Pharmaceutics*, *588*, 119736.

Li, R., He, M., Li, T., & Zhang, L. (2015). Preparation and properties of cellulose/silver nanocomposite fibers. *Carbohydrate Polymers*, *115*, 269—275.

CHAPTER 3

Nanomaterials as glucose sensors for diabetes monitoring

Introduction

Sensors and actuators are important inventions in the field of electronics as they are used in a wide variety of applications (Hassani et al., 2020). Biosensors represent a type of sensor used for detecting and measuring the activities and functions of living organisms either in vitro, in vivo, or in situ (Cooper, 2009; Cui et al., 2020). Glucose sensors are a type of biosensor used to measure blood glucose levels in the body (Zhang, Li, et al., 2020; Zhang, Chen, et al., 2020) in order to monitor the risk of unexpected rise (hyperglycemic) or fall (hypoglycemic) in glucose levels in diabetic patients (Sehit & Altintas, 2020). In addition, hyper and hypoglycemic alarms, self-blood glucose monitoring aid, and in vitro blood glucose analyzers are available for monitoring glucose levels in diabetic patients. Advancements in the biomedical field offer a wide variety of glucose-sensing systems that can monitor glucose levels of diabetic patients in real-time. Glucose sensors are one of the widely used health monitoring systems among other medical sensors (Juska & Pemble, 2020; Spichiger-Keller, 2008). However, the growth in the diabetic population globally and the complexity of biomolecular activities associated with diabetes call for the development of novel and high-precision glucose sensing systems to meet the market demand. The field of nanotechnology has found significant applications in the development and fabrication of advanced glucose monitoring systems. The unique properties of nanomaterials, their small size, and large surface-to-volume ratio play a significant role in the development of production of smart glucose sensors (Chen et al., 2013; Jeevanandam & Danquah, 2020). This chapter provides an overview of glucose sensors and discusses research advances in the development of novel nanomaterials for glucose sensing applications.

Developments in glucose sensors

A typical biosensor is an analytical unit or device, which consists of a bio or bioderived sensing element that is integrated with a physicochemical transducer (Ivars-Barceló et al., 2018; Turner, 2000). Biosensors recognize biological entities and differentiate target molecules such as microbes, enzymes, and nucleic acids; implement a magnetic, thermometric, electrochemical, optical, or piezoelectric transducer to convert

Emerging Nanomedicines for Diabetes Mellitus Theranostics
DOI: https://doi.org/10.1016/B978-0-323-85396-5.00004-X

recognized bioentity into a calculatable signal; and possess a system for processing signals into readable forms (Ahmed, 2020; Hiratsuka et al., 2008). Glucose sensors are a class of biosensors with majority of them based on electrochemical transducers for the detection of glucose (Thévenot et al., 2001; Wei et al., 2020). Generally, the glucose biosensor catalyzes β-D-glucose oxidation via glucose oxidase, hydrogen peroxidase, and molecular oxygen from gluconic acid (Okuda-Shimazaki et al., 2020; Weibel & Bright, 1971). Fig. 3.1 illustrates the general concepts of enzyme-based glucose biosensors. Similar concepts are employed in other types of glucose sensors with modifications in the transducer, molecular recognizer, and signal processors.

In 1965 the first blood glucose strip called Dextrostix was introduced to detect glucose level through color changes depending on glucose concentration (Clarke & Foster, 2012). Later, the first practical enzyme electrode was developed by Updike and Hicks in 1967 (Park et al., 2006). Based on their findings, the first research on the amperometric blood glucose determination via glucose oxidase catalysis and redox coupling was established in 1970 (Williams et al., 1970). In the year after (1971), Ames Reflectance Meter (ARM) was discovered by Miles Laboratory, United States, based on the concept of reflectometer and Dextrostix (Tonyushkina & Nichols, 2009).

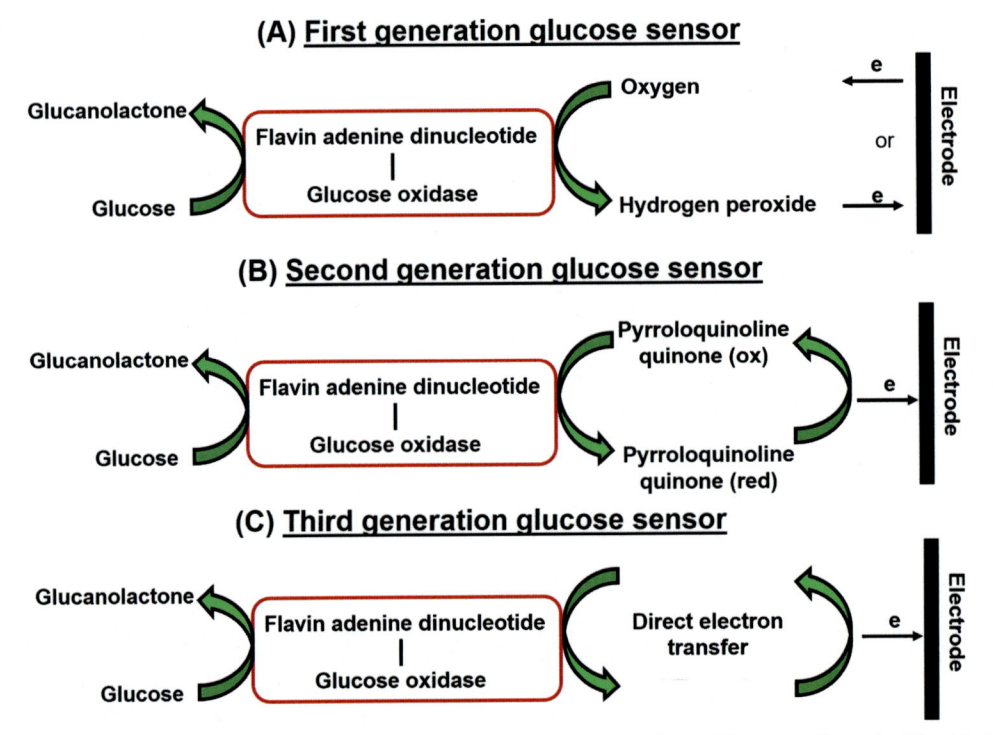

Figure 3.1 Developments in glucose sensors showing (A) first, (B) second and (C) third generations.

Some of the drawbacks in ARM, including high cost (~$650) and weight (~1 kg) (Tonyushkina & Nichols, 2009), led to the invention of Ames Eyetone glucose analyzer and the basic glucose enzyme electrode based on hydrogen peroxide detection (Guilbault & Lubrano, 1973). In 1975 the first commercial biosensor called Yellow Springs Instrument Company analyzer (Model 23A YSI analyzer) was launched (Burmeister & Arnold, 1995). The year after (1976), first bedside artificial pancreas was reported by Miles (Shichiri et al., 1983). In 1980 the second-generation glucose sensors became available on the market. These included screen-printed strips for self-monitoring of blood glucose and enhanced electrodes and membranes for improving sensor performance (Frew & Hill, 1987). The first needle-type enzyme electrode for subcutaneous implantation was introduced in 1982 (Shichiri et al., 1983) and the first ferrocene-based amperometric glucose biosensor was introduced in 1984 (Cass et al., 1984). In 1986 the American Diabetes Association, US Food and Drugs Administration (USFDA), and the US National Institutes of Health recommended self-monitoring of blood glucose in specific patients, and this was the major event that expanded the glucose sensor market throughout the world. In 1987 MediSense ExacTech blood glucose biosensor used for detecting blood glucose levels in real-time was launched (Turner, 2013). After a decade, the first commercial in vivo glucose sensor called MiniMed was launched in 1999 and a wearable noninvasive glucose monitoring system called GlucoWatch was also introduced in 2000 (Tierney et al., 2000; Yoo & Lee, 2010). Since the year 2000, nanoparticles and nanomaterials are widely explored in glucose sensors (Shipway et al., 2000). The Consensus error grid was proposed by Parkes et al. (2000) from the responses of 100 diabetologists to spot the errors associated with self-monitoring blood-glucose systems (Parkes et al., 2000). Similarly, the continuous glucose error grid analysis was demonstrated by Kovatchev et al. (2004) to determine the temporal continuous monitoring characteristics and the lag between continuous interstitial fluid sensor values and the blood reference values (Kovatchev et al., 2004; Oliver et al., 2009). Nanoparticles and nanomaterials are commonly utilized in glucose sensors introduced after 2000 (Sehit & Altintas, 2020). A timeline showing the developments in glucose sensors is described in Fig. 3.2. Nanomaterials have found significant applications in the development of novel glucose sensors for effective diabetes management.

Types of glucose sensors

Glucose sensors are mostly classified based on the method of detection, the mode of evaluation, and the nature of the output. Subclasses of glucose sensors are based on the type of equipment or technique used to detect glucose (Wang & Lee, 2015) as shown in Fig. 3.3.

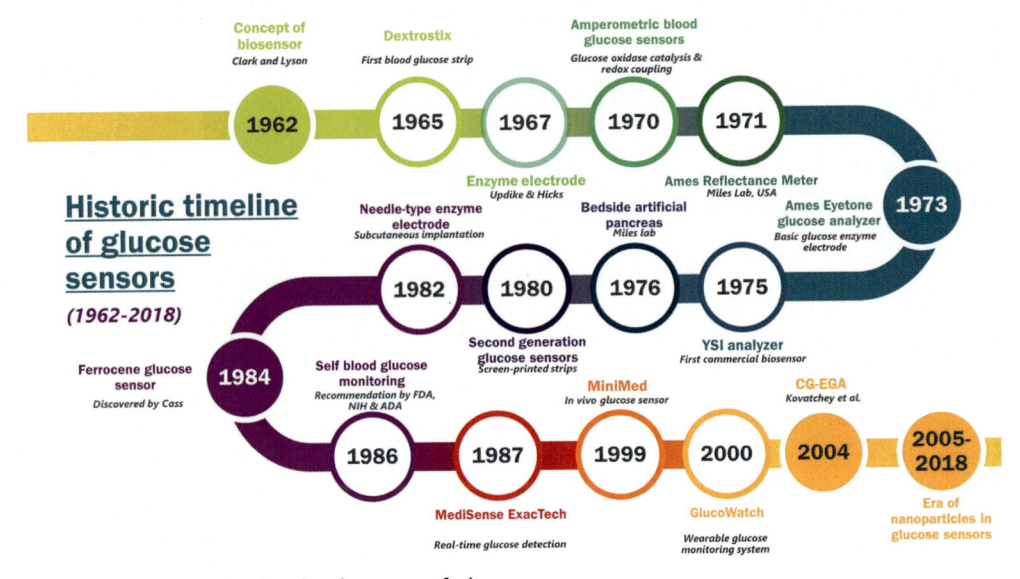

Figure 3.2 Progress in the development of glucose sensors.

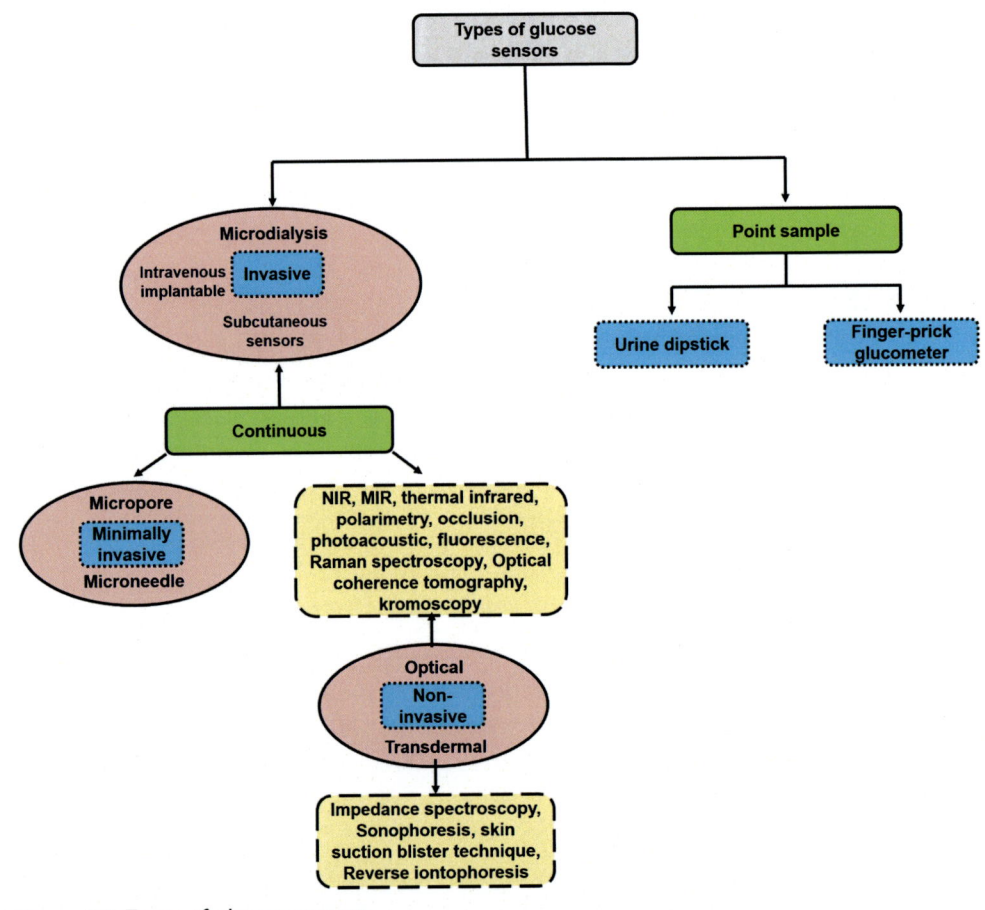

Figure 3.3 Types of glucose sensors.

Point-of-care glucose sensors

Point-of-care glucose sensors are used to monitor the level of blood glucose at the point of patient care, mostly outside the medical laboratory environment. It is usually based on electrochemical sensing, reflectance photometric measurements, and photometric strips. Some of the commercial glucose meters and test strips such as SureStep, Precision QID, Accu-Chek Advantage H, Precision G, AgaMatrix Presto, CareSens N, OneTouch Verio IQ, and Accu-Chek Comfort Curve belong to this class of glucose sensors (Lisi et al., 2020; Tang et al., 2000). Point-of-care glucose sensors can be further classified into finger-prick glucometer and urine dipstick depending on the sample collection point (Adepoyibi et al., 2013). Finger-prick glucometers are noninvasive in vivo glucose monitoring systems that use electrochemical sensing to detect glucose levels in the blood samples (Liakat et al., 2014). In recent times, glucometers that do not involve finger pricking to obtain blood are available and measure blood glucose level via the oxygen saturation level of the human skin (Thévenot et al., 2001; Wei et al., 2020). Urine dipsticks are not only used to monitor glucose variations in diabetic patients (Dinneen & Gerstein, 1997; Usui et al., 2020), but also to estimate proteinuria (Ginsberg et al., 1983), detect intestinal schistosomiasis (Standley et al., 2010), albuminuria (White et al., 2011), glomerular filtration rate (Johnson et al., 2004), epithelial dysplasia, and bladder tumor (Britton et al., 1989). Currently, automated urine dips are available commercially (Malia et al., 2018) to help in the early detection of chronic kidney diseases (Jonsson et al., 2020; Zečić et al., 2016) and multiple sclerosis—mediated urinary tract infections (Mahadeva et al., 2014) along with efficient monitoring of diabetes ketoacidosis (Misra & Oliver, 2015).

Continuous glucose sensors

Continuous glucose sensors are different from point sensors as they possess the ability to monitor glucose continuously in the human body. MiniMed is a perfect example of a continuous glucose monitoring system (Christoforidis et al., 2020; Mastrototaro, 2000). Continuous glucose sensors are subclassified into invasive (Chuang et al., 2004), minimally invasive, and noninvasive (Kwon & Burkoth, 2003; Ma et al., 2020) depending on the method of glucose detection.

Invasive glucose sensors

Invasive glucose sensors consist of implantable sensors that monitor glucose levels in diabetic patients. These sensors are subclassified into microdialysis, subcutaneous sensors, and intravenous implantable sensors (Kovatchev et al., 2004; Oliver et al., 2009). A hollow and fine fiber is implanted in the subcutaneous region where an isotonic fluid is perfused in an ex vivo reservoir to detect glucose electrochemically via microdialysis (Jaquins-Gerstl & Michael, 2020; Kubiak et al., 2006). A miniaturized glucose monitoring system has been fabricated using a combination of microdialysis and needle-type glucose sensing and has demonstrated promise for long-term subcutaneous glucose monitoring in ambulatory diabetic patients (Hashiguchi et al., 1994). Polyurethane membranes can be used to enhance the efficiency of implantable microdialysis systems (Tapsak et al., 2017) and for amperometric detection of glucose (Vargas et al., 2016).

Intravenous implantable sensors are implanted in the intravenous region of the body to evaluate the level of blood glucose invasively. Since the venous blood samples are not appropriate to calculate blood glucose level in diabetic patients, these sensors are not popular (Armour et al., 1990). Glucose sensors implanted in the subcutaneous region of the body are the most investigated and marketed implantable sensors for continuous detection of blood glucose levels in the interstitial body fluid (Bobbioni-Harsch et al., 1993). Conventional subcutaneous glucose sensors such as GlucoDay (Poscia et al., 2003), Medtronic CGMS system (Renard et al., 2010), DexCom STS (McGarraugh, 2009), and Freestyle Navigator (Weinstein et al., 2007) are available in the glucose sensor markets. However, the lack of functional biostability in subcutaneously implanted glucose sensors drives the need for improved invasive glucose sensors. The Bayesian Multiday Framework and the feasible factory calibration methods have introduced for the calibration of subcutaneous glucose sensors for enhanced blood glucose measurement (Acciaroli et al., 2018; Hoss et al., 2014). Six-day Enlite sensors (Bailey et al., 2014), dexamethasone-loaded polyurethane coatings (Vallejo-Heligon et al., 2016), and flexible sensors (Shah et al., 2015) are some of the fourth-generation subcutaneous continuous systems for enhanced glucose monitoring.

Minimally invasive glucose sensors

Micropores or microneedles belong to the group of minimally invasive glucose sensing techniques and combine invasive and noninvasive glucose sensors to collect samples from components implanted into the body to detect glucose in vivo (Wang & Lee, 2015). Micropore techniques are used to create tiny pores in the stratum corneum with complete skin penetration, eventually leading to the collection of interstitial fluid for monitoring glucose (Chuang et al., 2004). Several novel devices and methods have been introduced to enhance the skin perforation to create effective micropores for continuous glucose monitoring (Jina et al., 2017). Microneedles can be implanted in diabetic patients to collect glucose samples for the determination of their blood glucose levels (Smart & Subramanian, 2000). Dermal interstitial fluids are the best choice for minimally invasion extraction via microneedles and are highly beneficial for monitoring glucose levels in patients (Takeuchi et al., 2020; Wang et al., 2005). They are coupled with in-built glucose sensors for continuous glucose detection (Tiangco et al., 2018). Microneedles can be used along with insulin delivery systems (Jin et al., 2018), self-powered glucose biosensors (Strambini et al., 2015), optofluidic glucose biosensors (Ranamukhaarachchi et al., 2017), wearable noninvasive epidermal glucose sensors

(Kim et al., 2018), transdermal glucose biosensors (Ventrelli et al., 2015), and plasmon resonance sensors of glucose molecules (Li et al., 2017).

Noninvasive glucose sensors

Optical and transdermal glucose sensors are types of noninvasive sensors for monitoring glucose without implants or the collection of glucose samples with microneedles. Optical glucose sensors are beneficial in monitoring glucose levels in patients via several unique techniques such as Raman spectroscopy (Lyandres et al., 2008), photoacoustic (MacKenzie et al., 1999), infrared (Zeller et al., 1989), kromoscopy (Arnold, 1996), polarimetry (Coté et al., 1992), and optical coherence tomography (Arnold & Small, 2005). Raman spectroscopy uses weak Raman signals to assess the scattering of a single wavelength light depending on the vibrational or rotational state of energy within a molecule (Colthup, 2012). Surface-enhanced Raman spectroscopic techniques are used to monitor glucose noninvasively. Transcutaneous (Shih et al., 2015), boronate nanoprobes (Gu et al., 2016), bisboronic acid (Sharma et al., 2016), and femto-second stimulated (McAnally et al., 2017) Raman spectroscopic techniques are used to enhance the monitoring ability of glucose biosensors. Ultrasonic waves created by light absorption are the basic principle behind photoacoustic spectroscopy—mediated glucose sensors (Zhang, Li, et al., 2020; Zhang, Chen, et al., 2020). It is a surrogate glucose estimation technique and has been used in a prototype wristwatch to monitor glucose (Raz et al., 2003). The feasibility of photoacoustic spectroscopy has been investigated via in vitro and in vivo experiments for noninvasive blood glucose measurements and the results demonstrated promise for glucose detection even in the presence of other analytes (MacKenzie et al., 1999). The performance of noninvasive Aprise glucose sensor that detects blood glucose levels via photoacoustic spectroscopy was investigated based on 62 diabetic subjects and the study highly recommended this type of sensor for continuous and effective monitoring of glucose (Weiss et al., 2007). More recently, several types of noninvasive glucose sensors are used for glucose detection from human skin (Kottmann et al., 2016), glucose monitors that are invulnerable to the products of skin secretions (Sim et al., 2018), and in on-chip glucose sensing via highly sensitive saliva (Zhang, Chen, et al., 2020).

Infrared-based glucose monitoring systems have been widely available to diabetic patients for a long time. It can be categorized into thermal infrared, mid-infrared (MIR), and near-infrared (NIR) based glucose monitoring systems. Local glucose concentration analysis is based on cutaneous microcirculation, which is the principle of thermal infrared glucose sensors as a surrogate method of glucose measurement (Malchoff et al., 2002). NIR- and MIR-based glucose biosensors on the other hand are useful for continuous glucose monitoring. In these techniques, MIR with $2.5-50\ \mu m$ of wavelength and NIR with $0.7-1.4\ \mu m$ of wavelength are used for the detection of glucose (Nelson et al., 2006; Shen et al., 2003). However, drawbacks such as light scattering, which reduces signals-to-noise ratio and the need to overcome heterogenous extracellular glucose distribution exist with these glucose sensors (Heise et al., 1994). MIR sensing is used in label-free glucose sensor chips (Lin et al., 2016), fiber-optic glucose monitoring systems (Yu et al., 2014), and for sensing glucose in human saliva (Haas et al., 2018). NIR sensing is used in glucose monitoring from human serum (Goodarzi & Saeys, 2016), light emitting diodes (LEDs), holographs, and integrated circuits for detecting glucose in blood and body fluids (Song, Ha, et al., 2015; Song, Wei, et al., 2015; Vezouviou & Lowe, 2015; Yadav et al., 2014). Fluorescence resonance energy transfer (FRET) is also used in glucose analysis. Live imaging of glucose (Volkenhoff et al., 2018), real-time intravascular glucose measurement (Markle & Markle, 2017), and genetically encoded glucose sensors (Peroza et al., 2015) are possible due to recent enhancements in FRET-based sensing. Optical coherence tomography is used in multisport glucose monitors (Shakespeare et al., 2017), flowmetry for analyte level estimation (Reichgott et al., 2015), and glucose detection on human fingertip (Chen et al., 2013; Jeevanandam & Danquah, 2020) for rapid glucose estimation.

Impedance spectroscopy, sonophoresis, skin suction blister technique, and reverse iontophoresis are the subclasses of transdermal glucose sensors. Impedance or radio-wave spectroscopy detects glucose in the vascular compartment based on the dielectric properties of tissues (Pfützner et al., 2004). A commercial glucose sensor called Pendra device is developed based on this technique (DeVries et al., 2006). A wide variety of impedance spectroscopy that are coupled with other techniques for enhanced glucose monitoring (Daikuzono et al., 2016; Valiūnienė et al., 2017; Ward et al., 2018). Sonophoresis, particularly low-frequency sonophoresis, uses ultrasound for transdermal monitoring of glucose (Kost et al., 2000; Mitragotri & Kost, 2004). These unique sensors can be used to develop commercial air ultrasonic ceramic transducers to detect glucose concentration and for delivering transdermal insulin (Jabbari et al., 2015). The skin suction blister technique is used to collect fluid from the dermal—epidermal junction by creating a blister of a few millimeters in diameter via vacuum over the skin surface (Kiistala, 1968). Metabolomics have been used to determine glucose in skin suction blister fluids (Niedzwiecki et al., 2018) and the technique is used in mobile monitoring of blood glucose and bilirubin (Javid et al., 2018). Interstitial fluids are accessed via physical energy used in reverse iontophoresis to quantify glucose concentration in the transdermal region (Rao et al., 1993). This method possesses the ability to be an alternative to enzyme-based electrode sensor systems with low glucose concentration and reduced electrode fouling. However, drawbacks such as the requirement of longer warm-up and calibration time are associated with these sensors (Oliver et al., 2009). Reverse iontophoresis has been used in several skin-like biosensor systems (Kim et al., 2018) via electrochemical channel—based glucose sensors (Chen, Lu, et al., 2017) and also in novel tattoo-based glucose monitoring systems (Bandodkar et al., 2014). Fig. 3.3 presents a summary of different techniques used to monitor glucose levels in diabetic patients. The various drawbacks associated with all the different glucose monitoring techniques indicate the need for unique, novel, and efficient strategies for enhanced glucose monitoring.

Nanomaterials in glucose sensing

The use of nanomaterials offers opportunities to address some of the challenges associated with conventional glucose sensors. The unique functionalities offered by nanosized particles along with their enhanced surface-to-volume ratio can help in improving the efficacy of glucose sensors (Jeevanandam et al., 2020). Various nanomaterials have been explored or currently being explored for applications in glucose sensors (Li et al., 2017). Metal-based, carbon-based, dopants, polymers, thin films, and other novel nanosized particles and materials are widely investigated for glucose sensing applications (Thatikayala et al., 2020).

Metal-based nanoparticles and nanomaterials for glucose sensors

Gold nanoparticles represent one of the first widely used nanosized materials for sensor applications, including glucose monitoring systems (Zhou et al., 2020). Gold nanoparticles have been used in enzyme-free amperometric glucose sensors (Jena & Raj, 2006), the immobilization of glucose oxidase over gold electrode for biosensor fabrication (Zhang et al., 2005), calorimetric glucose detection (Jiang et al., 2010), FRET-based glucose sensors (Oh et al., 2005), and plasmon resonance—mediated glucose sensors (Aslan et al., 2004). In addition, various morphologies of nanosized gold such as rods, hexagons, and fibers have been used in glucose sensor fabrications. Gold nanorods with conducting polymers have been used in nonenzymatic glucose sensors (Çiftçi & Tamer, 2012). Gold nanowire array electrodes have been used for nonenzymatic voltametric and amperometric detection of glucose (Cherevko & Chung, 2009), electrospun gold nanofiber electrodes for glucose biosensors (Marx et al., 2011), gold nanodots for luminescent-based glucose sensors (Shiang et al., 2009), and horseradish peroxidase—functionalized gold nanoclusters for monitoring glucose (Wen et al., 2011). Gold nanoparticles have also been used in fabricating enzyme-free impedimetric glucose sensors (Ahammad et al., 2016), on-chip glucose monitoring from saliva (Zhang, Du, et al., 2015), electrochemical glucose detection (Thanh et al., 2016), surface-enhanced Raman scattering-based glucose sensors (Hu et al., 2017), and amperometric reagent-less glucose biosensors (German et al., 2015). Other morphologies of nanosized gold such as nanotubes (Chen et al., 2015), nanorods (Zhang, Chen, et al., 2017), nanowires (Muratova et al., 2016), nanodots (Ravindranath et al., 2016), nanocubes (Chu et al., 2015), and nanosphere (Zhong et al., 2014) have been explored for glucose monitoring applications.

Silver is another metal that is highly synthesized in nanoforms and is effectively utilized in glucose sensor applications (Deshmukh et al., 2020). Silver nanoparticles have been used in localized surface plasmon resonance—based glucose sensors (Serra et al., 2009) to enhance the response of glucose sensors (Ren et al., 2005), as a redox marker in potentiometric glucose biosensors (Ngeontae et al., 2009), in surface-enhanced Raman-based glucose sensors (Haynes et al., 2005), and in photometric glucose sensor based on nanoparticle decolorization (Tashkhourian et al., 2011). Distinct morphologies of silver nanomaterials include nanorod arrays for surface-enhanced Raman scattering-based glucose sensors (Sun et al., 2014), silver nanowires for amperometric glucose sensors (Wang et al., 2013), silver nanocubes for amperometric bienzyme glucose biosensor (Yang et al., 2014), fluorescent probe for sensitive glucose detection (Wen et al., 2011), silver nanoprism for calorimetric glucose visualization (Xia et al., 2013), and hexagonal silver nanoparticles for nonenzymatic glucose sensors (Kundu et al., 2015). Silver nanoparticles have also been used in U-shaped fiber-optic attenuated total reflectance (ATR) sensors for continuous monitoring of glucose (Li, Yu, et al., 2015), nonenzymatic glucose sensors (Baghayeri et al., 2016), antibacterial nonenzymatic glucose sensors (Hoa et al., 2017), and calorimetric glucose sensors (Nguyen et al., 2018). Silver nanowires have been used to fabricate glucose sensors with photonic crystal fiber (Yang et al., 2014), silver nanoclusters for precise determination of blood glucose concentration (Naaz et al., 2018), and for fluorometric "Turn-On" glucose sensors (Chen et al., 2013; Jeevanandam & Danquah, 2020). Nanostructured silver fabric as a freestanding Nanozyme for colorimetric detection of glucose in urine has also been used in glucose sensor applications (Karim et al., 2018). Other metal nanoparticles such as copper in the form of nanoparticles have been used to enhance nonenzymatic and electrochemical glucose sensors (Zhang, Chen, et al., 2020), nanorods and nanowires for nonenzymatic amperometric and voltametric glucose sensors (Zhang, Chen, et al., 2017), nanospheres for highly sensitive amperometric glucose sensors (Chen et al., 2016), and nanoclusters for fluorescent glucose biosensors (Wang, Shu, et al.,2015). Platinum (Sun et al., 2014), nickel (Ensafi et al., 2017), and silicon (Ding et al., 2017) nanoparticles have also been extensively investigated for glucose sensor applications.

Metal oxides are the most stable form of metal nanoparticles and have proven to possess glucose sensing properties. Zinc oxide (ZnO) nanoparticles (Dayakar et al., 2017a; Pan et al., 2020; Sodzel et al., 2015), nanorods (Ahmad et al., 2017; Marie et al., 2015), nanotubes (Zhou et al., 2016), nanoflowers (Muhimmah et al., 2018), and nanowires (Hsu et al., 2017) are highly useful in fabricating glucose monitoring systems. Copper oxides (CuO) are also highly utilized nanoparticles for glucose sensor applications (Zhang, Du, et al., 2015) and its various morphologies including nanorods (Gou et al., 2018), nanofibers (Lu et al., 2016, 2014), nanowires (Jiang et al., 2014; Mani et al., 2015; Zhang, Li, et al., 2020), and carnation-like hierarchical nanostructures (Zhang, Du, et al., 2015) can help to improve glucose sensing efficacy. Iron oxide (Fe_2O_4) nanoparticles are a unique set of magnetic metal oxides that are widely used in magnetic relaxation switching glucose sensors and in electrochemical blood glucose detection (Alcantara et al., 2016; Vallabani et al., 2017). Generally, these oxide nanoparticles are combined with carbon nanoparticles as composites for glucose sensing. Fe_2O_4 nanorods have been used in enzyme-free glucose detection (Zhang, Du, et al., 2015) and Fe_2O_4 nanofiber-based glucose sensors (Senthamizhan et al., 2016) have demonstrated to have glucose sensing properties. Nanoparticles of manganese (Yang, Liu, et al., 2015), silicon (Ranjani et al., 2015), nickel (Li et al., 2014), cerium (Zhou et al., 2020), and cobalt (Heidari & Habibi, 2016) are additional types of metal oxides that have been used in glucose sensing applications, mostly as composites. Besides, gold and copper nanoparticles for glucose sensing applications have been demonstrated commercially or used in clinical trials.

Carbon-based nanoparticles and nanomaterials in glucose sensing

Carbon-based nanoparticles are highly utilized in various components of glucose monitoring systems. Carbon nanoparticles have been used in nonenzymatic glucose sensors, especially in enhancing glucose detection ability in combination with metal oxide nanoparticles (Alizadeh et al., 2020; Karikalan et al., 2016). Other forms of carbon nanomaterials such as nanotubes, nanowires, nanofibers, and zero-dimensional dots have been used for the purpose of glucose detection, exploiting their high reactive surface bonds in forming nanocomposites to enhance glucose sensing (Yang, Denno, et al., 2015). It has been reported that both single-walled and multiwalled carbon nanotubes (CNTs) possess glucose detection ability as standalone sensors (Zhong et al., 2014). Carbon nanowires have been used to fabricate rapid label-free highly sensitive chemiresistive glucose biosensing platforms (Thiha et al., 2018), carbon nanofibers for developing glassy carbon electrodes for enhancing nonenzymatic glucose sensing (Lu et al., 2016), carbon nanospheres from cocoon silk as a metal-free electrocatalyst for monitoring glucose (Li, Li, et al., 2015), carbon dots for nonenzymatic amperometric sensor (Li, Zhong, et al., 2015), fluorescent blood glucose sensing (Nelson et al., 2006; Shen et al., 2003), mediator-less glucose biosensor (Zhao et al., 2015), and in glucose optical sensing applications (Wang, Shu, et al., 2015). In addition, other allotropes of carbon such as graphite, graphene, and fullerene have demonstrated effective glucose monitoring capability. Graphite is a three-dimensional carbon allotrope and its oxide nanoparticles have been utilized for field effect transistor (FET) fabrication of nonenzymatic glucose sensor (Said et al., 2017), graphite pencil electrode and graphite disks for enhancing nonenzymatic glucose sensing (Ju et al., 2016; Liu, Yang, et al., 2016; Zhang, Ding, et al., 2017). Nanosized oxide of graphene with a thin two-dimensional carbon allotrope (Farid et al., 2016), sheets (Hsieh et al., 2017), and nanostructures (Ngo et al., 2017) have been used to improve glucose sensing ability of metal oxides or composite nanostructures. Fullerenes (C60) are zero-dimensional allotropes of carbon that are highly beneficial as nanomediators for fabricating highly sensitive glucose biosensors (Afreen et al., 2015), nanostructured platforms for electrochemical glucose biosensors (Pilehvar & De Wael, 2015), and electrochemical catalytic and affinity glucose biosensor (Yáñez-Sedeño et al., 2017). Novel zero-dimensional fullerenes such as C70 have been used as a barrier layer in glucose sensors (Crane et al., 2018) and diamond nanowires have been used as a novel platform for electrochemical detection of glucose (Szunerits et al., 2015). Mostly, these carbon-based nanomaterials are used for point sample glucose sensing due to potential toxic reaction by the human body when used in continuous in vivo sensors.

Other novel nanomaterials for glucose sensing applications

Various novel nanomaterials that are fabricated via latest technologies are gaining interest among researchers for use in glucose monitoring systems (Chen et al., 2020). Glucose-sensitive polymeric nanoparticles such as egg phosphatidylcholine and dioleoylphos-phatidylethanolamine liposomes are utilized for producing self-regulated drug delivery systems based on glucose oxidase for glucose monitoring and diabetes treatment (Zhao et al., 2015). Further, cerium-coordinated adenosine triphosphate—based polymeric nanoparticles have been investigated for its glucose detecting property (Zeng et al., 2016). Flower-like copper cobaltite nanosheets mounted over graphite paper (Liu, Hui, et al., 2016), nickel cobaltite nanosheets (Naik et al., 2015), nickel hydroxide (Qian et al., 2018), and nickel subsulfide nanosheets (Huo et al., 2014) have been used as supercapacitor electrodes in enzyme-less glucose detection. Bimetallic nanoparticles such as platinum—gold (Singh & Dempsey, 2011), gold—ruthenium (Jo et al., 2014), platinum—lead (Guo et al., 2011), gold—nickel oxide (Ding et al., 2017), palladium—gold (Li & Du, 2017), copper—silver (Darabdhara et al., 2017), and gold—silver (Yu et al., 2014) have also been used in various glucose monitoring applications. Nanodendrites (Song, Ha, et al., 2015; Vezouviou & Lowe, 2015; Yadav et al., 2014), metal-doped nanoparticles (Muhr et al., 2017), and thin films (Rinaldi & Carballo, 2016) are other forms of nanomaterials that are beneficial for glucose monitoring. However, these novel nanomaterials still require extensive research efforts to support potential commercialization endeavors for glucose monitoring applications (Zaidi & Shin, 2016). Table 3.1 presents a summary of various nanomaterials used in glucose sensing applications.

Nanocomposites as glucose sensors

Nanocomposites are used in glucose sensing to incorporate enhanced sensing ability as mentioned earlier. Generally, nanocomposites are fabricated using metal oxides, carbon along with its associated materials, polymers, and other unique combination of materials to enhance glucose sensing ability (Deshmukh et al., 2020). In recent times, several metals-

Table 3.1 Nanomaterials in glucose sensing applications.

Nanoparticles or nanomaterials	Glucose sensors	References
Metal nanoparticles/materials		
Gold nanoparticles	Enzyme-free amperometric sensors	Jena and Raj (2006)
	Immobilization of glucose oxidase	Zhang et al. (2005)
	Calorimetric glucose detection	Jiang et al. (2010)
	FRET sensors	Oh et al. (2005)
	Third-generation horseradish peroxidase biosensor	Jia et al. (2002)
	Plasmon resonance—based sensors	Aslan et al. (2004)
	Enzyme-free impedimetric glucose sensor	Ahammad et al. (2016)
	Highly sensitive, on-chip glucose sensor	Zhang, Ni, et al. (2015), Zhang, Ma, et al. (2015)
	Electrochemical glucose detection	Thanh et al. (2016)
	Surface-enhanced Raman scattering—based glucose sensors	Hu et al. (2017)
	Amperometric reagent-less glucose biosensors	German et al. (2015)
	Enzymatic glucose sensor	Zhou et al. (2020)
Gold nanorods	Nonenzymatic glucose sensors	Çiftçi and Tamer (2012)
	Calorimetric glucose monitoring	Zhang et al. (2016)
	Highly sensitive on-site glucose detection in human urine	Zhang, Chen, et al. (2017)
Gold nanowires	Nonenzymatic voltametric and amperometric sensors	Cherevko and Chung (2009)
	Resistance over-potential for glucose detection	Muratova et al. (2016)
Gold nanofibers	Glucose biosensors	Marx et al. (2011)
Gold nanodots	Luminescent glucose sensors	Shiang et al. (2009), Ravindranath et al. (2016)
Gold nanoclusters	Fluorescent glucose sensors	Wen et al. (2011)
	Fluorescent ratiometric glucose detection	Meng et al. (2018)

(Continued)

Table 3.1 (Continued)

Nanoparticles or nanomaterials	Glucose sensors	References
Gold nanotubes	Ultrasensitive and stable nonenzymatic glucose sensors	Chen et al. (2015)
Gold nanocubes	Third-generation ultrasensitive glucose biosensor	Chu et al. (2015)
Gold nanosphere	Chemiluminescence glucose biosensor	Zhong et al. (2014)
Silver nanoparticles	Localized surface plasmon resonance—based glucose sensors	Serra et al. (2009)
	Potentiometric glucose sensor	Ngeontae et al. (2009)
	Surface-enhanced Raman glucose sensors	Haynes et al. (2005)
	Photometric glucose sensors	Tashkhourian et al. (2011)
	Fiber-optic attenuated total reflectance (ATR) glucose sensor	Li, Yu, et al. (2015), Li, Li, et al. (2015), Li, Zhong, et al. (2015)
	Nonenzymatic glucose sensor	Baghayeri et al. (2016), Deshmukh et al. (2020)
	Antibacterial, nonenzymatic glucose sensors	Hoa et al. (2017)
	Calorimetric glucose sensor	Nguyen et al. (2018)
Silver nanorods	Surface-enhanced Raman glucose sensors	Sun et al. (2014)
Silver nanowires	Amperometric glucose sensors	Wang et al. (2013)
	Glucose sensor with photonic crystal fiber	Yang et al. (2016)
Silver nanocubes	Amperometric bi-enzyme glucose biosensor	Yang et al. (2014)
Silver nanoprism	Calorimetric glucose sensing	Xia et al. (2013)
Hexagonal silver nanoparticles	Nonenzymatic glucose sensors	Kundu et al. (2015)
Silver nanoclusters		
	Fluorometric "Turn-On" glucose sensors	Chen, Zhao, et al. (2017)
Fabric with nanosilver	Freestanding Nanozyme for colorimetric detection of glucose in urine	Karim et al. (2018)
Copper nanoparticles	Nonenzymatic glucose sensors	Jiang et al. (2014), Zhang, Li, et al. (2020), Zhang, Chen, et al. (2020)
	Electrochemical glucose sensors	Mani et al. (2015)

(Continued)

Table 3.1 (Continued)

Nanoparticles or nanomaterials	Glucose sensors	References
Copper nanorods	Nonenzymatic amperometric glucose sensors	Liu, Hui, et al. (2016), Liu, Yang, et al. (2016)
Copper nanowires	Nonenzymatic amperometric glucose sensors	Ju et al. (2016)
	Nonenzymatic voltametric glucose sensors	Zhang, Chen, et al. (2017)
Copper nanospheres	Highly sensitive amperometric glucose sensor	Chen et al. (2016)
Copper nanoclusters	Fluorescent glucose biosensors	Wang, Yi, et al. (2015)
Platinum nanoparticles	Sensitive, nonenzymatic glucose sensor	Sun and Hur (2015)
Nickel nanoparticles	Enzyme-free electrochemical sensor	Ensafi et al. (2017)
Silicon nanoparticles	Fluorescent glucose monitor	Ding et al. (2017)
Metal oxide nanoparticles/materials		
ZnO nanoparticles	Nonenzymatic glucose sensor	Dayakar et al. (2017a)
	Continuous ultraviolet and visible luminescence sensors	Sodzel et al. (2015)
	Electrochemical sensors	Pan et al. (2020)
ZnO nanorods	Nonenzymatic glucose sensor	Marie et al. (2015; Ahmad et al. (2017)
ZnO nanotubes	Amperometric glucose sensors	Zhou et al. (2016)
ZnO nanowires	Illuminated ultraviolet/green LED-based nonenzymatic glucose biosensors	Hsu et al. (2017)
ZnO nanoflowers	Paper-based microfluidic blood glucose sensor	Muhimmah et al. (2018)
CuO nanoparticles	Nonenzymatic glucose sensor	Tian et al. (2015), Zhang, Ni, et al. (2015), Zhang, Ma, et al. (2015)
CuO nanorods	Nonenzymatic highly sensitive glucose sensor	Gou et al. (2018)
CuO nanofibers	Electrode material for glucose sensing	Lu et al. (2014)
CuO nanowires	Selective nonenzymatic glucose sensor	Zhang et al. (2014)

(*Continued*)

Table 3.1 (Continued)

Nanoparticles or nanomaterials	Glucose sensors	References
Iron oxide nanoparticles	Magnetic relaxation switching glucose sensors	Alcantara et al. (2016)
	Electrochemical blood glucose detection	Vallabani et al. (2017)
Iron oxide nanorods	Enzyme-free glucose sensors	Zhang, Ni, et al. (2015), Zhang, Ma, et al. (2015)
Manganese oxide nanoparticles	Selective nonenzymatic glucose sensor	Yang, Tong, et al. (2015)
Cerium oxide nanoparticles	Enzymatic glucose sensor	Zhou et al. (2017)
Silicon dioxide nanoparticles	Nonenzymatic glucose sensor	Ranjani et al. (2015)
Nickel oxide nanoparticles	Nonenzymatic glucose detection	Li et al. (2014)
Cobalt oxide nanoparticles	Enzyme-free amperometric glucose sensor	Heidari and Habibi (2016)

Carbon-based nanoparticles/materials

Nanoparticles or nanomaterials	Glucose sensors	References
Carbon nanoparticles	Nonenzymatic glucose detection	Karikalan et al. (2016)
	Optical glucose sensor	Alizadeh et al. (2020)
Single-walled CNTs	Electrochemical glucose sensor	Goornavar et al. (2014)
	Electrochemical paper-based nonenzymatic glucose sensor	Tran et al. (2018)
Multiwalled CNTs	Rapid, reusable, and sensitive glucose detector	Başkaya et al. (2017)
	Enzyme-free glucose sensors	Zhong et al. (2015)
Suspended carbon nanowires	Rapid, sensitive label-free chemiresistive biosensing platform	Thiha et al. (2018)
Carbon nanofibers	Fabrication of glassy carbon electrodes for enhanced nonenzymatic glucose sensing	Lu et al. (2016)
Carbon nanospheres	Metal-free electrocatalyst for glucose sensors	Li, Yu, et al. (2015), Li, Li, et al. (2015), Li, Zhong, et al. (2015)

(Continued)

Table 3.1 (Continued)

Nanoparticles or nanomaterials	Glucose sensors	References
Carbon nanodots	Nonenzymatic amperometric sensor	Li, Yu, et al. (2015), Li, Li, et al. (2015), Li, Zhong, et al. (2015)
	Fluorescent blood glucose sensor	Shen and Xia (2014)
	Mediator-less glucose biosensor	Zhao et al. (2015)
	Optical glucose sensor	Wang, Yi, et al. (2015)
Graphite oxide nanoparticles	FET for nonenzymatic glucose sensor	Said et al. (2017)
Graphite pencil electrode	Sensitive, nonenzymatic voltametric glucose detection	Kawde et al. (2017)
Graphite disks	Enhanced nonenzymatic glucose detection	Liu, Wu, et al. (2015)
Graphene oxide nanoparticles	Novel glucose sensor	Farid et al. (2016)
Graphene oxide sheets	Nonenzymatic glucose oxidation	Hsieh et al. (2017)
Fullerene (C60)	Nanomediator for highly sensitive glucose biosensor fabrication	Afreen et al. (2015)
	Nanostructured platforms for electrochemical glucose biosensors	Pilehvar and De Wael (2015)
	Electrochemical catalytic and affinity glucose biosensor	Yáñez-Sedeño et al. (2017)
C70	Barrier layer in glucose sensors	Crane et al. (2018)
Diamond nanowires	Novel electrochemical platform for glucose detection	Szunerits et al. (2015)

Novel nanoparticles/materials

Flower-like copper cobaltite nanosheets mounted over graphite paper	Enzyme-less glucose sensor	Liu, Hui, et al. (2016), Liu, Yang, et al. (2016)
Nickel cobaltite nanosheets	Glucose sensor	Naik et al. (2015)
Nickel hydroxide nanosheets	Sensitive fiber microelectrode for nonenzymatic glucose determination	Qian et al. (2018)
3D nickel subsulfide nanosheets	High-performance nonenzymatic glucose detection	Huo et al. (2014)

(Continued)

Table 3.1 (Continued)

Nanoparticles or nanomaterials	Glucose sensors	References
Platinum—gold nanoparticles	Amperometric nonenzymatic glucose sensor	Singh and Dempsey (2011), Zaidi and Shin (2016)
Gold—ruthenium nanoshells	Nonenzymatic glucose sensor	Jo et al. (2014)
Gold—nickel oxide nanobelts	Sensitive and selective glucose detection	Ding et al. (2011)
Gold—palladium nanoparticles	Nonenzymatic electrochemical glucose sensor	Li et al. (2017)
Gold—silver nanoparticles	Chemiluminescent glucose detection	Yu and He (2015)
Gold—platinum nanodendrites	Electrochemical nonenzymatic glucose sensor	Song, Ha, et al. (2015), Song, Wei, et al. (2015)
Europium-doped gadolinium vanadium oxide nanoparticles	Luminescent probe for enzymatic glucose sensing	Muhr et al. (2017)
Single-walled carbon nanotube-copper oxide-zinc oxide nanorods-graphene hybrid electrodes	Nonenzymatic glucose sensor	Chen et al. (2020)

based nanocomposites are fabricated for glucose monitoring applications. Gold—graphene (Shu et al., 2015), cobalt nanobeads—graphene oxides (Song, Wei, et al., 2015), anthill-like copper—carbon nanocomposites (Wei et al., 2014), core—shell iron—platinum (Mei et al., 2016), nickel nanoparticles-attapulgite-graphene oxide (Shen et al., 2016), and graphene oxide—copper oxide (Shen et al., 2016) are some of the important nanocomposites that have been used to improve glucose sensing. It can be noted that most metals are usually combined with carbon materials to form nanocomposites for enhanced glucose sensing due to the synergistic efficacy of the carbon structures in combination with the metals (Dag et al., 2005). Metals can also be combined with polymers or other materials as nanocomposites for glucose sensing. Examples include nickel oxide-copper oxide-polyaniline (Ghanbari & Babaei, 2016), nickel oxide-silicon carbide (Yang et al., 2014), mesoporous zinc sulfide-nickel sulfide (Wei et al., 2014), polypyrrole-chitosan-iron oxide nanocomposite films (AL-Mokaram et al., 2016), and gold-HS/SO_3H-PMOs, where HS is HS-CH_2-CH_2-CH_2 group, SO_3H is SO_3H-CH_2-CH_2-CH_2 group, and PMO is periodic mesoporous organosilic (Jia et al., 2015). Carbon-based nanocomposites with quantum dots (Chen et al., 2015; Maaoui et al., 2016), cadmium selenide quantum dots-graphene nanosheets (Zhang, Liu, et al., 2014), graphene oxide-copper (Wang, Wang, et al., 2015), gold-graphene (Turcheniuk et al., 2015), CNT-manganese oxide (Guo et al., 2011), multi-walled CNT-nickel oxide (Prasad & Bhat, 2015), and fullerene (Sutradhar & Patnaik, 2017) have widely been investigated for glucose sensing applications. Novel combinations of metals, polymers, and carbon have also been utilized to improve the efficiency of glucose monitoring systems (Liu, Guo, et al., 2015; Sedghi & Pezeshkian, 2015). Table 3.2 presents a summary of nanocomposites that have been investigated for glucose sensor applications. These nanocomposites are used in point sample, noninvasive, and minimally invasive glucose sensors. Even though, the incorporation of these nanocomposites in invasive glucose sensors may improve in vivo glucose sensing, cytotoxicity is a major concern. It is noteworthy that these nanocomposites are mostly synthesized using chemicals that are toxic to human cells (Kefeni et al., 2020; Pecoraro et al., 2018). However, these chemically fabricated nanocomposites can be used to initially design composite formulations with effective glucose sensing ability, and later replaced with biosynthesis methods based on the existing knowledge of bionanocomposite fabrication.

Biosynthesized nanoparticles as glucose sensors

Some of the challenges of chemically synthesized nanomaterials may be addressed via biosynthesis approaches relying on the use of benign biological materials. Biosynthesis approaches can be bacterial, fungal, and plant-mediated depending on the type of biological entity used for the synthesis process. Biosynthesized nanomaterials are mostly bioactive, biocompatible, and can be less toxic in vivo (Jeevanandam et al., 2016). It is possible to use biosynthesis approaches to synthesize a wide range of nanomaterials. However, research reports on biosynthesized nanoparticles for glucose sensing applications are limited. A novel nanobiocomplex of gold nanoparticles and Vmh2 hydrophobin (a unique amphiphilic fungal protein) was developed, and the complex demonstrated glucose sensing ability (Jane et al., 2016). Microbial glucose biosensor based on *Gluconobacter oxydans* and glassy carbon paste electrodes of graphene oxide nanoparticles and graphene-platinum hybrid nanoparticles has been developed for glucose sensing, and it showed a good glucose detecting property (Aslan et al., 2004). Also, bacterial cellulose from *Komeigatabacter xylinus* has been used as a base for the deposition of gold nanoparticles to fabricate highly flexible, self-powered biosensor for glucose detection (Lv et al., 2018). Viruses such as genetically engineered M13 bacteriophages have been used as templates to form manganese dioxide nanowires to enhance the efficacy of electrochemical glucose biosensors (Han et al., 2016). Plants are widely used as bioresource for the synthesis of various nanoparticles and nanocomposites. Copper, zinc oxide, and silver nanoparticles have been synthesized from *Ocimum tenuiflorum* leaf extract and demonstrated enhanced glucose detection properties (Dayakar et al., 2017b; Pan et al., 2020; Sodzel et al., 2015). The same leaf extract

Table 3.2 Nanocomposites for glucose sensing applications.

Nanocomposites	Glucose sensors	References
Gold-graphene	High-performance nonenzymatic glucose sensor	Shu et al. (2015)
Cobalt nanobeads-graphene oxides	Organic framework for glucose sensor	Song, Ha, et al. (2015), Song, Wei, et al. (2015)
Anthill-like copper-carbon nanocomposites	Nonenzymatic glucose sensor	Wei et al. (2014)
Core-shell iron-platinum	Nonenzymatic electrochemical glucose sensors	Mei et al. (2016)
Nickel nanoparticles-attapulgite-graphene oxide	Sensitive, nonenzymatic glucose sensor	Shen et al. (2016)
Hydroxide-bismuth oxynitrate	Nonenzymatic glucose sensor	Liu et al. (2017)
Nickel oxide-copper oxide-polyaniline	Nonenzymatic glucose sensor	Ghanbari and Babaei (2016)
Nickel oxide-silicon carbide	Electrocatalyst for nonenzymatic glucose sensor	Yang, Tong, et al. (2015)
Mesoporous zinc sulfide-nickel sulfide	Nonenzymatic electrochemical glucose sensors	Wei et al. (2015)
Polypyrrole-chitosan-iron oxide films	Nonenzymatic glucose sensor	AL-Mokaram et al. (2016)
Gold-HS/SO$_3$H-PMOs	Nonenzymatic electrochemiluminescence glucose sensor	Jia et al. (2015)
Carbon quantum dots-copper oxide	Nonenzymatic glucose sensor	Maaoui et al. (2016)
Silver-graphene quantum dots	Ultrasensitive colorimetric glucose detection	Chen et al. (2018)
Cadmium selenide-graphene nanosheets	Nonenzymatic photoelectrochemical glucose detection	Zhang, Xu, et al. (2014)
CNT-manganese dioxide	Composite electrode for glucose detection	Guo et al. (2015)
Multiwalled CNT-nickel oxide	Enzyme-free electrochemical glucose sensor	Prasad and Bhat (2015)
Fullerene (C60)-gold	Nonenzymatic glucose sensors	Sutradhar and Patnaik (2017)
Polyvinylpyrrolidone-graphene nanosheets-nickel nanoparticles-chitosan	High-performance nonenzymatic electrochemical glucose sensors	Liu, Wu, et al. (2015)
Multiwalled CNT-carboxyl group-poly (2-aminothiophenol)-gold nanoparticles	Nonenzymatic glucose sensors	Sedghi and Pezeshkian (2015)
Polyaniline-reduced graphene oxide-silver nanoparticles	Nonenzymatic electrochemical glucose sensors	Deshmukh et al. (2020)

has been used for the synthesis of silver–titanium dioxide and cerium oxide–copper oxide core–shell nanostructure for screen-printed electrode-based nonenzymatic glucose sensors (Dayakar, Rao, Vinod Kumar, et al., 2018; Dayakar, Rao, Bikshalu, et al., 2018). In addition, the gooseberry leaf extract–synthesized iron oxide nanoparticles (Rahman et al., 2017) and several other plant extracts–based nanoparticles used for glucose sensing have been reported (Bhattacharya et al., 2018). Biosynthesized nanomaterials have the potential to be less toxic to human cells and the environment, making them a good candidate for invasive glucose sensing. Table 3.3 presents a summary of different biosynthesized nanomaterials for glucose sensing applications.

Table 3.3 Biosynthesized nanomaterials for glucose sensing applications.

Biological entity	Nanomaterial	Glucose sensing	References
Fungal hydrophobin Vmh2	Polyethylene glycol-gold nanoparticle	Glucose monitoring systems	Jane et al. (2016)
Gluconobacter oxydans	Graphene oxide and graphene-platinum hybrid nanoparticles	Microbial glucose biosensor	Aslan and Anik (2016)
Komeigatabacter xylinus cellulose	Gold nanoparticles	Self-powered glucose sensor	Lv et al. (2018)
M13 bacteriophage (virus)	Manganese dioxide nanowires	Electrochemical glucose sensor	Han et al. (2016)
Ocimum tenuiflorum	Copper nanoparticles	Nonenzymatic glucose sensor	Dayakar et al. (2017a)
O. tenuiflorum	Silver nanoparticles	Nonenzymatic glucose biosensor	Dayakar, Venkateswara Rao, et al. (2018)
O. tenuiflorum	Zinc oxide nanoparticles	Nonenzymatic glucose biosensor	Dayakar et al. (2017a)
O. tenuiflorum	Cerium oxide-copper oxide core-shell nanostructure	Screen-printed electrode for nonenzymatic glucose biosensor	Dayakar, Rao, Vinod Kumar, et al. (2018), Dayakar, Rao, Bikshalu, et al. (2018)
O. tenuiflorum	Silver-titanium dioxide core-shell nanostructure	Nonenzymatic glucose biosensor	Dayakar, Rao, Vinod Kumar, et al. (2018), Dayakar, Rao, Bikshalu, et al. (2018)
Gooseberry leaf extract	Iron oxide nanoparticles	Glucose biosensor	Rahman et al. (2017)

Current trends in nanoglucose sensors

As discussed earlier, the advantages of nanotechnology have led to current and emerging innovations in glucose sensors. However, despite the availability of these nanomaterials, there is no standardized pipeline and benchmarks toward the development of nanomaterial-based glucose sensors, and this represents a significant challenge toward the commercialization of most of the currently developed technologies. Most of the developed nanomaterials-based glucose sensing technologies are used in minimally invasive and noninvasive continuous glucose monitoring systems or point sample glucose sensing. The current trend is shifting toward the fabrication of less toxic biosynthesized nanoparticles for use in continuous invasive glucose sensing. Breathalyzer nanosensors have been developed as novel glucose sensors for sensing glucose levels through breath (Banach et al., 2013). Nanotattoos are also becoming popular among young diabetic patients to monitor glucose level (Heo & Takeuchi, 2013). Incorporating glucose sensing capabilities into wearable devices have also been a major focus of recent technological advancements. This is a significant interest to develop multimodal glucose sensing platforms integrating bioaffinity systems, nanotechnology, data analytics, and signal processing to create high-precision glucose sensing platforms via wireless communication using mobile devices. There is a huge opportunity for further advancements in glucose sensing.

Conclusion

This chapter provides a summary of various glucose sensing technologies to support diabetes monitoring. In addition, it discusses various nanoparticles, nanomaterials, nanocomposites, and biosynthesized nanomaterials that have been used in glucose sensing applications. It demonstrates that nanosensors have the potential to revolutionize glucose sensing. However, more research advancements and standardized regulatory framework are needed to support the development and commercialization of nanomaterials-based glucose sensors for diabetes monitoring.

References

Acciaroli, G., Vettoretti, M., Facchinetti, A., Sparacino, G., & Cobelli, C. (2018). Reduction of blood glucose measurements to calibrate subcutaneous glucose sensors: A Bayesian Multiday Framework. *IEEE Transactions on Biomedical Engineering, 65*(3), 587–595.

Adepoyibi, T., Weigl, B., Greb, H., Neogi, T., & McGuire, H. (2013). New screening technologies for type 2 diabetes mellitus appropriate for use in tuberculosis patients. *Public Health Action, 3*(1), 10–17.

Afreen, S., Muthoosamy, K., Manickam, S., & Hashim, U. (2015). Functionalized fullerene (C60) as a potential nanomediator in the fabrication of highly sensitive biosensors. *Biosensors and Bioelectronics, 63*, 354–364.

Ahammad, A. S., Al Mamun, A., Akter, T., Mamun, M., Faraezi, S., & Monira, F. (2016). Enzyme-free impedimetric glucose sensor based on gold nanoparticles/polyaniline composite film. *Journal of Solid State Electrochemistry, 20*(7), 1933−1939.

Ahmad, R., Ahn, M.-S., & Hahn, Y.-B. (2017). Fabrication of a non-enzymatic glucose sensor field-effect transistor based on vertically-oriented ZnO nanorods modified with Fe_2O_3. *Electrochemistry Communications, 77*, 107−111.

Ahmed, J. (2020). Electrospinning for the manufacture of biosensor components: A mini-review. *Medical Devices & Sensors, 4*, e10136.

Alcantara, D., Lopez, S., García-Martin, M. L., & Pozo, D. (2016). Iron oxide nanoparticles as magnetic relaxation switching (MRSw) sensors: Current applications in nanomedicine. *Nanomedicine: Nanotechnology, Biology and Medicine, 12*(5), 1253−1262. Available from https://doi.org/10.1016/j.nano.2016.01.005.

Alizadeh, N., Salimi, A., & Hallaj, R. (2020). A strategy for visual optical determination of glucose based on a smartphone device using fluorescent boron-doped carbon nanoparticles as a light-up probe. *Microchimica Acta, 187*(1), 14.

AL-Mokaram, A. M. A. A., Yahya, R., Abdi, M. M., & Mahmud, H. N. M. E. (2016). One-step electrochemical deposition of Polypyrrole−Chitosan−Iron oxide nanocomposite films for non-enzymatic glucose biosensor. *Materials Letters, 183*, 90−93.

Armour, J. C., Lucisano, J. Y., McKean, B. D., & Gough, D. A. (1990). Application of chronic intravascular blood glucose sensor in dogs. *Diabetes, 39*(12), 1519−1526.

Arnold, M. A. (1996). Non-invasive glucose monitoring. *Current Opinion in Biotechnology, 7*(1), 46−49.

Arnold, M. A., & Small, G. W. (2005). Noninvasive glucose sensing. *Analytical Chemistry, 77*(17), 5429−5439.

Aslan, K., Lakowicz, J. R., & Geddes, C. D. (2004). Nanogold-plasmon-resonance-based glucose sensing. *Analytical Biochemistry, 330*(1), 145−155.

Aslan, S., & Anik, Ü. (2016). Microbial glucose biosensors based on glassy carbon paste electrodes modified with *Gluconobacter oxydans* and graphene oxide or graphene-platinum hybrid nanoparticles. *Microchimica Acta, 183*(1), 73−81. Available from https://doi.org/10.1007/s00604-015-1590-9.

Baghayeri, M., Amiri, A., & Farhadi, S. (2016). Development of non-enzymatic glucose sensor based on efficient loading Ag nanoparticles on functionalized carbon nanotubes. *Sensors and Actuators B: Chemical, 225*, 354−362. Available from https://doi.org/10.1016/j.snb.2015.11.003.

Bailey, T. S., Ahmann, A., Brazg, R., Christiansen, M., Garg, S., Watkins, E., Welsh, J. B., & Lee, S. W. (2014). Accuracy and acceptability of the 6-day Enlite continuous subcutaneous glucose sensor. *Diabetes Technology & Therapeutics, 16*(5), 277−283.

Banach, N. M., Rust, M. J., & Priefer, R. (2013). A portable spectrophotometer-based breathalyzer for point-of-care testing of diabetic patients. In *Proceedings of the thirty-ninth annual northeast bioengineering conference* (pp. 245−246). IEEE.

Bandodkar, A. J., Jia, W., Yardımcı, C., Wang, X., Ramirez, J., & Wang, J. (2014). Tattoo-based noninvasive glucose monitoring: A proof-of-concept study. *Analytical Chemistry, 87*(1), 394−398.

Başkaya, G., Yıldız, Y., Savk, A., Okyay, T. O., Eriş, S., Sert, H., & Şen, F. (2017). Rapid, sensitive, and reusable detection of glucose by highly monodisperse nickel nanoparticles decorated functionalized multi-walled carbon nanotubes. *Biosensors and Bioelectronics, 91*, 728−733.

Bhattacharya, S., Agarwal, A. K., Chanda, N., Pandey, A., Sen, A. K., Mandal, D., Mishra, S., & Singh, R. K. (2018). Green synthesized nanoparticles as potential nanosensors. In *Environmental, chemical and medical sensors* (pp. 137−164). Springer, Singapore. <https://doi.org/10.1007/978-981-10-7751-7_7>.

Bobbioni-Harsch, E., Rohner-Jeanrenaud, F., Koudelka, M., de Rooij, N., & Jeanrenaud, B. (1993). Lifespan of subcutaneous glucose sensors and their performances during dynamic glycaemia changes in rats. *Journal of Biomedical Engineering, 15*(6), 457−463.

Britton, J. P., Dowell, A. C., & Whelan, P. (1989). Dipstick haematuria and bladder cancer in men over 60: Results of a community study. *BMJ (Clinical Research ed.), 299*(6706), 1010−1012.

Burmeister, J. J., & Arnold, M. A. (1995). Accuracy of the YSI stat plus analyzer for glucose and lactate. *Analytical Letters, 28*(4), 581−592.

Cass, A. E., Davis, G., Francis, G. D., Hill, H. A. O., Aston, W. J., Higgins, I. J., Plotkin, E. V., Scott, L. D., & Turner, A. P. (1984). Ferrocene-mediated enzyme electrode for amperometric determination of glucose. *Analytical Chemistry, 56*(4), 667−671.

Chen, A., Ding, Y., Yang, Z., & Yang, S. (2015). Constructing heterostructure on highly roughened caterpillar-like gold nanotubes with cuprous oxide grains for ultrasensitive and stable nonenzymatic glucose sensor. *Biosensors and Bioelectronics*, 74, 967–973.

Chen, C., Xie, Q., Yang, D., Xiao, H., Fu, Y., Tan, Y., & Yao, S. (2013). Recent advances in electrochemical glucose biosensors: A review. *RSC Advances*, 3(14), 4473–4491.

Chen, C., Zhao, X.-L., Li, Z.-H., Zhu, Z.-G., Qian, S.-H., & Flewitt, A. J. (2017). Current and emerging technology for continuous glucose monitoring. *Sensors*, 17(1), 182.

Chen, H.-C., Su, W.-R., & Yeh, Y.-C. (2020). Functional channel of SWCNTs/Cu$_2$O/ZnO NRs/graphene hybrid electrodes for highly sensitive nonenzymatic glucose sensors. *ACS Applied Materials & Interfaces*, 12(29), 32905–32914.

Chen, S.-M., Devasenathipathy, R., Wang, S.-F., & Kohilarani, K. (2016). Highly sensitive amperometric sensor for the determination of glucose at histidine stabilized copper nanospheres decorated multiwalled carbon nanotubes. *International Journal of Electrochemical Science*, 11, 5416–5426.

Chen, T.-L., Lo, Y.-L., Liao, C.-C., & Phan, Q.-H. (2018). Noninvasive measurement of glucose concentration on human fingertip by optical coherence tomography. *Journal of Biomedical Optics*, 23(4), 047001.

Chen, Y., Lu, S., Zhang, S., Li, Y., Qu, Z., Chen, Y., Lu, B., Wang, X., & Feng, X. (2017). Skin-like biosensor system via electrochemical channels for noninvasive blood glucose monitoring. *Science Advances*, 3(12), e1701629.

Cherevko, S., & Chung, C.-H. (2009). Gold nanowire array electrode for non-enzymatic voltammetric and amperometric glucose detection. *Sensors and Actuators B: Chemical*, 142(1), 216–223.

Christoforidis, A., Kavoura, E., Nemtsa, A., Pappa, K., & Dimitriadou, M. (2020). Coronavirus lockdown effect on type 1 diabetes management on children wearing insulin pump equipped with continuous glucose monitoring system. *Diabetes Research and Clinical Practice*, 166, 108307.

Chu, Z., Liu, Y., Xu, Y., Shi, L., Peng, J., & Jin, W. (2015). In-situ fabrication of well-distributed gold nanocubes on thiol graphene as a third-generation biosensor for ultrasensitive glucose detection. *Electrochimica Acta*, 176, 162–171.

Chuang, H., Taylor, E., & Davison, T. W. (2004). Clinical evaluation of a continuous minimally invasive glucose flux sensor placed over ultrasonically permeated skin. *Diabetes Technology & Therapeutics*, 6(1), 21–30.

Çiftçi, H., & Tamer, U. (2012). Functional gold nanorod particles on conducting polymer poly (3-octylthiophene) as non-enzymatic glucose sensor. *Reactive and Functional Polymers*, 72(2), 127–132.

Clarke, S., & Foster, J. (2012). A history of blood glucose meters and their role in self-monitoring of diabetes mellitus. *British Journal of Biomedical Science*, 69(2), 83–93.

Colthup, N. (2012). *Introduction to infrared and Raman spectroscopy*. Elsevier.

Cooper, M. A. (2009). *Label-free biosensors: Techniques and applications*. Cambridge University Press.

Coté, G. L., Fox, M. D., & Northrop, R. B. (1992). Noninvasive optical polarimetric glucose sensing using a true phase measurement technique. *IEEE Transactions on Biomedical Engineering*, 39(7), 752–756.

Crane, B., Paterson, W., Barwell, N. P., & Culbert, B. (2018). *Barrier layer for glucose sensor*. Google Patents.

Cui, F., Yue, Y., Zhang, Y., Zhang, Z., & Zhou, H. S. (2020). Advancing biosensors with machine learning. *ACS Sensors*, 5(11), 3346–3364.

Dag, S., Ozturk, Y., Ciraci, S., & Yildirim, T. (2005). Adsorption and dissociation of hydrogen molecules on bare and functionalized carbon nanotubes. *Physical Review B*, 72(15), 155404.

Daikuzono, C., Florea, L., Delaney, C., Tesfay, H., Morrin, A., & Diamond, D. (2016). *Impedance spectroscopy for the detection of monosaccharides using functionalized carbon screen-printed electrodes on paper*.

Darabdhara, G., Sharma, B., Das, M. R., Boukherroub, R., & Szunerits, S. (2017). Cu-Ag bimetallic nanoparticles on reduced graphene oxide nanosheets as peroxidase mimic for glucose and ascorbic acid detection. *Sensors and Actuators B: Chemical*, 238, 842–851.

Dayakar, T., Rao, K. V., Bikshalu, K., Malapati, V., & Sadasivuni, K. K. (2018). Non-enzymatic sensing of glucose using screen-printed electrode modified with novel synthesized CeO$_2$@CuO core shell nanostructure. *Biosensors and Bioelectronics*, 111, 166–173. Available from https://doi.org/10.1016/j.bios.2018.03.063.

Dayakar, T., Rao, K. V., Bikshalu, K., Rajendar, V., & Park, S.-H. (2017a). Novel synthesis and characterization of pristine Cu nanoparticles for the non-enzymatic glucose biosensor. *Journal of Materials Science: Materials in Medicine*, 28(7), 109. Available from https://doi.org/10.1007/s10856-017-5907-6.

Dayakar, T., Rao, K. V., Bikshalu, K., Rajendar, V., & Park, S.-H. (2017b). Novel synthesis and structural analysis of zinc oxide nanoparticles for the non enzymatic glucose biosensor. *Materials Science and Engineering: C, 75*, 1472−1479.

Dayakar, T., Rao, K. V., Vinod Kumar, M., Bikshalu, K., Chakradhar, B., & Ramachandra Rao, K. (2018). Novel synthesis and characterization of Ag@TiO$_2$ core shell nanostructure for non-enzymatic glucose sensor. *Applied Surface Science, 435*, 216−224. Available from https://doi.org/10.1016/j.apsusc.2017.11.077.

Dayakar, T., Venkateswara Rao, K., Park, J., Sadasivuni, K. K., Ramachandra Rao, K., & Jayaram babu, N. (2018). Non-enzymatic biosensing of glucose based on silver nanoparticles synthesized from *Ocimum tenuiflorum* leaf extract and silver nitrate. *Materials Chemistry and Physics, 216*, 502−507. Available from https://doi.org/10.1016/j.matchemphys.2018.05.046.

Deshmukh, M. A., Kang, B.-C., & Ha, T.-J. (2020). Non-enzymatic electrochemical glucose sensors based on polyaniline/reduced-graphene-oxide nanocomposites functionalized with silver nanoparticles. *Journal of Materials Chemistry C, 8*(15), 5112−5123.

DeVries, J., Wentholt, I., Zwart, A., & Hoekstra, J. (2006). Pendra goes Dutch; lessons for the CE mark in Europe. *Diabetes Research and Clinical Practice, 74*, S93−S96.

Ding, L., Gong, Z., Yan, M., Yu, J., & Song, X. (2017). Determination of glucose by using fluorescent silicon nanoparticles and an inner filter caused by peroxidase-induced oxidation of o-phenylenediamine by hydrogen peroxide. *Microchimica Acta, 184*(11), 4531−4536.

Ding, Y., Liu, Y., Parisi, J., Zhang, L., & Lei, Y. (2011). A novel NiO−Au hybrid nanobelts based sensor for sensitive and selective glucose detection. *Biosensors and Bioelectronics, 28*(1), 393−398.

Dinneen, S. F., & Gerstein, H. C. (1997). The association of microalbuminuria and mortality in non-insulin-dependent diabetes mellitus: A systematic overview of the literature. *Archives of Internal Medicine, 157*(13), 1413−1418.

Ensafi, A. A., Ahmadi, N., & Rezaei, B. (2017). Nickel nanoparticles supported on porous silicon flour, application as a non-enzymatic electrochemical glucose sensor. *Sensors and Actuators B: Chemical, 239*, 807−815.

Farid, M. M., Goudini, L., Piri, F., Zamani, A., & Saadati, F. (2016). Molecular imprinting method for fabricating novel glucose sensor: Polyvinyl acetate electrode reinforced by MnO$_2$/CuO loaded on graphene oxide nanoparticles. *Food Chemistry, 194*, 61−67.

Frew, J. E., & Hill, H. A. O. (1987). Electrochemical biosensors. *Analytical Chemistry, 59*(15), 933A−944A.

German, N., Kausaite-Minkstimiene, A., Ramanavicius, A., Semashko, T., Mikhailova, R., & Ramanaviciene, A. (2015). The use of different glucose oxidases for the development of an amperometric reagentless glucose biosensor based on gold nanoparticles covered by polypyrrole. *Electrochimica Acta, 169*, 326−333.

Ghanbari, K., & Babaei, Z. (2016). Fabrication and characterization of non-enzymatic glucose sensor based on ternary NiO/CuO/polyaniline nanocomposite. *Analytical Biochemistry, 498*, 37−46.

Ginsberg, J. M., Chang, B. S., Matarese, R. A., & Garella, S. (1983). Use of single voided urine samples to estimate quantitative proteinuria. *New England Journal of Medicine, 309*(25), 1543−1546.

Goodarzi, M., & Saeys, W. (2016). Selection of the most informative near infrared spectroscopy wavebands for continuous glucose monitoring in human serum. *Talanta, 146*, 155−165.

Goornavar, V., Jeffers, R., Biradar, S., & Ramesh, G. T. (2014). Utilization of highly purified single wall carbon nanotubes dispersed in polymer thin films for an improved performance of an electrochemical glucose sensor. *Materials Science and Engineering: C, 40*, 299−307.

Gou, X., Sun, S., Yang, Q., Li, P., Liang, S., Zhang, X., & Yang, Z. (2018). A very facile strategy for the synthesis of ultrathin CuO nanorods towards non-enzymatic glucose sensing. *New Journal of Chemistry, 42*(8), 6364−6369. Available from https://doi.org/10.1039/C7NJ04717G.

Gu, X., Wang, H., Schultz, Z. D., & Camden, J. P. (2016). Sensing glucose in urine and serum and hydrogen peroxide in living cells by use of a novel boronate nanoprobe based on surface-enhanced raman spectroscopy. *Analytical Chemistry, 88*(14), 7191−7197.

Guilbault, G., & Lubrano, G. (1973). An enzyme electrode for the amperometric determination of glucose. *Analytica Chimica Acta, 64*(3), 439−455.

Guo, C., Li, H., Zhang, X., Huo, H., & Xu, C. (2015). 3D porous CNT/MnO$_2$ composite electrode for high-performance enzymeless glucose detection and supercapacitor application. *Sensors and Actuators B: Chemical, 206*, 407−414.

Guo, M., Wang, R., & Xu, X. (2011). Electrosynthesis of pinecone-shaped Pt—Pb nanostructures based on the application in glucose detection. *Materials Science and Engineering: C, 31*(8), 1700—1705.

Haas, J., Piron, P., Ernesto, V. C., Karlsson, M., Österlund, L., Nikolajeff, F., & Mizaikoff, B. (2018). Sensing glucose in human saliva with mid-infrared broadly tunable quantum cascade lasers and polycrystalline diamond waveguides. In *EUROPT (R) ODE XIV Naples 2018, Conference on optical chemical sensors and biosensors* (March 25—28, 2018, Naples, Italy).

Han, L., Shao, C., Liang, B., & Liu, A. (2016). Genetically engineered phage-templated MnO_2 nanowires: Synthesis and their application in electrochemical glucose biosensor operated at neutral pH condition. *ACS Applied Materials & Interfaces, 8*(22), 13768—13776. Available from https://doi.org/10.1021/acsami.6b03266.

Hashiguchi, Y., Sakakida, M., Nishida, K., Uemura, T., Kajiwara, K.-I., & Shichiri, M. (1994). Development of a miniaturized glucose monitoring system by combining a needle-type glucose sensor with microdialysis sampling method: Long-term subcutaneous tissue glucose monitoring in ambulatory diabetic patients. *Diabetes Care, 17*(5), 387—396.

Hassani, F. A., Shi, Q., Wen, F., He, T., Haroun, A., Yang, Y., Feng, Y., & Lee, C. (2020). Smart materials for smart healthcare—moving from sensors and actuators to self-sustained nanoenergy nanosystems. *Smart Materials in Medicine, 1.*

Haynes, C. L., Yonzon, C. R., Zhang, X., & Van Duyne, R. P. (2005). Surface-enhanced Raman sensors: Early history and the development of sensors for quantitative biowarfare agent and glucose detection. *Journal of Raman Spectroscopy, 36*(6-7), 471—484. Available from https://doi.org/10.1002/jrs.1376.

Heidari, H., & Habibi, E. (2016). Amperometric enzyme-free glucose sensor based on the use of a reduced graphene oxide paste electrode modified with electrodeposited cobalt oxide nanoparticles. *Microchimica Acta, 183*(7), 2259—2266. Available from https://doi.org/10.1007/s00604-016-1862-z.

Heise, H., Marbach, R., Koschinsky, T., & Gries, F. (1994). Noninvasive blood glucose sensors based on near-infrared spectroscopy. *Artificial Organs, 18*(6), 439—447.

Heo, Y. J., & Takeuchi, S. (2013). Towards smart tattoos: Implantable biosensors for continuous glucose monitoring. *Advanced Healthcare Materials, 2*(1), 43—56.

Hiratsuka, A., Fujisawa, K., & Muguruma, H. (2008). Amperometric biosensor based on glucose dehydrogenase and plasma-polymerized thin films. *Analytical Sciences, 24*(4), 483—486.

Hoa, L. T., Linh, N. T. Y., Chung, J. S., & Hur, S. H. (2017). Green synthesis of silver nanoparticle-decorated porous reduced graphene oxide for antibacterial non-enzymatic glucose sensors. *Ionics, 23*(6), 1525—1532. Available from https://doi.org/10.1007/s11581-016-1954-0.

Hoss, U., Budiman, E. S., Liu, H., & Christiansen, M. P. (2014). Feasibility of factory calibration for subcutaneous glucose sensors in subjects with diabetes. *Journal of Diabetes Science and Technology, 8*(1), 89—94.

Hsieh, C.-T., Lin, W.-H., Chen, Y.-F., Tzou, D.-Y., Chen, P.-Q., & Juang, R.-S. (2017). Microwave synthesis of copper catalysts onto reduced graphene oxide sheets for non-enzymatic glucose oxidation. *Journal of the Taiwan Institute of Chemical Engineers, 71*, 77—83.

Hsu, C.-L., Lin, J.-H., Hsu, D.-X., Wang, S.-H., Lin, S.-Y., & Hsueh, T.-J. (2017). Enhanced non-enzymatic glucose biosensor of ZnO nanowires via decorated Pt nanoparticles and illuminated with UV/green light emitting diodes. *Sensors and Actuators B: Chemical, 238*, 150—159.

Hu, Y., Cheng, H., Zhao, X., Wu, J., Muhammad, F., Lin, S., He, J., Zhou, L., Zhang, C., & Deng, Y. (2017). Surface-enhanced raman scattering active gold nanoparticles with enzyme-mimicking activities for measuring glucose and lactate in living tissues. *ACS Nano, 11*(6), 5558—5566.

Huo, H., Zhao, Y., & Xu, C. (2014). 3D Ni_3S_2 nanosheet arrays supported on Ni foam for high-performance supercapacitor and non-enzymatic glucose detection. *Journal of Materials Chemistry A, 2*(36), 15111—15117.

Ivars-Barceló, F., Zuliani, A., Fallah, M., Mashkour, M., Rahimnejad, M., & Luque, R. (2018). Novel applications of microbial fuel cells in sensors and biosensors. *Applied Sciences, 8*(7), 1184.

Jabbari, N., Asghari, M. H., Ahmadian, H., & Mikaili, P. (2015). Developing a commercial air ultrasonic ceramic transducer to transdermal insulin delivery. *Journal of Medical Signals and Sensors, 5*(2), 117.

Jane, P., Luca De, S., Ilaria, R., Alfredo Maria, G., Paola, G., Christophe, M., Sandra, C., & Jolanda, S. (2016). One-pot synthesis of a gold nanoparticle—Vmh2 hydrophobin nanobiocomplex for glucose monitoring. Nanotechnology, 27(19), 195701. <https://doi.org/10.1088/0957-4484/27/19/195701>.

Jaquins-Gerstl, A., & Michael, A. C. (2020). Dexamethasone-enhanced microdialysis and penetration injury. *Frontiers in Bioengineering and Biotechnology, 8*, 1386.

Javid, B., Fotouhi-Ghazvini, F., & Zakeri, F. S. (2018). Noninvasive optical diagnostic techniques for mobile blood glucose and bilirubin monitoring. *Journal of Medical Signals and Sensors, 8*(3), 125.

Jeevanandam, J., Chan, Y. S., & Danquah, M. K. (2016). Biosynthesis of metal and metal oxide nanoparticles. *ChemBioEng Reviews, 3*(2), 55–67.

Jeevanandam, J., & Danquah, M. K. (2020). Nanosensors for better diagnosis of health. *Nanofabrication for smart nanosensor applications* (Vol. 1, pp. 187–228). Elsevier.

Jeevanandam, J., Kaliyaperumal, A., Sundararam, M., & Danquah, M. K. (2020). Nanomaterials as toxic gas sensors and biosensors. *Nanosensor technologies for environmental monitoring* (pp. 389–430). Cham: Springer.

Jena, B. K., & Raj, C. R. (2006). Enzyme-free amperometric sensing of glucose by using gold nanoparticles. *Chemistry—A European Journal, 12*(10), 2702–2708.

Jia, J., Wang, B., Wu, A., Cheng, G., Li, Z., & Dong, S. (2002). A method to construct a third-generation horseradish peroxidase biosensor: Self-assembling gold nanoparticles to three-dimensional sol–gel network. *Analytical Chemistry, 74*(9), 2217–2223.

Jia, F., Zhong, H., Li, X., Zhu, F., Liu, G., Cheng, Z., Zhang, L., Yin, J., Sheng, Z., & Guo, L. (2015). Research on novel nonenzymatic ECL sensor using Au-HS/SO$_3$H-PMO (Et) nanocomposites for glucose detection. *Journal of Electroanalytical Chemistry, 758*, 93–99.

Jiang, D., Liu, Q., Wang, K., Qian, J., Dong, X., Yang, Z., Du, X., & Qiu, B. (2014). Enhanced non-enzymatic glucose sensing based on copper nanoparticles decorated nitrogen-doped graphene. *Biosensors and Bioelectronics, 54*, 273–278.

Jiang, Y., Zhao, H., Lin, Y., Zhu, N., Ma, Y., & Mao, L. (2010). Colorimetric detection of glucose in rat brain using gold nanoparticles. *Angewandte Chemie, 122*(28), 4910–4914.

Jin, X., Zhu, D. D., Chen, B. Z., Ashfaq, M., & Guo, X. D. (2018). Insulin delivery systems combined with microneedle technology. *Advanced Drug Delivery Reviews, 127*.

Jina, A. N., Tamada, J., Desai, S., & Lee, J. (2017). *Devices and methods for enhanced skin perforation for continuous glucose monitoring.* Google Patents.

Jo, A., Kang, M., Cha, A., Jang, H. S., Shim, J. H., Lee, N.-S., Kim, M. H., Lee, Y., & Lee, C. (2014). Nonenzymatic amperometric sensor for ascorbic acid based on hollow gold/ruthenium nanoshells. *Analytica Chimica Acta, 819*, 94–101.

Johnson, C. A., Levey, A. S., Coresh, J., Levin, A., Lau, J., & Eknoyan, G. (2004). Clinical practice guidelines for chronic kidney disease in adults: Part II. Glomerular filtration rate, proteinuria, and other markers. *American Family Physician, 70*(6), 1091–1097.

Jonsson, A. J., Lund, S. H., Eriksen, B. O., Palsson, R., & Indridason, O. S. (2020). The prevalence of chronic kidney disease in Iceland according to KDIGO criteria and age-adapted estimated glomerular filtration rate thresholds. *Kidney International, 98*(5), 1286–1295.

Ju, L., Wu, G., Lu, B., Li, X., Wu, H., & Liu, A. (2016). Non-enzymatic amperometric glucose sensor based on copper nanowires decorated reduced graphene oxide. *Electroanalysis, 28*(10), 2543–2551.

Juska, V. B., & Pemble, M. E. (2020). A critical review of electrochemical glucose sensing: Evolution of biosensor platforms based on advanced nanosystems. *Sensors, 20*(21), 6013.

Karikalan, N., Velmurugan, M., Chen, S.-M., & Karuppiah, C. (2016). Modern approach to the synthesis of Ni(OH)$_2$ decorated sulfur doped carbon nanoparticles for the nonenzymatic glucose sensor. *ACS Applied Materials & Interfaces, 8*(34), 22545–22553. Available from https://doi.org/10.1021/acsami.6b07260.

Karim, M. N., Anderson, S. R., Singh, S., Ramanathan, R., & Bansal, V. (2018). Nanostructured silver fabric as a free-standing NanoZyme for colorimetric detection of glucose in urine. *Biosensors and Bioelectronics, 110*, 8–15. Available from https://doi.org/10.1016/j.bios.2018.03.025.

Kawde, A., Aziz, M. A., El-Zohri, M., Baig, N., & Odewunmi, N. (2017). Cathodized gold nanoparticle-modified graphite pencil electrode for non-enzymatic sensitive voltammetric detection of glucose. *Electroanalysis, 29*(5), 1214–1221.

Kefeni, K. K., Msagati, T. A. M., Nkambule, T. T. I., & Mamba, B. B. (2020). Spinel ferrite nanoparticles and nanocomposites for biomedical applications and their toxicity. *Materials Science and Engineering: C, 107*, 110314.

Kiistala, U. (1968). Suction blister device for separation of viable epidermis from dermis. *The Journal of Investigative Dermatology, 50,* 129−137.

Kim, J., Campbell, A. S., & Wang, J. (2018). Wearable non-invasive epidermal glucose sensors: A review. *Talanta, 177,* 163−170.

Kost, J., Mitragotri, S., Gabbay, R. A., Pishko, M., & Langer, R. (2000). Transdermal monitoring of glucose and other analytes using ultrasound. *Nature Medicine, 6*(3), 347.

Kottmann, J., Rey, J. M., & Sigrist, M. W. (2016). Mid-Infrared photoacoustic detection of glucose in human skin: Towards non-invasive diagnostics. *Sensors, 16*(10), 1663.

Kovatchev, B. P., Gonder-Frederick, L. A., Cox, D. J., & Clarke, W. L. (2004). Evaluating the accuracy of continuous glucose-monitoring sensors: Continuous glucose−error grid analysis illustrated by TheraSense Freestyle Navigator data. *Diabetes Care, 27*(8), 1922−1928.

Kubiak, T., Wörle, B., Kuhr, B., Nied, I., Gläsner, G., Hermanns, N., Kulzer, B., & Haak, T. (2006). Microdialysis-based 48-hour continuous glucose monitoring with GlucoDay^{TM}: Clinical performance and patients' acceptance. *Diabetes Technology & Therapeutics, 8*(5), 570−575.

Kundu, M. K., Sadhukhan, M., & Barman, S. (2015). Ordered assemblies of silver nanoparticles on carbon nitride sheets and their application in the non-enzymatic sensing of hydrogen peroxide and glucose. *Journal of Materials Chemistry B, 3*(7), 1289−1300. Available from https://doi.org/10.1039/C4TB01740D.

Kwon, S. Y., & Burkoth, T. L. (2003). *Non-or minimally invasive monitoring methods.* Google Patents.

Li, D., Su, J., Yang, J., Yu, S., Zhang, J., Xu, K., & Yu, H. (2017). Optical surface plasmon resonance sensor modified by mutant glucose/galactose-binding protein for affinity detection of glucose molecules. *Biomedical Optics Express, 8*(11), 5206−5217.

Li, D., Yu, S., Sun, C., Zou, C., Yu, H., & Xu, K. (2015). U-shaped fiber-optic ATR sensor enhanced by silver nanoparticles for continuous glucose monitoring. *Biosensors and Bioelectronics, 72,* 370−375. Available from https://doi.org/10.1016/j.bios.2015.05.023.

Li, M., Bo, X., Mu, Z., Zhang, Y., & Guo, L. (2014). Electrodeposition of nickel oxide and platinum nanoparticles on electrochemically reduced graphene oxide film as a nonenzymatic glucose sensor. *Sensors and Actuators B: Chemical, 192,* 261−268. Available from https://doi.org/10.1016/j.snb.2013.10.140.

Li, T., Li, Y., Wang, C., Gao, Z.-D., & Song, Y.-Y. (2015). Nitrogen-doped carbon nanospheres derived from cocoon silk as metal-free electrocatalyst for glucose sensing. *Talanta, 144,* 1245−1251.

Li, X., & Du, X. (2017). Molybdenum disulfide nanosheets supported Au-Pd bimetallic nanoparticles for non-enzymatic electrochemical sensing of hydrogen peroxide and glucose. *Sensors and Actuators B: Chemical, 239,* 536−543.

Li, Y., Zhong, Y., Zhang, Y., Weng, W., & Li, S. (2015). Carbon quantum dots/octahedral Cu_2O nanocomposites for non-enzymatic glucose and hydrogen peroxide amperometric sensor. *Sensors and Actuators B: Chemical, 206,* 735−743.

Liakat, S., Bors, K. A., Xu, L., Woods, C. M., Doyle, J., & Gmachl, C. F. (2014). Noninvasive in vivo glucose sensing on human subjects using mid-infrared light. *Biomedical Optics Express, 5*(7), 2397−2404.

Lin, P. T., Lin, H. G., Han, Z., Jin, T., Millender, R., Kimerling, L. C., & Agarwal, A. (2016). Label-free glucose sensing using chip-scale mid-infrared integrated photonics. *Advanced Optical Materials, 4*(11), 1755−1759.

Lisi, F., Peterson, J. R., & Gooding, J. J. (2020). The application of personal glucose meters as universal point-of-care diagnostic tools. *Biosensors and Bioelectronics, 148,* 111835. Available from https://doi.org/10.1016/j.bios.2019.111835.

Liu, H., Wu, X., Yang, B., Li, Z., Lei, L., & Zhang, X. (2015). Three-dimensional porous NiO nanosheets vertically grown on graphite disks for enhanced performance non-enzymatic glucose sensor. *Electrochimica Acta, 174,* 745−752.

Liu, S., Hui, K., & Hui, K. (2016). Flower-like copper cobaltite nanosheets on graphite paper as high-performance supercapacitor electrodes and enzymeless glucose sensors. *ACS Applied Materials & Interfaces, 8*(5), 3258−3267.

Liu, X., Yang, W., Chen, L., & Jia, J. (2016). Synthesis of copper nanorods for non-enzymatic amperometric sensing of glucose. *Microchimica Acta, 183*(8), 2369−2375.

Liu, Z., Guo, Y., & Dong, C. (2015). A high performance nonenzymatic electrochemical glucose sensor based on polyvinylpyrrolidone—graphene nanosheets—nickel nanoparticles—chitosan nanocomposite. *Talanta, 137*, 87—93.

Liu, G. Q., Zhong, H., Li, X. R., Yang, K., Jia, F. F., Cheng, Z. P., & Qian, H. Y. (2017). Research on nonenzymatic electrochemical sensor using HO-BiONO3 nanocomposites for glucose detection. *Sensors and Actuators B: Chemical, 242*, 484—491.

Lu, N., Shao, C., Li, X., Miao, F., Wang, K., & Liu, Y. (2016). CuO nanoparticles/nitrogen-doped carbon nanofibers modified glassy carbon electrodes for non-enzymatic glucose sensors with improved sensitivity. *Ceramics International, 42*(9), 11285—11293.

Lu, N., Shao, C., Li, X., Shen, T., Zhang, M., Miao, F., Zhang, P., Zhang, X., Wang, K., Zhang, Y., & Liu, Y. (2014). CuO/Cu$_2$O nanofibers as electrode materials for non-enzymatic glucose sensors with improved sensitivity. *RSC Advances, 4*(59), 31056—31061. Available from https://doi.org/10.1039/C4RA03258F.

Lv, P., Zhou, H., Mensah, A., Feng, Q., Wang, D., Hu, X., Cai, Y., AmerigoLucia, L., Li, D., & Wei, Q. (2018). A highly flexible self-powered biosensor for glucose detection by epitaxial deposition of gold nanoparticles on conductive bacterial cellulose. *Chemical Engineering Journal, 351*, 177—188. Available from https://doi.org/10.1016/j.cej.2018.06.098.

Lyandres, O., Yuen, J. M., Shah, N. C., Van Duyne, R. P., Walsh, J. T., Jr, & Glucksberg, M. R. (2008). Progress toward an in vivo surface-enhanced Raman spectroscopy glucose sensor. *Diabetes Technology & Therapeutics, 10*(4), 257—265.

Ma, M., Zhou, Y., Li, J., Ge, Z., He, H., Tao, T., Cai, Z., Wang, X., Chang, G., & He, Y. (2020). Non-invasive detection of glucose via a solution-gated graphene transistor. *Analyst, 145*(3), 887—896.

Maaoui, H., Teodoresu, F., Wang, Q., Pan, G.-H., Addad, A., Chtourou, R., Szunerits, S., & Boukherroub, R. (2016). Non-enzymatic glucose sensing using carbon quantum dots decorated with copper oxide nanoparticles. *Sensors, 16*(10), 1720.

MacKenzie, H. A., Ashton, H. S., Spiers, S., Shen, Y., Freeborn, S. S., Hannigan, J., Lindberg, J., & Rae, P. (1999). Advances in photoacoustic noninvasive glucose testing. *Clinical Chemistry, 45*(9), 1587—1595.

Mahadeva, A., Tanasescu, R., & Gran, B. (2014). Urinary tract infections in multiple sclerosis: Under-diagnosed and under-treated? A clinical audit at a large University Hospital. *American Journal of Clinical and Experimental Immunology, 3*(1), 57.

Malchoff, C. D., Shoukri, K., Landau, J. I., & Buchert, J. M. (2002). A novel noninvasive blood glucose monitor. *Diabetes Care, 25*(12), 2268—2275.

Malia, L., Strumph, K., Brancato, J., Johnson, S. T., Mittal, A., Chicaiza, H., & Smith, S. (2018). Fast and sensitive: Automated point-of-care urine dips. *American Academy of Pediatrics, 141*.

Mani, V., Devasenathipathy, R., Chen, S.-M., Wang, S.-F., Devi, P., & Tai, Y. (2015). Electrodeposition of copper nanoparticles using pectin scaffold at graphene nanosheets for electrochemical sensing of glucose and hydrogen peroxide. *Electrochimica Acta, 176*, 804—810.

Marie, M., Mandal, S., & Manasreh, O. (2015). An electrochemical glucose sensor based on zinc oxide nanorods. *Sensors, 15*(8), 18714—18723.

Markle, D. R., & Markle, W. (2017). *Equilibrium non-consuming fluorescence sensor for real time intravascular glucose measurement*. Google Patents.

Marx, S., Jose, M. V., Andersen, J. D., & Russell, A. J. (2011). Electrospun gold nanofiber electrodes for biosensors. *Biosensors and Bioelectronics, 26*(6), 2981—2986.

Mastrototaro, J. J. (2000). The MiniMed continuous glucose monitoring system. *Diabetes Technology & Therapeutics, 2*(1, Suppl. 1), 13—18.

McAnally, M. O., Phelan, B. T., Young, R. M., Wasielewski, M. R., Schatz, G. C., & Van Duyne, R. P. (2017). Quantitative determination of the differential Raman scattering cross sections of glucose by femtosecond stimulated Raman scattering. *Analytical Chemistry, 89*(13), 6931—6935.

McGarraugh, G. (2009). The chemistry of commercial continuous glucose monitors. *Diabetes Technology & Therapeutics, 11*(S1), S-17—S-24.

Mei, H., Wu, W., Yu, B., Wu, H., Wang, S., & Xia, Q. (2016). Nonenzymatic electrochemical sensor based on Fe@ Pt core—shell nanoparticles for hydrogen peroxide, glucose and formaldehyde. *Sensors and Actuators B: Chemical, 223*, 68—75.

Meng, F., Yin, H., Li, Y., Zheng, S., Gan, F., & Ye, G. (2018). One-step synthesis of enzyme-stabilized gold nanoclusters for fluorescent ratiometric detection of hydrogen peroxide, glucose and uric acid. *Microchemical Journal, 141,* 431−437.

Misra, S., & Oliver, N. (2015). Utility of ketone measurement in the prevention, diagnosis and management of diabetic ketoacidosis. *Diabetic Medicine, 32*(1), 14−23.

Mitragotri, S., & Kost, J. (2004). Low-frequency sonophoresis: A review. *Advanced Drug Delivery Reviews, 56*(5), 589−601.

Muhimmah, L. C., Roekmono, Hadi, H., Yuwono, R. A., & Wahyuono, R. A. (2018). Blood plasma separation in ZnO nanoflowers-supported paper based microfluidic for glucose sensing. In *Proceedings of the AIP conference* (Vol. 1945, p. 020006). AIP Publishing.

Muhr, V., Buchner, M., Hirsch, T., Jovanović, D. J., Dolić, S. D., Dramićanin, M. D., & Wolfbeis, O. S. (2017). Europium-doped GdVO$_4$ nanocrystals as a luminescent probe for hydrogen peroxide and for enzymatic sensing of glucose. *Sensors and Actuators B: Chemical, 241,* 349−356. Available from https://doi.org/10.1016/j.snb.2016.10.090.

Muratova, I., Mikhelson, K., Ermolenko, Y., Offenhäusser, A., & Mourzina, Y. (2016). On "resistance overpotential" caused by a potential drop along the ultrathin high aspect ratio gold nanowire electrodes in cyclic voltammetry. *Journal of Solid State Electrochemistry, 20*(12), 3359−3365.

Naaz, S., Poddar, S., Bayen, S. P., Mondal, M. K., Roy, D., Mondal, S. K., Chowdhury, P., & Saha, S. K. (2018). Tenfold enhancement of fluorescence quantum yield of water soluble silver nanoclusters for nano-molar level glucose sensing and precise determination of blood glucose level. *Sensors and Actuators B: Chemical, 255,* 332−340. Available from https://doi.org/10.1016/j.snb.2017.07.143.

Naik, K. K., Kumar, S., & Rout, C. S. (2015). Electrodeposited spinel NiCo$_2$O$_4$ nanosheet arrays for glucose sensing application. *RSC Advances, 5*(91), 74585−74591.

Nelson, L. A., McCann, J. C., Loepke, A. W., Wu, J., Ben-Dor, B., & Kurth, D. C. (2006). Development and validation of a multiwavelength spatial domain near-infrared oximeter to detect cerebral hypoxia-ischemia. *Journal of Biomedical Optics, 11*(6), 064022.

Ngeontae, W., Janrungroatsakul, W., Maneewattanapinyo, P., Ekgasit, S., Aeungmaitrepirom, W., & Tuntulani, T. (2009). Novel potentiometric approach in glucose biosensor using silver nanoparticles as redox marker. *Sensors and Actuators B: Chemical, 137*(1), 320−326.

Ngo, Y.-L. T., Chung, J. S., & Hur, S. H. (2017). Multi-dimensional Ag/NiO/reduced graphene oxide nanostructures for a highly sensitive non-enzymatic glucose sensor. *Journal of Alloys and Compounds, 712,* 742−751.

Nguyen, N. D., Nguyen, T. V., Chu, A. D., Tran, H. V., Tran, L. T., & Huynh, C. D. (2018). A label-free colorimetric sensor based on silver nanoparticles directed to hydrogen peroxide and glucose. *Arabian Journal of Chemistry.* Available from https://doi.org/10.1016/j.arabjc.2017.12.035.

Niedzwiecki, M. M., Samant, P., Walker, D. I., Tran, V., Jones, D. P., Prausnitz, M. R., & Miller, G. W. (2018). Human suction blister fluid composition determined using high-resolution metabolomics. *Analytical Chemistry, 90*(6), 3786−3792.

Oh, E., Hong, M.-Y., Lee, D., Nam, S.-H., Yoon, H. C., & Kim, H.-S. (2005). Inhibition assay of biomolecules based on fluorescence resonance energy transfer (FRET) between quantum dots and gold nanoparticles. *Journal of the American Chemical Society, 127*(10), 3270−3271.

Okuda-Shimazaki, J., Yoshida, H., & Sode, K. (2020). FAD dependent glucose dehydrogenases−Discovery and engineering of representative glucose sensing enzymes. *Bioelectrochemistry (Amsterdam, Netherlands), 132,* 107414.

Oliver, N., Toumazou, C., Cass, A., & Johnston, D. (2009). Glucose sensors: A review of current and emerging technology. *Diabetic Medicine, 26*(3), 197−210.

Pan, Y., Zuo, J., Hou, Z., Huang, Y., & Huang, C. (2020). Preparation of electrochemical sensor based on zinc oxide nanoparticles for simultaneous determination of AA, DA, and UA. *Frontiers in Chemistry, 8.*

Park, S., Boo, H., & Chung, T. D. (2006). Electrochemical non-enzymatic glucose sensors. *Analytica Chimica Acta, 556*(1), 46−57.

Parkes, J. L., Slatin, S. L., Pardo, S., & Ginsberg, B. H. (2000). A new consensus error grid to evaluate the clinical significance of inaccuracies in the measurement of blood glucose. *Diabetes Care, 23*(8), 1143−1148.

Pecoraro, R., D'Angelo, D., Filice, S., Scalese, S., Capparucci, F., Marino, F., Iaria, C., Guerriero, G., Tibullo, D., & Scalisi, E. M. (2018). Toxicity evaluation of graphene oxide and titania loaded nafion membranes in zebrafish. *Frontiers in Physiology, 8*, 1039.

Peroza, E. A., Boumezbeur, A.-H., & Zamboni, N. (2015). Rapid, randomized development of genetically encoded FRET sensors for small molecules. *Analyst, 140*(13), 4540–4548.

Pfützner, A., Caduff, A., Larbig, M., Schrepfer, T., & Forst, T. (2004). Impact of posture and fixation technique on impedance spectroscopy used for continuous and noninvasive glucose monitoring. *Diabetes Technology & Therapeutics, 6*(4), 435–441.

Pilehvar, S., & De Wael, K. (2015). Recent advances in electrochemical biosensors based on fullerene-C60 nano-structured platforms. *Biosensors, 5*(4), 712–735.

Poscia, A., Mascini, M., Moscone, D., Luzzana, M., Caramenti, G., Cremonesi, P., Valgimigli, F., Bongiovanni, C., & Varalli, M. (2003). A microdialysis technique for continuous subcutaneous glucose monitoring in diabetic patients (part 1). *Biosensors and Bioelectronics, 18*(7), 891–898.

Prasad, R., & Bhat, B. R. (2015). Multi-wall carbon nanotube–NiO nanoparticle composite as enzyme-free electrochemical glucose sensor. *Sensors and Actuators B: Chemical, 220*, 81–90.

Qian, Q., Hu, Q., Li, L., Shi, P., Zhou, J., Kong, J., Zhang, X., Sun, G., & Huang, W. (2018). Sensitive fiber microelectrode made of nickel hydroxide nanosheets embedded in highly-aligned carbon nanotube scaffold for nonenzymatic glucose determination. *Sensors and Actuators B: Chemical, 257*, 23–28.

Rahman, S. S. U., Qureshi, M. T., Sultana, K., Rehman, W., Khan, M. Y., Asif, M. H., Farooq, M., & Sultana, N. (2017). Single step growth of iron oxide nanoparticles and their use as glucose biosensor. *Results in Physics, 7*, 4451–4456. Available from https://doi.org/10.1016/j.rinp.2017.11.001.

Ranamukhaarachchi, S. A., Padeste, C., Häfeli, U. O., Stoeber, B., & Cadarso, V. J. (2017). Design considerations of a hollow microneedle-optofluidic biosensing platform incorporating enzyme-linked assays. *Journal of Micromechanics and Microengineering, 28*(2), 024002.

Ranjani, M., Sathishkumar, Y., Lee, Y. S., JinYoo, D., Kim, A. R., & Gnana kumar, G. (2015). Ni–Co alloy nanostructures anchored on mesoporous silica nanoparticles for non-enzymatic glucose sensor applications. *RSC Advances, 5*(71), 57804–57814. Available from https://doi.org/10.1039/C5RA08471G.

Rao, G., Glikfeld, P., & Guy, R. H. (1993). Reverse iontophoresis: Development of a noninvasive approach for glucose monitoring. *Pharmaceutical Research, 10*(12), 1751–1755.

Ravindranath, R., Roy, P., & Chang, H. (2016). Synthesis, optical properties, and sensing applications of gold nanodots. *The Chemical Record, 16*(3), 1664–1675.

Raz, I., Wainstein, J., & Argaman, D. (2003). Continuous non invasive venous blood glucose monitor. *Diabetes, 52*, A98.

Reichgott, S., Shakespeare, W. J., Kechter, G., Wallace, P. W., & Schurman, M. J. (2015). *Flowometry in optical coherence tomography for analyte level estimation*. Google Patents.

Ren, X., Meng, X., Chen, D., Tang, F., & Jiao, J. (2005). Using silver nanoparticle to enhance current response of biosensor. *Biosensors and Bioelectronics, 21*(3), 433–437.

Renard, E., Place, J., Cantwell, M., Chevassus, H., & Palerm, C. C. (2010). Closed-loop insulin delivery using a subcutaneous glucose sensor and intraperitoneal insulin delivery: Feasibility study testing a new model for the artificial pancreas. *Diabetes Care, 33*(1), 121–127.

Rinaldi, A. L., & Carballo, R. (2016). Impedimetric non-enzymatic glucose sensor based on nickel hydroxide thin film onto gold electrode. *Sensors and Actuators B: Chemical, 228*, 43–52.

Said, K., Ayesh, A. I., Qamhieh, N. N., Awwad, F., Mahmoud, S. T., & Hisaindee, S. (2017). Fabrication and characterization of graphite oxide–nanoparticle composite based field effect transistors for non-enzymatic glucose sensor applications. *Journal of Alloys and Compounds, 694*, 1061–1066.

Sedghi, R., & Pezeshkian, Z. (2015). Fabrication of non-enzymatic glucose sensor based on nanocomposite of MWCNTs-COOH-Poly (2-aminothiophenol)-Au NPs. *Sensors and Actuators B: Chemical, 219*, 119–124.

Sehit, E., & Altintas, Z. (2020). Significance of nanomaterials in electrochemical glucose sensors: An updated review (2016-2020). *Biosensors and Bioelectronics*, 112165.

Senthamizhan, A., Balusamy, B., & Uyar, T. (2016). Glucose sensors based on electrospun nanofibers: A review. *Analytical and Bioanalytical Chemistry, 408*(5), 1285–1306. Available from https://doi.org/10.1007/s00216-015-9152-x.

Serra, A., Filippo, E., Re, M., Palmisano, M., Vittori-Antisari, M., Buccolieri, A., & Manno, D. (2009). Non-functionalized silver nanoparticles for a localized surface plasmon resonance-based glucose sensor. *Nanotechnology*, *20*(16), 165501.

Shah, R., Gottlieb, R. K., & Larson, E. A. (2015). *Flexible sensor apparatus*. Google Patents.

Shakespeare, W. J., Bennett, W. H., Iceman, J. T., Apple, H. P., Wallace, P. W., & Schurman, M. J. (2017). *Multispot monitoring for use in optical coherence tomography*. Google Patents.

Sharma, B., Bugga, P., Madison, L. R., Henry, A.-I., Blaber, M. G., Greeneltch, N. G., Chiang, N., Mrksich, M., Schatz, G. C., & Van Duyne, R. P. (2016). Bisboronic acids for selective, physiologically relevant direct glucose sensing with surface-enhanced Raman spectroscopy. *Journal of the American Chemical Society*, *138*(42), 13952−13959.

Shen, P., & Xia, Y. (2014). Synthesis-modification integration: One-step fabrication of boronic acid functionalized carbon dots for fluorescent blood sugar sensing. *Analytical Chemistry*, *86*(11), 5323−5329.

Shen, Y., Davies, A., Linfield, E., Elsey, T., Taday, P., & Arnone, D. (2003). The use of Fourier-transform infrared spectroscopy for the quantitative determination of glucose concentration in whole blood. *Physics in Medicine & Biology*, *48*(13), 2023.

Shen, Z., Gao, W., Li, P., Wang, X., Zheng, Q., Wu, H., Ma, Y., Guan, W., Wu, S., & Yu, Y. (2016). Highly sensitive nonenzymatic glucose sensor based on Nickel nanoparticle−attapulgite-reduced graphene oxide-modified glassy carbon electrode. *Talanta*, *159*, 194−199.

Shiang, Y.-C., Huang, C.-C., & Chang, H.-T. (2009). Gold nanodot-based luminescent sensor for the detection of hydrogen peroxide and glucose. *Chemical Communications*, *23*, 3437−3439.

Shichiri, M., Kawamori, R., Goriya, Y., Yamasaki, Y., Nomura, M., Hakui, N., & Abe, H. (1983). Glycaemic control in pancreatectomized dogs with a wearable artificial endocrine pancreas. *Diabetologia*, *24*(3), 179−184.

Shih, W.-C., Bechtel, K. L., & Rebec, M. V. (2015). Noninvasive glucose sensing by transcutaneous Raman spectroscopy. *Journal of Biomedical Optics*, *20*(5), 051036.

Shipway, A. N., Katz, E., & Willner, I. (2000). Nanoparticle arrays on surfaces for electronic, optical, and sensor applications. *Chemphyschem: A European Journal of Chemical Physics and Physical Chemistry*, *1*(1), 18−52.

Shu, H., Chang, G., Su, J., Cao, L., Huang, Q., Zhang, Y., Xia, T., & He, Y. (2015). Single-step electrochemical deposition of high performance Au-graphene nanocomposites for nonenzymatic glucose sensing. *Sensors and Actuators B: Chemical*, *220*, 331−339.

Sim, J. Y., Ahn, C.-G., Jeong, E.-J., & Kim, B. K. (2018). In vivo microscopic photoacoustic spectroscopy for non-invasive glucose monitoring invulnerable to skin secretion products. *Scientific Reports*, *8*(1), 1059.

Singh, B., & Dempsey, E. (2011). Pt based nanocomposites for direct glucose determination sensitive nonenzymatic amperometric glucose detection by carbon supported PtAu based bimetallic nanomaterials. *ECS Transactions*, *35*(25), 75−91.

Smart, W. H., & Subramanian, K. (2000). The use of silicon microfabrication technology in painless blood glucose monitoring. *Diabetes Technology & Therapeutics*, *2*(4), 549−559.

Sodzel, D., Khranovskyy, V., Beni, V., Turner, A. P., Viter, R., Eriksson, M. O., Holtz, P.-O., Janot, J.-M., Bechelany, M., & Balme, S. (2015). Continuous sensing of hydrogen peroxide and glucose via quenching of the UV and visible luminescence of ZnO nanoparticles. *Microchimica Acta*, *182*(9−10), 1819−1826.

Song, K., Ha, U., Park, S., Bae, J., & Yoo, H.-J. (2015). An impedance and multi-wavelength near-infrared spectroscopy IC for non-invasive blood glucose estimation. *IEEE Journal of Solid-State Circuits*, *50*(4), 1025−1037.

Song, Y., Wei, C., He, J., Li, X., Lu, X., & Wang, L. (2015). Porous Co nanobeads/rGO nanocomposites derived from rGO/Co-metal organic frameworks for glucose sensing. *Sensors and Actuators B: Chemical*, *220*, 1056−1063.

Spichiger-Keller, U. E. (2008). *Chemical sensors and biosensors for medical and biological applications*. John Wiley & Sons.

Standley, C., Lwambo, N., Lange, C., Kariuki, H., Adriko, M., & Stothard, J. (2010). Performance of circulating cathodic antigen (CCA) urine-dipsticks for rapid detection of intestinal schistosomiasis in schoolchildren from shoreline communities of Lake Victoria. *Parasites & Vectors*, *3*(1), 7.

Strambini, L., Longo, A., Scarano, S., Prescimone, T., Palchetti, I., Minunni, M., Giannessi, D., & Barillaro, G. (2015). Self-powered microneedle-based biosensors for pain-free high-accuracy measurement of glycaemia in interstitial fluid. *Biosensors and Bioelectronics, 66*, 162–168.

Sun, K. G., & Hur, S. H. (2015). Highly sensitive non-enzymatic glucose sensor based on Pt nanoparticle decorated graphene oxide hydrogel. *Sensors and Actuators B: Chemical, 210*, 618–623.

Sun, X., Stagon, S., Huang, H., Chen, J., & Lei, Y. (2014). Functionalized aligned silver nanorod arrays for glucose sensing through surface enhanced Raman scattering. *RSC Advances, 4*(45), 23382–23388. Available from https://doi.org/10.1039/C4RA02423K.

Sutradhar, S., & Patnaik, A. (2017). A new fullerene-C60–Nanogold composite for non-enzymatic glucose sensing. *Sensors and Actuators B: Chemical, 241*, 681–689.

Szunerits, S., Coffinier, Y., & Boukherroub, R. (2015). Diamond nanowires: A novel platform for electrochemistry and matrix-free mass spectrometry. *Sensors, 15*(6), 12573–12593.

Takeuchi, K., Takama, N., Kinoshita, R., Okitsu, T., & Kim, B. (2020). Flexible and porous microneedles of PDMS for continuous glucose monitoring. *Biomedical Microdevices, 22*(4), 1–12.

Tang, Z., Lee, J. H., Louie, R. F., & Kost, G. J. (2000). Effects of different hematocrit levels on glucose measurements with handheld meters for point-of-care testing. *Archives of Pathology & Laboratory Medicine, 124*(8), 1135–1140.

Tapsak, M. A., Rhodes, R. K., Shults, M. C., & McClure, J. D. (2017). *Techniques to improve polyurethane membranes for implantable glucose sensors.* Google Patents.

Tashkhourian, J., Hormozi-Nezhad, M. R., Khodaveisi, J., & Dashti, R. (2011). A novel photometric glucose biosensor based on decolorizing of silver nanoparticles. *Sensors and Actuators B: Chemical, 158*(1), 185–189. Available from https://doi.org/10.1016/j.snb.2011.06.002.

Thanh, T. D., Balamurugan, J., Lee, S. H., Kim, N. H., & Lee, J. H. (2016). Effective seed-assisted synthesis of gold nanoparticles anchored nitrogen-doped graphene for electrochemical detection of glucose and dopamine. *Biosensors and Bioelectronics, 81*, 259–267.

Thatikayala, D., Ponnamma, D., Sadasivuni, K. K., Cabibihan, J.-J., Al-Ali, A. K., Malik, R. A., & Min, B. (2020). Progress of advanced nanomaterials in the non-enzymatic electrochemical sensing of glucose and H_2O_2. *Biosensors, 10*(11), 151.

Thiha, A., Ibrahim, F., Muniandy, S., Dinshaw, I. J., Teh, S. J., Thong, K. L., Leo, B. F., & Madou, M. (2018). All-carbon suspended nanowire sensors as a rapid highly-sensitive label-free chemiresistive biosensing platform. *Biosensors and Bioelectronics, 107*, 145–152.

Thévenot, D. R., Toth, K., Durst, R. A., & Wilson, G. S. (2001). Electrochemical biosensors: Recommended definitions and classification. *Analytical Letters, 34*(5), 635–659.

Tian, Y., Liu, Y., Wang, W., Zhang, X., & Peng, W. (2015). CuO nanoparticles on sulfur-doped graphene for nonenzymatic glucose sensing. *Electrochimica Acta, 156*, 244–251.

Tiangco, C., Andar, A., Sevilla, F., III, Rao, G., & Tolosa, L. (2018). Transdermal glucose monitoring using glucose binding protein-based fiber optic biosensor coupled with microneedles. *Sensors & Transducers, 28*, 2–6.

Tierney, M. J., Tamada, J. A., Potts, R. O., Eastman, R. C., Pitzer, K., Ackerman, N. R., & Fermi, S. J. (2000). *The GlucoWatch® biographer: A frequent, automatic and noninvasive glucose monitor.* Taylor & Francis.

Tonyushkina, K., & Nichols, J. H. (2009). Glucose meters: A review of technical challenges to obtaining accurate results. *Journal of Diabetes Science and Technology, 3*(4), 971–980. Available from https://doi.org/10.1177/193229680900300446.

Tran, V.-K., Ko, E., Geng, Y., Kim, M. K., Jin, G. H., Son, S. E., Hur, W., & Seong, G. H. (2018). Micro-patterning of single-walled carbon nanotubes and its surface modification with gold nanoparticles for electrochemical paper-based non-enzymatic glucose sensor. *Journal of Electroanalytical Chemistry, 826*, 29–37.

Turcheniuk, K., Boukherroub, R., & Szunerits, S. (2015). Gold–graphene nanocomposites for sensing and biomedical applications. *Journal of Materials Chemistry B, 3*(21), 4301–4324.

Turner, A. P. (2000). Biosensors—sense and sensitivity. *Science (New York, N.Y.), 290*(5495), 1315–1317.

Turner, A. P. (2013). Biosensors: Sense and sensibility. *Chemical Society Reviews, 42*(8), 3184–3196.

Usui, T., Yoshida, Y., Nishi, H., Yanagimoto, S., Matsuyama, Y., & Nangaku, M. (2020). Diagnostic accuracy of urine dipstick for proteinuria category in Japanese workers. *Clinical and Experimental Nephrology, 24*(2), 151−156.

Valiūnienė, A., Rekertaitė, A. I., Ramanavičienė, A., Mikoliūnaitė, L., & Ramanavičius, A. (2017). Fast Fourier transformation electrochemical impedance spectroscopy for the investigation of inactivation of glucose biosensor based on graphite electrode modified by Prussian blue, polypyrrole and glucose oxidase. *Colloids and Surfaces A: Physicochemical and Engineering Aspects, 532*, 165−171.

Vallabani, N. V. S., Karakoti, A. S., & Singh, S. (2017). ATP-mediated intrinsic peroxidase-like activity of Fe3O4-based nanozyme: One step detection of blood glucose at physiological pH. *Colloids and Surfaces B: Biointerfaces, 153*, 52−60. Available from https://doi.org/10.1016/j.colsurfb.2017.02.004.

Vallejo-Heligon, S. G., Brown, N. L., Reichert, W. M., & Klitzman, B. (2016). Porous, Dexamethasone-loaded polyurethane coatings extend performance window of implantable glucose sensors in vivo. *Acta Biomaterialia, 30*, 106−115.

Vargas, E., Ruiz, M., Campuzano, S., Reviejo, A., & Pingarrón, J. (2016). Non-invasive determination of glucose directly in raw fruits using a continuous flow system based on microdialysis sampling and amperometric detection at an integrated enzymatic biosensor. *Analytica Chimica Acta, 914*, 53−61.

Ventrelli, L., Marsilio Strambini, L., & Barillaro, G. (2015). Microneedles for transdermal biosensing: Current picture and future direction. *Advanced Healthcare Materials, 4*(17), 2606−2640.

Vezouviou, E., & Lowe, C. R. (2015). A near infrared holographic glucose sensor. *Biosensors and Bioelectronics, 68*, 371−381.

Volkenhoff, A., Hirrlinger, J., Kappel, J. M., Klämbt, C., & Schirmeier, S. (2018). Live imaging using a FRET glucose sensor reveals glucose delivery to all cell types in the Drosophila brain. *Journal of Insect Physiology, 106*, 55−64.

Wang, C., Shu, S., Yao, Y., & Song, Q. (2015). A fluorescent biosensor of lysozyme-stabilized copper nanoclusters for the selective detection of glucose. *RSC Advances, 5*(123), 101599−101606.

Wang, H., Yi, J., Velado, D., Yu, Y., & Zhou, S. (2015). Immobilization of carbon dots in molecularly imprinted microgels for optical sensing of glucose at physiological pH. *ACS Applied Materials & Interfaces, 7*(29), 15735−15745.

Wang, H.-C., & Lee, A.-R. (2015). Recent developments in blood glucose sensors. *Journal of Food and Drug Analysis, 23*(2), 191−200.

Wang, L., Gao, X., Jin, L., Wu, Q., Chen, Z., & Lin, X. (2013). Amperometric glucose biosensor based on silver nanowires and glucose oxidase. *Sensors and Actuators B: Chemical, 176*, 9−14. Available from https://doi.org/10.1016/j.snb.2012.08.077.

Wang, P. M., Cornwell, M., & Prausnitz, M. R. (2005). Minimally invasive extraction of dermal interstitial fluid for glucose monitoring using microneedles. *Diabetes Technology & Therapeutics, 7*(1), 131−141.

Wang, Q., Wang, Q., Li, M., Szunerits, S., & Boukherroub, R. (2015). Preparation of reduced graphene oxide/Cu nanoparticle composites through electrophoretic deposition: Application for nonenzymatic glucose sensing. *RSC Advances, 5*(21), 15861−15869.

Ward, A., Hannah, A., Kendrick, S., Tucker, N., MacGregor, G., & Connolly, P. (2018). Identification and characterisation of *Staphylococcus aureus* on low cost screen printed carbon electrodes using impedance spectroscopy. *Biosensors and Bioelectronics, 110*, 65−70.

Wei, C., Cheng, C., Zhao, J., Wang, Z., Wu, H., Gu, K., Du, W., & Pang, H. (2015). Mesoporous ZnS−NiS nanocomposites for nonenzymatic electrochemical glucose sensors. *ChemistryOpen, 4*(1), 32−38.

Wei, C., Li, X., Xu, F., Tan, H., Li, Z., Sun, L., & Song, Y. (2014). Metal organic framework−derived anthill-like Cu@ carbon nanocomposites for nonenzymatic glucose sensor. *Analytical Methods, 6*(5), 1550−1557.

Wei, S., Hao, Y., Ying, Z., Xu, C., Wei, Q., Xue, S., Cheng, H.-M., Ren, W., Ma, L.-P., & Zeng, Y. (2020). Transfer-free CVD graphene for highly sensitive glucose sensors. *Journal of Materials Science & Technology, 37*, 71−76.

Weibel, M. K., & Bright, H. J. (1971). The glucose oxidase mechanism interpretation of the pH dependence. *Journal of Biological Chemistry, 246*(9), 2734−2744.

Weinstein, R. L., Schwartz, S. L., Brazg, R. L., Bugler, J. R., Peyser, T. A., & McGarraugh, G. V. (2007). Accuracy of the 5-day FreeStyle Navigator Continuous Glucose Monitoring System: Comparison with frequent laboratory reference measurements. *Diabetes Care*, *30*(5), 1125−1130.

Weiss, R., Yegorchikov, Y., Shusterman, A., & Raz, I. (2007). Noninvasive continuous glucose monitoring using photoacoustic technology—results from the first 62 subjects. *Diabetes Technology & Therapeutics*, *9*(1), 68−74.

Wen, F., Dong, Y., Feng, L., Wang, S., Zhang, S., & Zhang, X. (2011). Horseradish peroxidase functionalized fluorescent gold nanoclusters for hydrogen peroxide sensing. *Analytical Chemistry*, *83*(4), 1193−1196.

White, S. L., Yu, R., Craig, J. C., Polkinghorne, K. R., Atkins, R. C., & Chadban, S. J. (2011). Diagnostic accuracy of urine dipsticks for detection of albuminuria in the general community. *American Journal of Kidney Diseases*, *58*(1), 19−28.

Williams, D. L., Doig, A. R., & Korosi, A. (1970). Electrochemical-enzymatic analysis of blood glucose and lactate. *Analytical Chemistry*, *42*(1), 118−121.

Xia, Y., Ye, J., Tan, K., Wang, J., & Yang, G. (2013). Colorimetric visualization of glucose at the submicromole level in serum by a homogenous silver nanoprism−glucose oxidase system. *Analytical Chemistry*, *85*(13), 6241−6247. Available from https://doi.org/10.1021/ac303591n.

Yadav, J., Rani, A., Singh, V., & Murari, B. M. (2014). Near-infrared LED based non-invasive blood glucose sensor. In *Proceedings of the international conference on signal processing and integrated networks (SPIN)*, (pp. 591−594). IEEE.

Yang, C., Denno, M. E., Pyakurel, P., & Venton, B. J. (2015). Recent trends in carbon nanomaterial-based electrochemical sensors for biomolecules: A review. *Analytica Chimica Acta*, *887*, 17−37.

Yang, P., Tong, X., Wang, G., Gao, Z., Guo, X., & Qin, Y. (2015). NiO/SiC nanocomposite prepared by atomic layer deposition used as a novel electrocatalyst for nonenzymatic glucose sensing. *ACS Applied Materials & Interfaces*, *7*(8), 4772−4777.

Yang, P., Wang, L., Wu, Q., Chen, Z., & Lin, X. (2014). A method for determination of glucose by an amperometric bienzyme biosensor based on silver nanocubes modified Au electrode. *Sensors and Actuators B: Chemical*, *194*, 71−78. Available from https://doi.org/10.1016/j.snb.2013.12.074.

Yang, S., Liu, L., Wang, G., Li, G., Deng, D., & Qu, L. (2015). One-pot synthesis of Mn_3O_4 nanoparticles decorated with nitrogen-doped reduced graphene oxide for sensitive nonenzymatic glucose sensing. *Journal of Electroanalytical Chemistry*, *755*, 15−21. Available from https://doi.org/10.1016/j.jelechem.2015.07.021.

Yang, X. C., Lu, Y., Wang, M. T., & Yao, J. Q. (2016). A photonic crystal fiber glucose sensor filled with silver nanowires. *Optics Communications*, *359*, 279−284. Available from https://doi.org/10.1016/j.optcom.2015.09.102.

Yoo, E.-H., & Lee, S.-Y. (2010). Glucose biosensors: An overview of use in clinical practice. *Sensors*, *10*(5), 4558−4576.

Yu, H., & He, Y. (2015). Seed-assisted synthesis of dendritic Au−Ag bimetallic nanoparticles with chemiluminescence activity and their application in glucose detection. *Sensors and Actuators B: Chemical*, *209*, 877−882.

Yu, S., Li, D., Chong, H., Sun, C., Yu, H., & Xu, K. (2014). In vitro glucose measurement using tunable mid-infrared laser spectroscopy combined with fiber-optic sensor. *Biomedical Optics Express*, *5*(1), 275−286.

Yáñez-Sedeño, P., Campuzano, S., & Pingarrón, J. M. (2017). Fullerenes in electrochemical catalytic and affinity biosensing: A review. *C*, *3*(3), 21.

Zaidi, S. A., & Shin, J. H. (2016). Recent developments in nanostructure based electrochemical glucose sensors. *Talanta*, *149*, 30−42.

Zeller, H., Novak, P., & Landgraf, R. (1989). Blood glucose measurement by infrared spectroscopy. *The International Journal of Artificial Organs*, *12*(2), 129−135.

Zeng, H.-H., Qiu, W.-B., Zhang, L., Liang, R.-P., & Qiu, J.-D. (2016). Lanthanide coordination polymer nanoparticles as an excellent artificial peroxidase for hydrogen peroxide detection. *Analytical Chemistry*, *88*(12), 6342−6348. Available from https://doi.org/10.1021/acs.analchem.6b00630.

Zečić, S., Bašić, E., & Begićević, N. (2016). Advantage of urine dipsticks for determination of ALB/CRE ratio in early detection of chronic kidney disease. *European Urology Supplements*, *15*(10), e1299.

Zhang, C., Ni, H., Chen, R., Zhan, W., Zhang, B., Lei, R., Xiao, T., & Zha, Y. (2015). Enzyme-free glucose sensing based on Fe_3O_4 nanorod arrays. *Microchimica Acta, 182*(9), 1811−1818. Available from https://doi.org/10.1007/s00604-015-1511-y.

Zhang, J., Ma, J., Zhang, S., Wang, W., & Chen, Z. (2015). A highly sensitive nonenzymatic glucose sensor based on CuO nanoparticles decorated carbon spheres. *Sensors and Actuators B: Chemical, 211*, 385−391.

Zhang, L., Ding, Y., Li, R., Ye, C., Zhao, G., & Wang, Y. (2017). Electrodeposition of ultra-long copper nanowires on a titanium foil electrode for nonenzymatic voltammetric sensing of glucose. *Microchimica Acta, 184*(8), 2837−2843.

Zhang, S., Wang, N., Niu, Y., & Sun, C. (2005). Immobilization of glucose oxidase on gold nanoparticles modified Au electrode for the construction of biosensor. *Sensors and Actuators B: Chemical, 109*(2), 367−374.

Zhang, W., Du, Y., & Wang, M. L. (2015). On-chip highly sensitive saliva glucose sensing using multilayer films composed of single-walled carbon nanotubes, gold nanoparticles, and glucose oxidase. *Sensing and Bio-Sensing Research, 4*, 96−102.

Zhang, W., Li, R., Xing, L., Wang, X., & Gou, X. (2016). Carnation-like CuO hierarchical nanostructures assembled by porous nanosheets for nonenzymatic glucose sensing. *Electroanalysis, 28*(9), 2214−2221. Available from https://doi.org/10.1002/elan.201600132.

Zhang, X., Xu, F., Zhao, B., Ji, X., Yao, Y., Wu, D., Gao, Z., & Jiang, K. (2014). Synthesis of CdS quantum dots decorated graphene nanosheets and non-enzymatic photoelectrochemical detection of glucose. *Electrochimica Acta, 133*, 615−622.

Zhang, Y., Li, N., Xiang, Y., Wang, D., Zhang, P., Wang, Y., Lu, S., Xu, R., & Zhao, J. (2020). A flexible non-enzymatic glucose sensor based on copper nanoparticles anchored on laser-induced graphene. *Carbon, 156*, 506−513.

Zhang, Y., Liu, Y., Su, L., Zhang, Z., Huo, D., Hou, C., & Lei, Y. (2014). CuO nanowires based sensitive and selective non-enzymatic glucose detection. *Sensors and Actuators B: Chemical, 191*, 86−93. Available from https://doi.org/10.1016/j.snb.2013.08.096.

Zhang, Y.-J., Chen, S., Yu, Y.-L., & Wang, J.-H. (2020). A miniaturized photoacoustic device with laptop readout for point-of-care testing of blood glucose. *Talanta, 209*, 120527.

Zhang, Z., Chen, Z., Cheng, F., Zhang, Y., & Chen, L. (2017). Highly sensitive on-site detection of glucose in human urine with naked eye based on enzymatic-like reaction mediated etching of gold nanorods. *Biosensors and Bioelectronics, 89*, 932−936.

Zhao, M., Gao, Y., Sun, J., & Gao, F. (2015). Mediatorless glucose biosensor and direct electron transfer type glucose/air biofuel cell enabled with carbon nanodots. *Analytical Chemistry, 87*(5), 2615−2622.

Zhong, A., Luo, X., Chen, L., Wei, S., Liang, Y., & Li, X. (2015). Enzyme-free sensing of glucose on a copper electrode modified with nickel nanoparticles and multiwalled carbon nanotubes. *Microchimica Acta, 182*(5−6), 1197−1204.

Zhong, X., Chai, Y.-Q., & Yuan, R. (2014). A novel strategy for synthesis of hollow gold nanosphere and its application in electrogenerated chemiluminescence glucose biosensor. *Talanta, 128*, 9−14.

Zhou, F., Jing, W., Liu, S., Mao, Q., Xu, Y., Han, F., Wei, Z., & Jiang, Z. (2020). Electrodeposition of gold nanoparticles on ZnO nanorods for improved performance of enzymatic glucose sensors. *Materials Science in Semiconductor Processing, 105*, 104708.

Zhou, F., Jing, W., Wu, Q., Gao, W., Jiang, Z., Shi, J., & Cui, Q. (2016). Effects of the surface morphologies of ZnO nanotube arrays on the performance of amperometric glucose sensors. *Materials Science in Semiconductor Processing, 56*, 137−144.

Zhou, Y., Uzun, S. D., Briseno, A. L., Carter, K. R., & Watkins, J. J. (2017). Fabrication of three-dimensional woodpile cerium oxide nanostructure via solution-based soft nanoimprint lithography for enzymatic glucose sensor. In *Meeting abstracts* (pp. 1926−1926). The Electrochemical Society.

Further reading

Kim, G. J., & Kim, K. O. (2020). Novel glucose-responsive of the transparent nanofiber hydrogel patches as a wearable biosensor via electrospinning. *Scientific Reports, 10*(1), 1−12.

Li, G., & Wen, D. (2020). Sensing nanomaterials of wearable glucose sensors. *Chinese Chemical Letters.*

Shichiri, M. (1987). Needle-type glucose sensor and its clinical applications. *Biosensors: Fundamentals and Applications*, 409−424.

Wei, T.-T., Tsai, H.-Y., Yang, C.-C., Hsiao, W.-T., & Huang, K.-C. (2016). Noninvasive glucose evaluation by human skin oxygen saturation level. In *Proceedings of the IEEE conference on international instrumentation and measurement technology (I2MTC)*, (pp. 1−5). IEEE.

Wen, T., Qu, F., Li, N. B., & Luo, H. Q. (2012). Polyethyleneimine-capped silver nanoclusters as a fluorescence probe for sensitive detection of hydrogen peroxide and glucose. *Analytica Chimica Acta, 749*, 56−62. Available from https://doi.org/10.1016/j.aca.2012.08.048.

Zhao, L., Xiao, C., wang, L., Gai, G., & Ding, J. (2016). Glucose-sensitive polymer nanoparticles for self-regulated drug delivery. *Chemical Communications, 52*(49), 7633−7652. Available from https://doi.org/10.1039/C6CC02202B.

CHAPTER 4

Nano-tattoos—a novel approach for glucose monitoring and diabetes management

Introduction

Nano-tattoos are a unique array of biosensors embedded intradermally in patients and are functionalized to change colors depending on the target markers in the bloodstream (Najigivi et al., 2020). Since its discovery, these novel tattoos are used in the detection of glucose and have become beneficial in recent times among youngsters for diabetes management (Bennett & Naranja, 2013). Conventionally, the level of blood glucose is determined through electrochemical reactions by evaluating a small fingerstick blood sample in a sensor strip, which is later inserted into a portable monitor (Wang, 2008). Electrochemical sensors require a blood sample extracted from the patient usually by a finger–prick method. The collection of blood samples requires preparation and several pre-processing steps (Newman & Turner, 2005) that affected the turnaround times for results. Implantable glucose sensors were later discovered to address some of these challenges. Implantable glucose sensors have a chip or transducer that is injected into a diabetic patient's body (Rodrigues et al., 2020). The sensor detects the blood glucose level and sends signals to an external monitor, a computer, mobile device, or watch, for continuous glucose detection (Teymourian et al., 2020). This method also possesses drawbacks such as the biocompatibility of implantable sensors and reliability of the glucose detection (Gough et al., 2010).

Nano-tattoos or smart tattoos use optical-based fluorescent (or even visible colors) biosensing for continuous monitoring of glucose (Srivastava et al., 2011). Advancements in the fabrication of nano-sized transducers and molecular detectors have facilitated the fabrication of these conceptual smart tattoos into real practical sensors (McShane, 1999, 2002). Further, developments in the fabrication and characterization of glucose sensing nanoparticles, such as carbon nanotubes (CNT), have catalyzed the emergence of novel tattoos (Jia et al., 2013; Sehit & Altintas, 2020). Whilst wearable glucose sensors, such as glucose monitoring smartwatches, are now available on the market (Garg et al., 1999; Li & Wen, 2020), nanotattoo-based glucose sensors are popular among young diabetic patients, as it is embedded in the skin and uses color changes to indicate blood glucose levels (Pickup, 2015). Thus, nano-tattoos are considered as a new generation, personalized,

Emerging Nanomedicines for Diabetes Mellitus Theranostics
DOI: https://doi.org/10.1016/B978-0-323-85396-5.00003-8

and continuous glucose sensors. This chapter discusses nano-tattoos, their method of fabrication, characteristics, mode of mechanism, and performance.

Origin of nanotattoo-based glucose sensors

The concept of smart tattoos emerged before the introduction of smartwatches for glucose sensors (Russell et al., 1999). In the year between 1990 and 1999, several concepts and experimental designs for the fabrication of smart tattoos for the detection of blood glucose levels were reported (Beastall et al., 1995; Geers, 1994; McShane, 1999, 2002). Later, the idea of fabricating smart tattoos into nano-tattoos using nanoparticles was introduced. Nano-tattoos are embedded into the skin and emit fluorescent light depending on the changes in blood glucose levels (Brown & McShane, 2006). After 2010, various types of nanomaterials started to occupy the market of biomedical sensors, and these materials found application in glucose monitoring systems. The development of nanomaterials, such as CNT, nanocomposites, and biocompatible nanostructures, supported the fabrication of nano-tattoos for detecting blood glucose (Saxl et al., 2011). In the same time period, nano-tattoos that change color with blood glucose concentration was investigated (Kellogg, 2010). The concept of biomedical tattoos, especially nano-tattoos for glucose sensing, became common in recent times, particularly after 2015 (Ly & Lee, 2017). Several nanoparticle-based inks that change color with glucose concentration were used as paints for nano-tattoos in diabetic patients. In countries such as the United States and Russia, people use these types of nano-inks as personal glucose monitoring systems (Beans, 2018). Circuit-enabled tattoos were also introduced to detect glucose levels, signal color changes, and convert signals to digital analog for display on wearable devices (Vega et al., 2017). Since young people prefer these modernized and fashionable tattoo-based glucose sensors, it is believed that the pharmaceutical companies will soon obtain the Food and Drug Administration approvals for glucose sensing tattoos (Allon et al., 2017).

Fabrication of nano-tattoos

Nano-tattoos for glucose sensing applications is fabricated mostly by two different approaches. One approach utilizes nanomaterials in the form of microelectrode or intradermal fluorescent implants as glucose sensors incorporated into patients. For example, a polymer-based hydrogel coating has been used as a drug-eluting glucose microelectrode sensor in diabetic rats to detect blood glucose for a month. The sensor microelectrodes were coated with biocompatible hydrogels to increase glucose detection efficiency and tissue-device interface for increasing the lifetime of the biosensor (Wang et al., 2013). Further, nanosensors that are fabricated based on the nano-electro-mechanical system (NEMS) concepts are also used (Faintuch et al., 2020) for nano-tattoo glucose detection applications (Coffel & Nuxoll, 2018). Optical air mirrors for fluorescence detection and photonic nanofences for optical sensors can be fabricated using NEMS technology for glucose monitoring applications (Cadarso et al., 2016; Llobera et al., 2007). Carbon-based nanomaterials, including CNT, have been introduced for the fabrication of NEMS for glucose sensor applications (Joshi et al., 2017; Paranjape & Zhou, 2016). Nanofibers synthesized via electrospinning approaches can be fabricated into nanocomposite biosensors via layer-by-layer deposition to serve as novel smart tattoos (Sibuyi et al., 2019). It has been demonstrated that the layer-by-layer technique can lead to the formation of nanofibers with a thickness of 10 nm, enabling implantation in the subcutaneous region of diabetic patients for enhanced monitoring of glucose (Meetoo et al., 2019; Speer et al., 2020).

Another approach is to use nanomaterial as inks for tattooing glucose sensors in the intradermal skin of diabetic patients. A temporary tattoo-based, all printed glucose sensor has been fabricated using this approach (Bandodkar et al., 2013, 2014a, 2014b). The tattoo was fabricated using stainless steel, a Papilio temporary tattoo paper, Prussian blue conductive carbon ink as the working electrode, insulator ink, semiautomatic screen printer, silver/silver chloride ink as reverse iontophoresis electrodes, and pseudo counter (Bandodkar et al., 2013, 2014a, 2014b). The electrodes in the tattoo sensor were covered by an agarose hydrogel before epidermal biosensing. The in vitro characterization results revealed that these tattoos possess the ability to detect glucose at the micromolar level without the interference of chemical species. Similarly, Dermal Abyss (d–abyss) was innovated with an interactive display of patterned biosensors that are placed in the skin as tattoos to change colors in response to variations in the level of biomarker present in the interstitial fluid (Vega et al., 2017). In this unique nano-tattoo, traditional tattoo inks were replaced with fluorescent and colorimetric sensors that can detect variations in glucose, sodium levels, and pH of interstitial fluids. Colorimetric glucose-sensing tattoos act as a skin-embedded wearable device and the metabolism of the diabetic patient acts as an input for the biosensor (Vega et al., 2017).

The lateral approach of using nanoparticles as tattoo inks to detect glucose levels is widely explored for nano-tattoos due to their unique color-changing or fluorescence emitting capability in response to glucose concentration in blood or body fluids. Several glucose-detecting nanoparticles, nanocomposites, and nanoformulations can be fabricated for use in medical tattoos. Carbon, quantum dots, and certain polymer nanoparticles have been investigated to possess the ability to change color depending on glucose levels in body fluids when incorporated as inks and tattooed on the skin of diabetic animal models (Khan & Pickup, 2013; Russell et al., 1999; Polak et al., 2009). The cytotoxicity of these embedded electrodes and inks for tattoos that change color or emit fluorescence forms a crucial part of the evaluation process in utilizing them as smart medical tattoos among patients. Table 4.1 presents a summary of nano-tattoos that are used for the detection of glucose in body fluids.

Table 4.1 Nano-tattoos used in personalized glucose monitoring systems.

Glucose sensor	Type and uniqueness	Reference
Implantable glucose sensor	Real-time glucose monitoring. Drug-eluting biocompatible hydrogel coating for implantable electrode type biosensors.	Wang et al. (2013)
BioMEMS based glucose sensors	Fluorescent based glucose sensors implanted in patients.	Coffel et al. (2018)
NEMS based glucose sensors	Optical air mirrors for glucose detection by emitting fluorescence.	Llobera et al. (2007)
NEMS based glucose sensor	Photonic nanofences for optical glucose sensing.	Cadarso et al. (2016)
Field effect transistors (FET) based sensors	Multi-walled carbon nanotubes as FET for skin embedded glucose sensors.	Paranjape et al. (2016)
Resonance frequency shift-based glucose sensors	Zinc oxide thin film based piezoelectric micro-cantilever for the detection of blood glucose.	Joshi et al. (2017)
Tattoo-based glucose sensor	Solid-contact ion-selective electrodes as tattoo for noninvasive potentiometric glucose monitoring system	Bandodkar et al. (2013)
Temporary tattoo-based sweat sensors	Noninvasive epidermal tattoo-based sensor coupled with a miniature wearable wireless transceiver	Bandodkar et al. (2014a)
Epidermal glucose detector	Flexible, easy-to-wear electrodes covered by hydrogels as tattoos for glucose sensing.	Bandodkar et al. (2014b)
Dermal Abyss medical tattoos	Colorimetric, wearable device as tattoos for sensing glucose.	Vega et al. (2017)

Amyloid aggregation

In type 2 diabetes, the aggregation of proteins and peptides via amyloid is the most common pathology, where islet amyloid polypeptide of human (IAPP) cosecreted by pancreatic beta-cells with insulin is essential for enhanced glycemic control (Ke et al., 2017). The polypeptide can aggregate together and transform from monomer to oligomer, amyloid fibrils, and protofibrils, which eventually exhibits toxicity towards beta cells and leads to diabetic condition through various factors such as alterations in cell pH (4.5—5), existence of physiological metals namely zinc or calcium, homeostasis of glucose, chaperones, insulin, and chaperone-like proteins (Sudhakar et al., 2017; Zhang et al., 2017). Several nanoparticles have been identified to reduce protein or peptide aggregation mediated by amyloids, thereby controlling glucose levels in cells. Faridi et al. (2019) used graphene quantum dots to manage the dysregulation of protein in pancreatic beta-cells exposed to human islet amyloid polypeptide. The study showed that the quantum dots possessed an enhanced ability to regulate 29 proteins from IAPP species from aberrant expression via hydrogen bonding as well as hydrophobic interactions (Faridi et al., 2019). Nanoparticles have also been used as tattoos or implantable electrodes for the detection of amyloid aggregation for enhanced monitoring of glucose. Kwon et al. (2020) fabricated a novel vertical nanowire electrode array that generates local electric fields in the event of amyloid-beta disaggregation causing plaques. The study showed that these electrode arrays were better than the flat-type electrodes with a 2.7-fold high charge capacity to distinguish the transition of amyloid-beta from beta-sheet to alpha-helix, which is crucial in the detection of amyloid aggregation and its associated complications (Kwon et al., 2020). Also, Li et al. (2017) monitored the aggregation of beta-amyloid by using gold nanorods conjugated by multifunctional peptides. In this study, the authors combined the near-infrared absorption property of gold nanorods with amyloid-beta-15−20 and polyoxome-talates as the known amyloid-beta inhibitors. The study revealed that the nanorods possess an improved ability to inhibit the aggregation of amyloid-beta and disintegrate the deposits of amyloid with infrared radiation to protect the cells from amyloid-beta toxicity in mice (Li et al., 2017). However, this method is considered as an indirect approach to monitor glucose homeostasis in diabetic cells or patients, thus to date, the application of nano-tattoos for detecting amyloid aggregation is rare.

Recent updates in nano-tattoos

Many research studies have been reported using distinct nanomaterials as novel medical tattoos for glucose detection in diabetic patients as shown in Table 4.2. Graphene-based electronic tattoo sensors have been introduced as a new-age glucose monitoring system (Kabiri Ameri et al., 2017). These sensors are about ~ 500 nm thickness, $\sim 40\%$ stretchable, $\sim 85\%$ optically transparent, and are designed into filamentary serpentine. These sensors can be incorporated into the human skin, similar to a temporary tattoo, to measure the target molecules such as glucose (Kabiri Ameri et al., 2017). Likewise, electrodes in the form of films have been fabricated using carbon nanotube paste modified film paper as medical skin tattoos for real-time glucose monitoring (Ly & Lee, 2017). For example, lactate oxidase mod-ified silver nanoparticle electrodes have been fabricated as a mechanically flexible biosensor in the cross-serpentine fashion for detecting lactate concentration. This noninvasive, flexible electrochemical sensor can be merged with near-field communication-based wireless data trans-mission to measure analytes such as glucose and lactates (Abrar et al., 2016). Another study reported the fabrication of smart diagnostic tattoos using glucose oxidase labeled with fluores-cein or ruthenium chelate with magnetic nanoparticles (del Barrio et al., 2017). It was dem-onstrated that the enzyme-based tattoo was effective for ontinuous glucose monitoring with a sensor lifetime of less than a week (del Barrio et al., 2017). In addition, a stretchable, skin

Table 4.2 Summary of smart and nano tattoo technologies with the potential for glucose monitoring.

Smart / nano-tattoos	Special features	Reference
Tattoo-like epidermal sensors	Graphene-based, multimodal, sub-micron thick electronic wearable sensors.	Kabiri Ameri et al. (2017)
Medical skin tattoos with real-time glucose sensors	Carbon nanotube paste film paper as electrodes.	Ly et al. (2017)
Silver nanoparticles smart tattoos	Lactate oxidase modified silver nanoparticle electrodes as noninvasive, flexible electrochemical sensors.	Abrar et al. (2016)
Enzyme-based tattoo	Fabricated from a mixture of glucose oxidase, fluorescein, ruthenium chelates, and magnetic nanoparticles for glucose detection.	del Barrio et al. (2017)
Electrochemical sweat based skin-attachable tattoo as glucose sensor	Carbon nanotube (CNT) patterned in gold nanosheets with a mixture of cobalt tungstate and polyaniline polymer as a temporary tattoo that can detect glucose from patients' sweat.	Oh et al. (2018)
Electrode-based tattoos as glucose sensors	Fabricated by lancet-free, label-free, flexible, nano-porous, zinc oxide thin film for glucose detection from human sweat.	Bhide et al. (2018)
Visible color-changing tattoo	Short-term temporary tattoo with electrodes that can detect blood glucose beneath the skin.	Vieira (2018)
DuoSkin	Gold-leaf based tattoo that changes color depending on body temperature	Kao et al. (2016)
Epidermal electronic system	Ultrathin temporary tattoos with all electronic components embedded within for detecting heart, muscle and brain activity.	Kim et al. (2011)
My UV Patch	Tattoo with photosensitive dye that changes color when exposed to ultraviolet (UV) rays.	Beans (2018)
Living tattoos	Tattoo that uses bacteria printed in a flexible hydrogel and emits light in response to chemical compounds in human skin.	Liu et al. (2018)

attachable, electrochemical sensor (Oh et al., 2018) and a lancet-free, label-free, flexible, nano-porous, zinc oxide thin film electrode-based biosensors (Bhide et al., 2018) have been fabricated for the detection of glucose from human sweat.

Researchers from UC San Diego Jacobs School of Engineering have also fabricated a visible color-changing tattoo for the detecting of glucose levels, similar to Dermal Abyss. This is a temporary tattoo that can be placed on the skin for glucose detection. It has two electrodes that conduct electricity through the skin and force glucose molecules to reach the surface for monitoring (Diego & Vieira, 2018). Other wearable tattoo-based sensors, such as DuoSkin, fabricated via gold leaf, act as temporary tattoos and change color with changes in the body temperature (Kao et al., 2016). Researchers from Northwestern University fabricated an epidermal electronic system into an ultrathin, stretchable, silicone-based membrane that can withhold electrodes, sensors, power supplies, electronics, and communication components, and can stick to the skin as a temporary tattoo (Kim et al., 2011, 2020). This smart tattoo can be used to detect and measure the activity of the brain, muscle, and heart. This tattoo concept was used to fabricate novel medical tattoo products such as BioStamp, My UV Patch, LogicInk, living tattoos, and graphene tattoos to detect and monitor various target molecules (Beans, 2018). Further, German scientists have developed a novel intradermal tattoo that changes color in response to changes in the albumin, glucose, and pH levels in the blood. The color change was monitored with a smartphone camera and a mobile application, which was powered by an algorithm for the determination of the biomarker concentration. The researchers used dyes, such as methyl red, bromothymol blue, and phenolphthalein, as inks for the tattoo, and were administered via nano-sized injections into the skin without affecting the integrity of the epidermal layer (Yetisen et al., 2019). It is anticipated that smart and nano tattoos will become more available on the market in the future due to the rapid developments in the field and the patronization of young diabetic patients.

Mechanism of nano-tattoos as glucose sensors

Smart/nano-tattoos use electrodes fabricated with nanomaterials or colloidal nanoparticles that can conduct electricity. A common mechanism of smart/nano-tattoos for detecting glucose levels in diabetic patients is illustrated in Fig. 4.1. The colloidal nanoparticles are inks that act as the cathode and anode. In other instances, an actual micro-electrode or transducer is attached to the skin like a tattoo, which possesses a cathode and an anode. When a patient eats, the process of digestion leads to an increase in blood glucose levels. It is noteworthy that the glucose in human sweat is directly proportional to the blood glucose level (Cho et al., 2017; Cui et al., 2020).

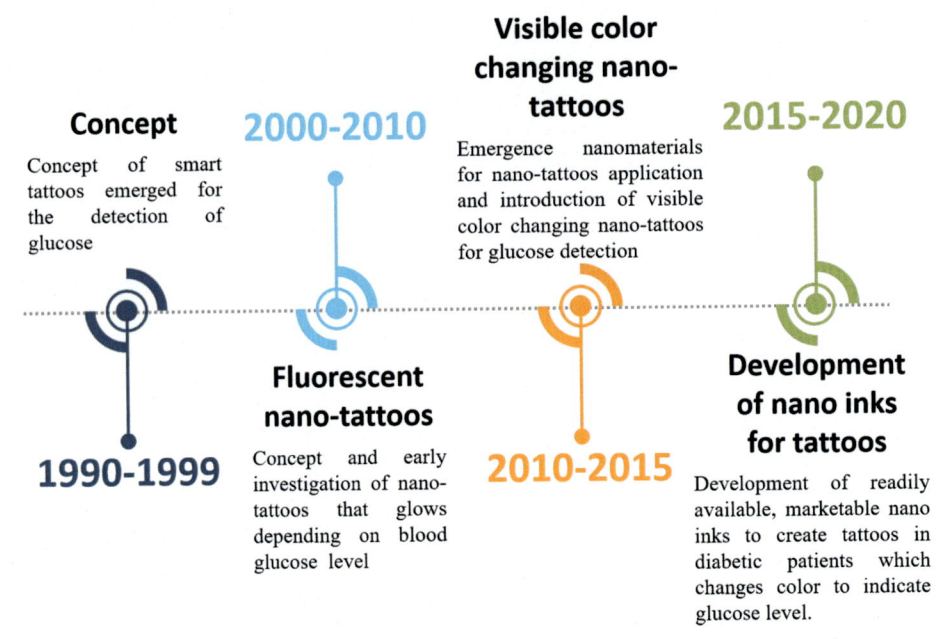

Figure 4.1 A schematic of the mechanism of glucose detection by smart/nano-tattoos.

Thus, the sweat from the patient is used to detect blood glucose concentrations. Once the sweat is in contact with the tattoo, the glucose oxidase converts the glucose to gluconic acid, which is detected by the cathode. The color of the tattoo changes to a visible color in the case of a colorimetric sensor and emits fluorescence in the case of a fluorescent sensor, immediately after the cathodic detection of gluconic acid. This color change informs the patient to take prescribed medications to manage the blood glucose level. The color changes back to the original to indicate the blood glucose level is normal. A typical rapid nano–tattoo detects increases in glucose levels in 20 min, consisting of 5 min waiting time, 10 min reaction time, and 5 min detection time (Bandodkar et al., 2013, 2014a, 2014b). There are several modifications that are being introduced with advancements in technology to reduce the time of detection. It is anticipated that real-time detection of glucose using nano-tattoos would be possible in the future.

Advantages of nano-tattoos

In recent times, there is an increasing trend among young diabetic patients to reveal their disease status via tattoos as shown in Fig. 4.2. These tattoos are inked to inform the public that they are diagnosed with diabetes and fighting against the disorder. An advantage of nano-tattoos is their optical color-changing ability, depending on glucose

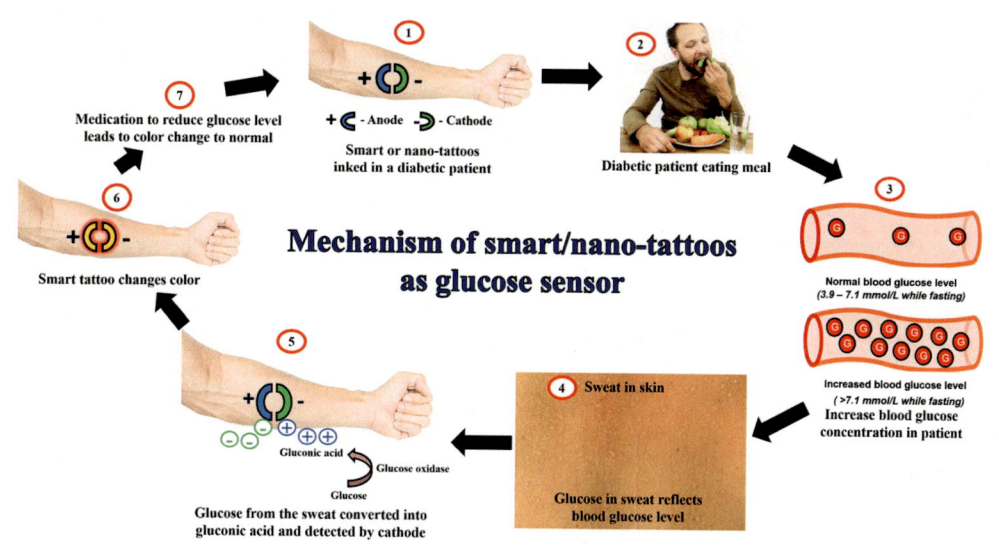

Figure 4.2 Different tattoos inked by young diabetic patients to reveal their disease status. *Image courtesy: Diabetic ink (#diabeticink) from http://www.tumblr.com.*

levels. Hence, there is no need for an additional device or equipment for glucose detection. It is a promising approach for personalized diabetes management. For example, during travels, color-changing nano-tattoos can be highly beneficial to indicate blood glucose levels without any external device. In recent advancements, nano-tattoos are investigated to transfer measured glucose levels directly to the doctor via a wireless transmitter (Bandodkar & Wang, 2014c; Mostafalu et al., 2016). Such smart tattoos will be highly useful, not only in monitoring glucose but also to alert doctors about severe conditions of elevated glucose levels.

As smart tattoos are similar to regular tattoos, young diabetes patients use them without any awkward feeling, unlike wearable technologies that monitor glucose, such as glucose meters. As diabetes in children under 15 years is expected to increase to a significant population in developed, developing, and third world countries (Weisman et al., 2017), nano-tattoos as glucose sensors are expected to thrive on the pharmaceutical companies to support diabetic youngsters.

Challenges in using nano-tattoos as glucose sensors

The prime challenge in utilizing nano-tattoos as glucose sensors is the potential toxicity associated with colloidal nanoparticles used as inks to create the tattoo. Researchers are interested in understanding specific toxic reactions triggered by nanoparticles based on their morphology, size, and surface properties (Chrishtop et al., 2021; Pogribna & Hammons, 2021). Evaluation of colloidal nanoparticles as inks for tattoos in animal models is essential before introducing them to humans. Another limitation of nano-tattoos is the lack of quantification of blood glucose levels. Thus it is difficult to correlate the blood glucose concentration to the level of medication required. However, micro-chips with wireless transmitters are being incorporated into nano-tattoos to transfer blood glucose signals to mobile devices for correlation to actual blood glucose outputs. More advanced nano-tattoos will not only change colors but also transfer data to an external device that inject drugs to manage blood glucose level. The reliability of nano-tattoos is also a major concern. Nonspecific detection of analytes can generate false results as the blood contains other compounds and sugar molecules or glucose enantiomers. These challenges need to be addressed to support the widespread use of nano-tattoos for personalized diabetes management.

Future perspective

Smart or nano-tattoos are considered the initial stage for the development of highly efficient biomedical cyber-physical tools for monitoring target molecules in the bloodstream. The concept of smart tattoo is applied in contact lenses that can detect glucose and change color depending on the glucose level. Such holographic contact lenses for glucose detection have been clinically tested (Domschke et al., 2006). Also, contact lenses with real-time colorimetric glucose detection ability (Bruen et al., 2015) and wearable contact lenses for continuous glucose monitoring using smartphones have been developed (Elsherif et al., 2018; Kim et al., 2011, 2020). These devices will be useful as dual-purpose sensors along with prosthesis applications. Bioluminescent enzymes are useful in smart tattoos as they are biologically derived, maybe less toxic towards humans, and can emit fluorescence or visible light. These bioluminescent enzymes along with nanoformulation technology can enhance the efficiency of smart tattoos in the future. It is evident from recent advancements that smart/nano-tattoos can have an increasing presence on the market to support diabetes management.

Conclusion

This chapter provides an overview of smart/nano-tattoos as glucose monitoring systems for diabetes management. The recent updates will provide scientists with information about the emerging trends in smart/nano-tattoos to create opportunities for further developments to meet specific user requirements. The chapter also provides a simplified understanding of the working mechanism of nano-tattoos for glucose detection and establishes the need for advanced technologies to support the development of nano-tattoos for real-time glucose monitoring. The chapter provides an account of the immediate challenges of nano-tattoos and proposes potential solutions to facilitate the development of novel, efficient smart/nano-tattoos to support personalized diabetes management (Fig. 4.3).

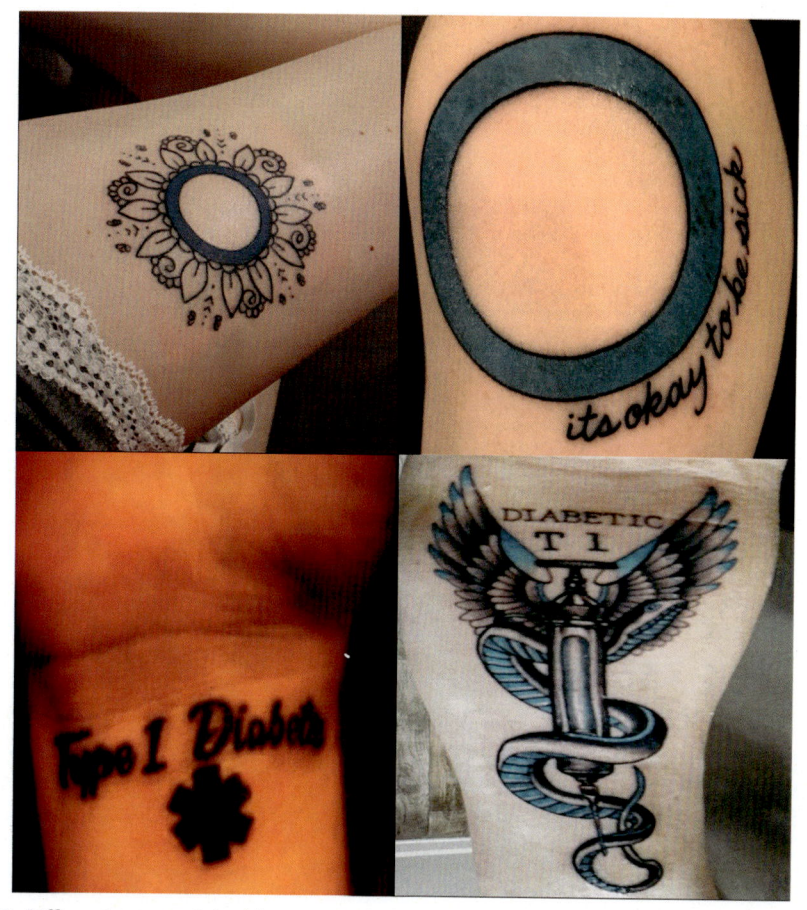

Figure 4.3 Different tattoos inked by young diabetic patients to reveal their disease status. Image courtesy — Diabetic ink (#diabeticink) from www.tumblr.com.

References

Abrar, M. A., Dong, Y., Lee, P. K., & Kim, W. S. (2016). Bendable electro-chemical lactate sensor printed with silver nano-particles. *Scientific Reports*, *6*(1), 1—9.

Allon, I., Ben-Yehudah, A., Dekel, R., Solbakk, J.-H., Weltring, K.-M., & Siegal, G. (2017). Ethical issues in nanomedicine: Tempest in a teapot? *Medicine, Health Care and Philosophy*, *20*(1), 3—11.

Bandodkar, A. J., Hung, V. W., Jia, W., Valdés-Ramírez, G., Windmiller, J. R., Martinez, A. G., & Wang, J. (2013). Tattoo-based potentiometric ion-selective sensors for epidermal pH monitoring. *Analyst*, *138*(1), 123—128.

Bandodkar, A. J., Jia, W., Yardımcı, C., Wang, X., Ramirez, J., & Wang, J. (2014a). Tattoo-based noninvasive glucose monitoring: A proof-of-concept study. *Analytical Chemistry*, *87*(1), 394—398.

Bandodkar, A. J., Molinnus, D., Mirza, O., Guinovart, T., Windmiller, J. R., Valdés-Ramírez, G., & Wang, J. (2014b). Epidermal tattoo potentiometric sodium sensors with wireless signal transduction for continuous non-invasive sweat monitoring. *Biosensors and Bioelectronics*, *54*, 603—609. Available from https://doi.org/10.1016/j.bios.2013.11.039.

Bandodkar, A. J., & Wang, J. (2014c). Non-invasive wearable electrochemical sensors: a review. *Trends in Biotechnology, 32*(7), 363–371.

Beans, C. (2018). Science and Culture: Wearable tech meets tattoo art in a bid to revolutionize both. *Proceedings of the National Academy of Sciences of the United States of America, 115*(14), 3504–3506. Available from https://doi.org/10.1073/pnas.1803214115.

Beastall, G. H., Gibson, I. H., & Martin, J. (1995). Successful Suicide by Insulin Injection in a Non-Diabetic. *Medicine, Science and the Law, 35*(1), 79–85.

Bennett, M. G., & Naranja, R. J., Jr. (2013). Getting nano tattoos right—A checklist of legal and ethical hurdles for an emerging nanomedical technology. *Nanomedicine: Nanotechnology, Biology and Medicine, 9*(6), 729–731.

Bhide, A., Muthukumar, S., Saini, A., & Prasad, S. (2018). Simultaneous lancet-free monitoring of alcohol and glucose from low-volumes of perspired human sweat. *Scientific Reports, 8*(1), 1–11.

Brown, J. Q., & McShane, M. J. (2006). Modeling of spherical fluorescent glucose microsensor systems: Design of enzymatic smart tattoos. *Biosensors and Bioelectronics, 21*(9), 1760–1769.

Bruen, D., Albatal, R., Florea, L., & Diamond, D. (2015). Contact lenses for real-time colorimetric sensing of glucose. In: Advanced Materials World Congress 2015, 23-26 Aug 2015, Stockholm, Sweden.

Cadarso, V. J., Llobera, A., Puyol, M., & Schift, H. (2016). Integrated photonic nanofences: Combining subwavelength waveguides with an enhanced evanescent field for sensing applications. *ACS Nano, 10*(1), 778–785.

Cho, E., Mohammadifar, M., & Choi, S. (2017). *A self-powered sensor patch for glucose monitoring in sweat* (pp. 366–369). IEEE.

Chrishtop, V. V., Mironov, V. A., Prilepskii, A. Y., Nikonorova, V. G., & Vinogradov, V. V. (2021). Organ-specific toxicity of magnetic iron oxide-based nanoparticles. *Nanotoxicology, 15*(2), 167–204.

Coffel, J., & Nuxoll, E. (2018). BioMEMS for biosensors and closed-loop drug delivery. *International Journal of Pharmaceutics, 544*(2), 335–349.

Cui, Y., Duan, W., Jin, Y., Wo, F., Xi, F., & Wu, J. (2020). Ratiometric fluorescent Nanohybrid for noninvasive and visual monitoring of sweat glucose. *ACS Sensors, 5*(7), 2096–2105.

del Barrio, M., Moros, M., Puertas, S., de la Fuente, J. M., Grazú, V., Cebolla, V., & Galbán, J. (2017) Glucose oxidase immobilized on magnetic nanoparticles: Nanobiosensors for fluorescent glucose monitoring. *Microchimica Acta, 184*(5), 1325–1333. Available from https://doi.org/10.1007/s00604-017-2120-8.

Diego, U.S., & Vieira, G. (2018). Available from https://www.healthline.com/health-news/needle-free-tattoo-may-help-make-diabetes-pain-free#1.

Domschke, A., March, W. F., Kabilan, S., & Lowe, C. (2006). Initial clinical testing of a holographic non-invasive contact lens glucose sensor. *Diabetes Technology & Therapeutics, 8*(1), 89–93.

Elsherif, M., Hassan, M. U., Yetisen, A. K., & Butt, H. (2018). Wearable Contact Lens Biosensors for Continuous Glucose Monitoring using Smartphones. *ACS Nano*.

Faintuch, J., Faintuch, S., Speer, A. M., & Choudhury, M. (2020). Nanotechnology: Can It Be a Crusader in Diabesity? In *Obesity and Diabetes: Scientific Advances and Best Practice*, (pp. 947–955). Springer International Publishing. Available from https://doi.org/10.1007/978-3-030-53370-0_70.

Faridi, A., Sun, Y., Mortimer, M., Aranha, R. R., Nandakumar, A., Li, Y., & Ke, P. C. (2019). Graphene quantum dots rescue protein dysregulation of pancreatic β-cells exposed to human islet amyloid polypeptide. *Nano Research, 12*(11), 2827–2834. Available from https://doi.org/10.1007/s12274-019-2520-7.

Garg, S. K., Potts, R. O., Ackerman, N. R., Fermi, S. J., Tamada, J. A., & Chase, H. P. (1999). Correlation of fingerstick blood glucose measurements with GlucoWatch biographer glucose results in young subjects with type 1 diabetes. *Diabetes Care, 22*(10), 1708–1714.

Geers, R. (1994). Electronic monitoring of farm animals: a review of research and development requirements and expected benefits. *Computers and Electronics in Agriculture, 10*(1), 1–9.

Gough, D. A., Kumosa, L. S., Routh, T. L., Lin, J. T., & Lucisano, J. Y. (2010). Function of an implanted tissue glucose sensor for more than 1 year in animals. *Science Translational Medicine, 2*(42), 42ra53–42ra53.

Jia, W., Bandodkar, A. J., Valdés-Ramírez, G., Windmiller, J. R., Yang, Z., Ramírez, J., & Wang, J. (2013). Electrochemical tattoo biosensors for real-time noninvasive lactate monitoring in human perspiration. *Analytical Chemistry*, *85*(14), 6553−6560.

Joshi, P., Singh, J., Jain, V., & Akhtar, J. (2017). Design and Simulation of Micro-cantilever Based Sensor for Glucose Detection. *Mody University International Journal of Computing and Engineering Research*, *1*(2), 104−108.

Kabiri Ameri, S., Ho, R., Jang, H., Tao, L., Wang, Y., Wang, L., & Lu, N. (2017). Graphene electronic tattoo sensors. *ACS Nano*, *11*(8), 7634−7641.

Kao, H.-L. C., Holz, C., Roseway, A., Calvo, A., & Schmandt, C. (2016). *DuoSkin: Rapidly prototyping on-skin user interfaces using skin-friendly materials. In Proceedings of the 2016 ACM International Symposium on Wearable Computers* (pp. 16−23).

Ke, P. C., Sani, M.-A., Ding, F., Kakinen, A., Javed, I., Separovic, F., & Mezzenga, R. (2017). Implications of peptide assemblies in amyloid diseases. *Chemical Society Reviews*, *46*(21), 6492−6531.

Kellogg, S. (2010). Nanotechnology: Small Science, Huge Hurdles. *Wash. Law.*, *25*, 23.

Khan, F., & Pickup, J. C. (2013). Near-infrared fluorescence glucose sensing based on glucose/galactose-binding protein coupled to 651-Blue Oxazine. *Biochemical and Biophysical Research Communications*, *438*(3), 488−492.

Kim, S., Jeon, H.-J., Park, S., Lee, D. Y., & Chung, E. (2020). Tear Glucose Measurement by Reflectance Spectrum of a Nanoparticle Embedded Contact Lens. *Scientific Reports*, *10*(1), 1−8.

Kim, D. H., Lu, N., Ma, R., Kim, Y. S., Kim, R. H., Wang, S., & Rogers, J. A. (2011). Epidermal electronics. *Science*, *333*(6044), 838−843. Available from https://doi.org/10.1126/science.1206157.

Kwon, J., Choi, J. S., Lee, J., Na, J., Sung, J., Lee, H.-J., & Choi, H.-J. (2020). Disaggregation of Amyloid-β Plaques by a Local Electric Field Generated by a Vertical Nanowire Electrode Array. *ACS Applied Materials & Interfaces*, *12*(50), 55596−55604. Available from https://doi.org/10.1021/acsami.0c16000.

Li, M., Guan, Y., Zhao, A., Ren, J., & Qu, X. (2017). Using Multifunctional Peptide Conjugated Au Nanorods for Monitoring β-amyloid Aggregation and Chemo-Photothermal Treatment of Alzheimer's Disease. *Theranostics*, *7*(12), 2996−3006. Available from https://doi.org/10.7150/thno.18459.

Li, G., & Wen, D. (2020). Sensing nanomaterials of wearable glucose sensors. *Chinese Chemical Letters*.

Liu, X., Yuk, H., Lin, S., Parada, G. A., Tang, T. C., Tham, E., & Zhao, X. (2018). 3D printing of living responsive materials and devices. *Advanced Materials*, *30*(4), 1704821.

Llobera, A., Demming, S., Wilke, R., & Büttgenbach, S. (2007). Multiple internal reflection poly (dimethylsiloxane) systems for optical sensing. *Lab on a Chip*, *7*(11), 1560−1566.

Ly, S. Y., & Lee, C. H. (2017). Clinical In Vivo Bio Assay of Glucose in Human Skin by a Tattoo Film Carbon Nano Tube Sensor. *Journal of the Korean Applied Science and Technology*, *34*(3), 595−601.

McShane, M. J. (1999). *Design of an optical probe and signal processing for an implantable fluorescence-based glucose sensor*. Texas A&M University.

McShane, M. J. (2002). Potential for glucose monitoring with nanoengineered fluorescent biosensors. *Diabetes Technology & Therapeutics*, *4*(4), 533−538.

Meetoo, D., Wong, L., & Ochieng, B. (2019). Smart tattoo: Technology for monitoring blood glucose in the future. *British Journal of Nursing*, *28*(2), 110−115.

Mostafalu, P., Akbari, M., Alberti, K. A., Xu, Q., Khademhosseini, A., & Sonkusale, S. R. (2016). A toolkit of thread-based microfluidics, sensors, and electronics for 3D tissue embedding for medical diagnostics. *Microsystems & Nanoengineering*, *2*(1), 1−10.

Najigivi, I. D. S., Mirmotallebi, S., & Najigivi, A. (2020). Contribution of nanotechnology and nanomaterials to the treatment of diabetic patients by aid of novel inventions. *Journal of American Science*, *16*(5).

Newman, J. D., & Turner, A. P. (2005). Home blood glucose biosensors: A commercial perspective. *Biosensors and Bioelectronics*, *20*(12), 2435−2453.

Oh, S. Y., Hong, S. Y., Jeong, Y. R., Yun, J., Park, H., Jin, S. W., & Ha, J. S. (2018). Skin-Attachable, Stretchable Electrochemical Sweat Sensor for Glucose and pH Detection. *ACS Applied Materials & Interfaces*, *10*(16), 13729−13740. Available from https://doi.org/10.1021/acsami.8b03342.

Paranjape, M., & Zhou, J. (2016). Systems and process for forming carbon nanotube sensors. *U.S. Patent No. 9, 302, 908.*

Pickup, J. (2015). Banting Memorial Lecture 2014 Technology and diabetes care: appropriate and personalized. *Diabetic Medicine, 32*(1), 3—13.

Pogribna, M., & Hammons, G. (2021). Epigenetic Effects of nanomaterials and nanoparticles. *Journal of Nanobiotechnology, 19*(1), 1—18.

Polak, A.J., Ballerstadt, R., Beuhler, A., & Gamboa, C. (2009). Sensor device and methods for manufacture. Google Patents.

Rodrigues, D., Barbosa, A. I., Rebelo, R., Kwon, I. K., Reis, R. L., & Correlo, V. M. (2020). Skin-integrated wearable systems and implantable biosensors: A comprehensive review. *Biosensors, 10*(7), 79.

Russell, R. J., Pishko, M. V., Gefrides, C. C., McShane, M. J., & Cote, G. L. (1999). A fluorescence-based glucose biosensor using concanavalin A and dextran encapsulated in a poly (ethylene glycol) hydrogel. *Analytical Chemistry, 71*(15), 3126—3132.

Saxl, T., Khan, F., Ferla, M., Birch, D., & Pickup, J. (2011). A fluorescence lifetime-based fibre-optic glucose sensor using glucose/galactose-binding protein. *Analyst, 136*(5), 968—972.

Sehit, E., & Altintas, Z. (2020). Significance of nanomaterials in electrochemical glucose sensors: An updated review (2016-2020). *Biosensors and Bioelectronics, 159*, 112165.

Sibuyi, N. R. S., Moabelo, K. L., Meyer, M., Onani, M. O., Dube, A., & Madiehe, A. M. (2019). Nanotechnology advances towards development of targeted-treatment for obesity. *Journal of Nanobiotechnology, 17*(1), 1—21.

Speer, A. M., & Choudhury, M. (2020). *Nanotechnology: Can it be a crusader in diabesity? In* Obesity and diabetes (pp. 947—955). Cham: Springer.

Srivastava, R., Jayant, R. D., Chaudhary, A., & McShane, M. J. (2011). Smart tattoo" glucose biosensors and effect of coencapsulated anti-inflammatory agents. *Journal of Diabetes Science and Technology, 5*(1), 76—85.

Sudhakar, S., Kalipillai, P., Santhosh, P. B., & Mani, E. (2017). Role of surface charge of inhibitors on amyloid beta fibrillation. *The Journal of Physical Chemistry C, 121*(11), 6339—6348.

Teymourian, H., Barfidokht, A., & Wang, J. (2020). Electrochemical glucose sensors in diabetes management: an updated review (2010—2020). *Chemical Society Reviews*.

Vega, K., Jiang, N., Liu, X., Kan, V., Barry, N., Maes, P., Yetisen, A., & Paradiso, J. (2017). The dermal abyss: Interfacing with the skin by tattooing biosensors. In *Proceedings of the 2017 ACM International Symposium on Wearable Computers*, (pp. 138—145).

Vieira, G. (2018). Needle-free 'Tattoo' may help make diabetes pain-free. https://www.healthline.com/health-news/needle-free-tattoo-may-help-make-diabetes-pain-free#1.

Wang, J. (2008). Electrochemical glucose biosensors. *Chemical Reviews, 108*(2), 814—825.

Wang, Y., Papadimitrakopoulos, F., & Burgess, D. J. (2013). Polymeric "smart" coatings to prevent foreign body response to implantable biosensors. *Journal of Controlled Release, 169*(3), 341—347. Available from https://doi.org/10.1016/j.jconrel.2012.12.028.

Weisman, A., Bai, J.-W., Cardinez, M., Kramer, C. K., & Perkins, B. A. (2017). Effect of artificial pancreas systems on glycaemic control in patients with type 1 diabetes: A systematic review and *meta*-analysis of outpatient randomised controlled trials. *The Lancet Diabetes & Endocrinology, 5*(7), 501—512.

Yetisen, A. K., Moreddu, R., Seifi, S., Jiang, N., Vega, K., Dong, X., & Elsner, M. (2019). Dermal tattoo biosensors for colorimetric metabolite detection. *Angewandte Chemie, 131*(31), 10616—10623.

Zhang, X., St. Clair, J. R., London, E., & Raleigh, D. P. (2017). Islet amyloid polypeptide membrane interactions: effects of membrane composition. *Biochemistry, 56*(2), 376—390.

CHAPTER 5

Metal and metal oxide nanoparticles: synthesis, properties, and applications as nanomedicines for diabetes treatment

Introduction

The conversion of bulk metals into nanosized particles leads to drastic alterations in their properties. For example, the boiling point of bulk gold is 2700°C whilst nanosized gold has a boiling point of <700°C (Hayat, 2012). The strong extinction band of surface plasmon resonance properties possessed by metal nanoparticles in the visible spectrum facilitates a wide range of industrial applications (Fedlheim & Foss, 2001). Oxides of metals are a subclass of metal nanoparticles used extensively in various applications, including electronics and biomedicine. Metal oxide nanoparticles are considered highly stable metallic nanoparticles due to the strong energy of bonds between the oxygen and the metal atom (Jeevanandam et al., 2016, 2017; Jeevanandam, Danquah, et al., 2015; Jeevanandam, Pal, et al., 2019; Jeevanandam, San Chan, et al., 2016). It is noteworthy that the properties of both metal and metal oxide nanoparticles can be modified via the synthesis process; thus the synthesis method is critical in defining the properties of the nanoparticles for a specific application (Jeevanandam et al., 2016, 2017; Jeevanandam, Danquah, et al., 2015; Jeevanandam, Pal, et al., 2019; Jeevanandam, San Chan, et al., 2016).

Metal and metal oxide nanoparticles can possess catalytic, thermal, electronic, electrical, mechanical, optical, and magnetic properties (Schmid, 2005). Nonlinear optics (Stepanov, 2016), localized surface plasmon resonance (Linic et al., 2015), heat generation (Govorov & Richardson, 2007), thermal stability (Zhao et al., 2010), chemo-electronic (Yan et al., 2016), electrical conductivity (Lin et al., 2015), high mechanical stiffness (Huang et al., 2017), and magneto-elastic properties are some reported properties of metal nanoparticles. Metal oxides also have unique properties, such as high conductivity in semiconductors (Franke et al., 2006), photocatalytic activity (Katwal et al., 2015), superparamagnetism (Ling et al., 2015), minimal thermal contact resistance (Sarafraz & Hormozi, 2016), and enhanced optical absorptivity (Milanese et al., 2016). These exclusive properties are valuable in biomedical and pharmaceutical applications. Metal nanoparticles have been used as biomaterials to fabricate stents (Schmid et al., 2017) and implants (GhavamiNejad, Aguilar, et al., 2015; GhavamiNejad, Park, et al., 2016), for targeted drug delivery (Tiwari et al., 2011), and as nutraceuticals (Mazumder et al., 2019). Metal oxide nanoparticles, including

Emerging Nanomedicines for Diabetes Mellitus Theranostics
DOI: https://doi.org/10.1016/B978-0-323-85396-5.00001-4

111

iron oxide, copper oxide, and zinc oxide are also used as nanomedicines to treat various diseases (Pavlatou & Gorgoulis, 2021; Sahoo et al., 2017). These metal/metal oxide nanoparticles can be used as photocatalytic antimicrobial agents, magnetic resonance contrast imaging agents, cell labeling, and as controlled drug delivery systems (Matijević et al., 2012). This chapter discusses various metal and metal oxide nanoparticles and appraises their potential for diabetes treatment.

Classification of metal and metal oxide nanoparticles

Metal and metal oxide nanoparticles are classified according to their types, synthesis procedures, and properties as shown in Fig. 5.1. They can also be classified based on morphology, dimensionality, and geometry. In this chapter, the classification is based on types, synthesis methods, and properties.

Types

Metal nanoparticles are classified, based on the type of metals and their novel combinations, into nanosized pure metals, rare earth elements, and dopant incorporated and composite particles. Metal oxide nanoparticles are oxygen-bound metal nanoparticles; hence, they can also be similarly classified.

Pure metals

Pure metal nanoparticles are made of single metals formed as colloidal or powdered nanosized particles. Gold nanoparticles are an example of pure metal nanoparticles. Gold nanoparticles have been applied in various disease treatment and biosensing applications (Carabineiro, 2017; Daraee et al., 2016; Yadav et al., 2020). Silver, copper, and iron are other pure metal nanoparticles that are available and are under extensive research for biomedical and pharmaceutical applications (Babushkina et al., 2017; Brennan et al., 2015; Franci et al., 2015; GhavamiNejad, Aguilar, et al., 2015; GhavamiNejad, Park, et al.,

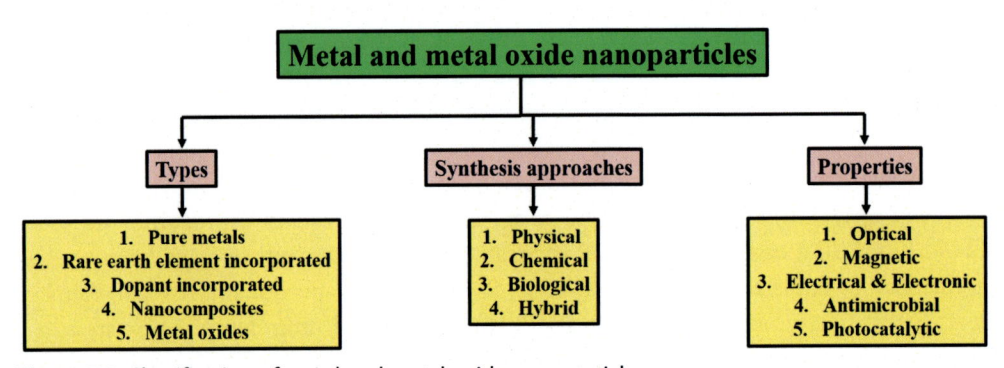

Figure 5.1 Classification of metal and metal oxide nanoparticles.

2016; Huber, 2005; Kumar & Poornachandra, 2015; Rafique et al., 2017; Rubilar et al., 2013; Vincent et al., 2016; Wei et al., 2015; Xiao et al., 2018).

Rare earth elements

Rare earth metals, including neodymium, yttrium, ytterbium, cerium, lanthanum, europium, and gadolinium, can be combined with metal nanoparticles to enhance their properties (Nethi et al., 2020). Cerium, yttrium (Guzman et al., 2005), ytterbium (Zhang, Li, et al., 2017; Zhang, Zhao, et al., 2016), erbium (Min et al., 2009), and lanthanum (Wang et al., 2014) have been combined or functionalized with gold nanoparticles to enhance pharmaceutical value. Also, the biomedical properties of nanosized silver particles can be improved via the addition of rare earth elements such as lanthanum, terbium (Piasecki et al., 2010), praseodymium (Naranjo et al., 2005), erbium (Singh, Mondal, et al., 2015; Singh, Giri, et al., 2010), and europium (Zmojda et al., 2017).

Dopants

Doping is the process of including an additional ion to a crystal structure to enhance its properties. In metal nanoparticles, carbon (He et al., 2016), nitrogen (Wu, Hiltunen, et al., 2011; Wu, Wu, et al., 2015), and phosphorus (Yang et al., 2012) are commonly used as dopants to improve specific properties. Metal ions such as barium, calcium, magnesium (Santra et al., 2013), palladium (Jin & Nobusada, 2014), nickel (Munir et al., 2020), and nematic liquid crystals (Qi & Hegmann, 2006) have been added to gold nanoparticles to improve their biological properties. Mercury, lead, zinc, copper (Lee & Hyeon, 2012; Li, Xu, et al., 2013; Li, Yan, et al., 2015; Li, Shang, et al., 2005; Li, Wei, et al., 2016; Li, Zhang, et al., 2017; Ma, Li, et al., 2017; Ma, Tang, et al., 2008; Meyer et al., 2007) have been doped into silver, copper, and iron nanoparticles, respectively, to enhance their pharmaceutical properties for various applications.

Composites

Nanosized composites are formed from the combination of two or more metals, metal-metal oxides, metal-carbon materials, metal-polymer, and complex metals. Silver-gold (Dutta Choudhury et al., 2012), gold-copper (Neaţu et al., 2014), silver-copper (Beyerlein et al., 2011), gold-silicon (Mohapatra et al., 2008), and gold-iron (Carpenter et al., 2000) are examples of common metal-metal nanocomposites used in biomedical applications. Also, metal-metal oxide nanocomposites such as gold-iron oxides (Leung et al., 2012), silver-iron oxides (Leung et al., 2012), zinc oxide (ZnO)-copper (Habibi & Karimi, 2015), and titanium dioxide (TiO_2)-zinc (Ehsan et al., 2015) are commonly used in biomedical applications. Metal-carbon nanocomposites including gold-carbon nanotubes (CNT) (Alagiri et al., 2017), silver-CNT (Chaudhari et al., 2016), platinum-graphene oxide (Zhang, Li, et al., 2017; Zhang, Zhao, et al., 2016), gold-carbon dots (Priyadarshini et al., 2018), metal-polymer nanocomposites such as gold-molecularly imprinted polymers

(Ahmad, Senapati, et al., 2003; Ahmad, Griffete, et al., 2015; Ahmad, Srivastava, et al., 2021), copper-polymer (Tamayo et al., 2016), and silver-polymer (Palza, 2015) are unique composites that are under research for biomedical applications. Novel metal-dendrimer (Varnavski et al., 2001), metal matrix-CNT-graphene nanosheets (Tjong, 2013), chitosan-g-poly(acrylamide)-zinc (Pathania et al., 2016), block copolymer-metal complex (Wakayama & Yonekura, 2017), gold-sulfur-doped graphene quantum dot (Anh & Doong, 2018), porous silica nanocarriers with gold-carbon quantum dots (Chen et al., 2020), and hydrophilic graphene oxide-iron oxide-gold-polyethylene glycol (PEG) (Jiang et al., 2016) are examples of complex nanocomposites used in biomedical research.

Metal oxide nanoparticles

Metal oxide nanoparticles represent a large subclass of metal nanoparticles as most metals form oxides naturally to be stable in the environment. Similar to metals, nanosized metal oxides can be classified into pure metal oxides, rare earth elements, dopants incorporated, and nanocomposites. Zinc (Mishra et al., 2017; Rajeshkumar & Sandhiya, 2020), iron (Wu, Hiltunen, et al., 2011; Wu, Wu, et al., 2015), copper (Katwal et al., 2015), aluminum (Rajeshwari et al., 2015), magnesium (Paucar Álvarez et al., 2017), manganese (Song et al., 2016), titanium (Sherin et al., 2017), and silicon oxides (Sabziparvar et al., 2018) are examples of common metal oxide nanoparticles used in biomedical and pharmaceutical applications. Yttrium-doped iron oxide (Aghazadeh et al., 2018), erbium-doped cerium oxide (Maria Magdalane et al., 2018), europium doped gadolinium oxide (Chaudhary et al., 2017), tantalum-doped zinc oxide (Guo et al., 2015), and lanthanum doped copper oxide (Sasikala et al., 2016) nanoparticles are examples of rare earth element−doped metal oxide nanoparticles. Additionally, magnesium, iron-doped zinc oxide (Sharma et al., 2016), zinc oxide−doped titanium dioxide (Kaviyarasu et al., 2017), and cobalt-doped zinc oxide (Oves et al., 2015) are some metal-doped metal oxide nanoparticles. Zinc oxide-iron oxide-graphene oxide (Ojha et al., 2017), zinc oxide-silicon dioxide-silver (Kokate et al., 2016), chitosan−carboxymethyl cellulose-zinc oxide (Youssef et al., 2016), and cobalt-doped iron oxide-spider silk (Singh, Mondal, et al., 2015; Singh, Giri, et al., 2010) are some novel metal oxide nanocomposites for biomedical applications. The synthesis methods used for these nanoparticles influence their stability along with size, morphology, surface charge, and crystal structure. Thus the synthesis approach and parameters are highly essential in the generation of nanoparticles with distinct properties.

Synthesis approaches

Metal and metal oxide nanoparticles are synthesized via three different approaches, namely physical, chemical, and biological. It is noteworthy that every nanoparticle fabrication approach possesses merits and demerits. Researchers continue to develop hybrid methods to generate metal and metal oxide nanoparticles with optimal properties for tailored applications.

Physical methods

Evaporation-condensation, arc discharge, laser ablation method, physical deposition, sputtering, and ball milling are some physical methods used to produce nanosized metal particles. The evaporation-condensation technique consists of the thermal-mediated decomposition of pure metals as precursors in distinct decomposition chambers to generate pure metal nanoparticles (Förster et al., 2012). The method of removing metals from the solid precursor via laser beam irradiation and the settlement of metal ions into nanosized particles is called laser ablation. The capacity to produce metal nanoparticles with unique morphologies under minimal heat transfer is a major advantage of this method (Dolgaev et al., 2002; Yang, 2007). Arc discharge method is used for metal nanoparticles synthesis via voltage application across direct-current arc between inert gas immersed metal electrodes (Teng et al., 2017). The high yield of crystalline metal nanoparticles with great mechanical strength is an important advantage of this method (Maria & Mieno, 2015; Tharchanaa et al., 2021). The physical deposition method is similar to laser ablation, where a series of processes is used to deposit thin layers of metals to form thin metal films and coatings. Ultra-thin coatings and thin films with highly crystalline, mechanically strong metal nanoparticles can be produced via the physical deposition method (Richter et al., 2010). Sputtering is a physical deposition method in which metal nanoparticles are formed as self-sustained plasma by deposition over a substrate with controlled inert gas into a vacuum chamber and electrically energizing a cathode (Kashin & Ananikov, 2011). Various metals can be coated over a substrate with high purity and crystallinity and this is a major advantage of this technique. Mechanical ball milling is a top-down method used to produce different sizes of spherical-shaped metal (Ma, Li, et al., 2016; Ma, Tang, et al., 2008; Rak et al., 2014) and metal oxide nanoparticles (Arbain et al., 2011; Damonte et al., 2004). Gold nanoparticles in different sizes (Swihart, 2003), nanowires (Feng et al., 2009), and morphologically different nanosized gold can be developed via the physical method (Pazos-Pérez et al., 2008). The evaporation-condensation method has been used to synthesize spherical silver nanoparticles, silver nanospheriods have been synthesized by laser ablation (Iravani et al., 2014), and the solid-liquid phase arc discharge method has been used to synthesize silver nanowires (Zhou et al., 1999). Other silver nanorods and nanowires can be synthesized using physical deposition and sputtering methods (Aslan et al., 2005; Lee et al., 2012). Other metal nanoparticles such as copper (Förster et al., 2012), iron (Muñoz et al., 2007), and platinum (Caillard et al., 2005) can also be synthesized using these physical methods. Further, metal oxide nanoparticles such as zinc oxide (Thareja & Shukla, 2007), iron oxides (Hassanjani-Roshan et al., 2011), rare earth metals including cerium oxide (Kim et al., 2012), doped metal oxides (Ederth et al., 2003; Liu et al., 2010), and nanocomposites (Qiang et al., 2006; Yu et al., 2005) have also been synthesized using physical methods. The capacity to synthesize monodispersed and uniform nanoparticles with no solvent contamination is a major advantage of physical methods. However, the requirement of costly equipment, high energy consumption, and delayed thermal stability are some challenges of physical methods (Iravani et al., 2014).

Chemical methods

There are several chemical-based techniques that are available for the fabrication of metal nanoparticles. The sol–gel approach is a common chemical synthesis method used to fabricate metal nanoparticles from smaller molecules (bottom-up). In this method, monomers are converted into a colloidal solution named sol which later forms into an integrated network called gel to form metal or metal oxide nanoparticles (Xu et al., 2007). The use of low temperatures for nanoparticle formation, compared to other methods, and the formation of very fine powders and composites are some main advantages of the sol–gel method (Carter & Norton, 2013). Chemical reduction, precipitation, and coprecipitation methods are the widely used chemical synthesis techniques to fabricate nanosized metal particles. The reduction of bulk metals into nanosized metal particles with the help of chemicals and solvents as reducing agents is called chemical reduction. This is a simple synthesis method and requires capping or surfactant agents to prevent the agglomeration of particles (Dang et al., 2011). Precipitation is the method of forming metal nanoparticles from chemically precipitated metal hydroxides and their calcination at high temperatures (Ghorbani et al., 2015). Coprecipitation is used to synthesize nanosized metal oxide particles from solutions of metal salts and bases under an inert atmosphere, mostly at high temperatures (Ahmed et al., 2020; Rai et al., 2015). Precipitation and coprecipitation methods are highly beneficial in yielding nanoparticles with distinct sizes and morphology via alterations in the temperature and time of calcination (Jeevanandam et al., 2016, 2017; Jeevanandam, Danquah, et al., 2015; Jeevanandam, Pal, et al., 2019; Jeevanandam, San Chan, et al., 2016). The polyol method is used to form high-quality metal nanoparticles in high-boiling and multivalent alcohol solutions (Dong et al., 2015). Minimizing surface oxidation and reductive ability at elevated temperatures toward the formation of metal nanoparticles are the advantages of the polyol method (Ariga et al., 2017). Hydro- and solvothermal synthesis is based on crystallizing metal nanoparticles from high-temperature aqueous and solvent solutions at high pressure (Darr et al., 2017). This method is significant in fabricating metal nanoparticles that are not stable at elevated temperatures and high pressure (Hayashi & Hakuta, 2010). Electrochemical synthesis is a highly efficient chemical method to fabricate metal nanoparticles of different sizes and shapes using electrical voltage to drive the deposit of nanoparticles over a substrate. Low cost, simplicity, high purity, and low synthesis temperature are the advantages of electrochemical synthesis methods (Li, Xu, et al., 2013; Li, Yan, et al., 2015; Li, Shang, et al., 2005; Li, Wei, et al., 2016; Li, Zhang, et al., 2017). Even though chemical synthesis methods possess several advantages and yield stable metal nanoparticles, the utilization of toxic chemicals is a major drawback

(Jeevanandam et al., 2016, 2017; Jeevanandam, Danquah, et al., 2015; Jeevanandam, Pal, et al., 2019; Jeevanandam, San Chan, et al., 2016).

Biological methods

Biological methods for metal nanoparticles synthesis seek to address the drawbacks of physical and chemical approaches to yield less toxic nanoparticles. Various species of bacteria, algae, fungi, viruses, and plants have been utilized for nanoparticles synthesis. The bacterial synthesis method involves intracellular synthesis with the bacteria (Mohanpuria et al., 2008), as well as synthesis via bacterial metabolites and enzymes (Durán et al., 2011). Gold, silver, copper, and iron are some intracellularly synthesized metal nanoparticles (Ahmad, Senapati, et al., 2003; Ahmad, Griffete, et al., 2015; Ahmad, Srivastava, et al., 2021). The potential for large-scale production without hazardous and expensive chemicals are some advantages of using bacteria for metal nanoparticles synthesis (Jeevanandam et al., 2016, 2017; Jeevanandam, Danquah, et al., 2015; Jeevanandam, Pal, et al., 2019; Jeevanandam, San Chan, et al., 2016). Similar to bacteria, fungi and algae have been used to fabricate metal and metal oxide nanoparticles via both intracellular mechanisms (Senapati et al., 2012; Vanaja et al., 2015) and extracellular metabolites (Cuevas et al., 2015; Kathiraven et al., 2015). Viruses are a unique set of microorganisms used for metal nanoparticle synthesis, especially in template-assisted synthesis to yield specific morphologies of nanoparticles (Dujardin et al., 2003). Viruses are natural nanosized materials and can multiply in favorable environments (Jeevanandam et al., 2016, 2017; Jeevanandam, Danquah, et al., 2015; Jeevanandam, Pal, et al., 2019; Jeevanandam, San Chan, et al., 2016). Plants are widely used for the biosynthesis of nanoparticles and the presence of phytochemicals and biomolecules, such as coenzymes and secondary metabolites, facilitates the reduction of metals into ions and nanoparticles (Jeevanandam et al., 2020; Saratale et al., 2018). Plant-based synthesis of metal nanoparticles presents difficulties associated with size control, morphology, and stability compared to other methods (Jeevanandam et al., 2016, 2017; Jeevanandam, Danquah, et al., 2015; Jeevanandam, Pal, et al., 2019; Jeevanandam, San Chan, et al., 2016).

Hybrid methods

Hybrid methods rely on a perfect blend of the advantages of different nanoparticles synthesis methods to obtain nanoparticles with desired size, morphology, and geometry for specific applications. For example, the replacement of chemicals with biomolecules and microbial extracts with similar properties are commonplace in hybrid nanoparticles synthesis methods. Chemical synthesis methods are also coupled with ultrasonication, photoinduced reduction, ultraviolet-initiated photoreduction, laser, and different light irradiation methods, and microwave-assisted synthesis to enhance the efficiency of metal nanoparticles synthesis to yield the desired properties (Iravani et al., 2014; Siaw et al., 2020). Biosynthesis methods usually use chemical precursors for nanoparticles formation, and biochemicals are utilized as capping and reducing agents in chemical methods (Mohanpuria et al., 2008; Samat & Nor, 2013). Hybrid synthesis methods combining two or more approaches are gaining importance in nanoparticles production.

Properties of metal and metal oxide nanoparticles

The types of nanoparticles and synthesis approaches are major factors that affect the properties of nanoparticles. Metal and metal oxide nanoparticles possess unique optical, magnetic, electrical, and electronics, antimicrobial, and photocatalytic properties that are highly significant in biomedical and pharmaceutical applications.

Optical properties

Some metal and metal oxide nanoparticles possess optical properties including nonlinear absorption, refraction optical limitation in Z-scan (Stepanov, 2016), localized surface plasmon resonance (Tang et al., 2021), and color change due to red and blue shifts (Zaleska-Medynska et al., 2016). Metal nanoparticles can possess distinct optical absorption, extinction, and scattering properties, depending on their size, morphology, and crystal structures (Sosa et al., 2003).

Magnetic properties

The magnetic properties of metal nanoparticles are based on their sizes. It is possible to improve the magnetic properties of bulk metals by reducing their size to the nano level (Li, Xu, et al., 2013; Li, Yan, et al., 2015; Li, Shang, et al., 2005; Li, Wei, et al., 2016; Li, Zhang, et al., 2017). Iron oxide is a common magnetic metal nanoparticle (Shi et al., 2015). Magnetic properties, such as permanent magnetism, magnetic anisotropy, hysteresis (Crespo et al., 2004), paramagnetism (Peng et al., 2008), and superparamagnetism are exhibited by some magnetic metal and metal oxide nanoparticles (Garfinkel et al., 2020; Graczyk et al., 2015).

Electrical and electronic properties

The electronic and electrical properties of nanosized metal and metal oxide—based particles are dependent on their crystal structures. Metals usually act both as conductors and semiconductors, whereas metal oxides behave as conductors, semiconductors, and sometimes insulators and superconductors (Zhu et al., 2017). Electrical conductivity, resistivity, temperature coefficient of resistance, permittivity, and thermoelectricity are some electrical properties of nanosized metal particles (Kelly et al., 2003; Tommalieh et al., 2020). Further, metal nanoparticles possess electronic properties including electrostatic equilibrium, an enhanced field of charged conductors, and dielectric potential (Kreibig, 1974; Naito et al., 2008).

Antimicrobial properties

Some metal and metal oxide nanoparticles possess enhanced antimicrobial properties to inhibit the growth of disease-causing microbes such as bacteria, fungi, and viruses (Gold et al., 2018). The size of metal nanoparticles plays a significant role in their antimicrobial activity (Daniel et al., 2018). Metal nanoparticles possess the potential to penetrate cells and break down microbial DNA to cause cell death. Metal ions can also disintegrate and inhibit the secretion of certain enzymes, eventually leading to microbial cell death. Metal oxide nanoparticles can release reactive oxygen species, leading to the inhibition of microbial growth (Grumezescu et al., 2016; Sánchez-López et al., 2020).

Photocatalytic properties

The narrow bandgap of semiconductor metal oxide nanoparticles results in unique photocatalytic properties that are highly beneficial in solar cells, wastewater treatment, and light-mediated antimicrobial applications (Ji et al., 2015). The size of the nanoparticles determines the bandgap and also photocatalytic activity. Zinc oxide (Bhuyan et al., 2015), titanium dioxide (Li, Xu, et al., 2013; Li, Yan, et al., 2015; Li, Shang, et al., 2005; Li, Wei, et al., 2016; Li, Zhang, et al., 2017; Ren, et al., 2020), doped metal oxide nanoparticles (Yousefi et al., 2015), and metal oxide nanocomposites (Magdalane et al., 2016) are some nanoparticles with enhanced photocatalytic activity.

Metal nanoparticles as nanomedicines for diabetes treatment

Amongst the various metal nanoparticles, only a few have been explored as nanomedicines for the treatment of diabetes. Alkaladi et al. (2014) showed that silver nanoparticles possess antidiabetic properties by reducing blood glucose levels, increasing insulin concentration in the serum, and enhancing glucokinase activity (Alkaladi et al., 2014). Ghosh et al. (2015) showed that spherical monodispersed 86 nm copper nanoparticles synthesized via aqueous tuber extract of *Dioscorea bulbifera* possess antidiabetic activity. The copper nanoparticles inhibited α-amylase and α-glucosidase enzymes to reduce hyperglycemic conditions for the treatment of diabetes patients (Ghosh et al., 2015). Further, spherical-shaped silver nanoparticles synthesized from the leaf extract of *Ficus palmata* exhibited enhanced antidiabetic activity (Sati et al., 2020). The results showed that the silver nanoparticles possess improved ability for the inhibition of α-amylase and α-glucosidase activity as a potential antidiabetic agent. Other metals such as selenium, palladium, and platinum nanoparticles have been demonstrated to possess antihyperglycemic and antidiabetic activity (Li, Xu, et al., 2013; Li, Yan, et al., 2015; Li, Shang, et al., 2005; Li, Wei, et al., 2016; Li, Zhang, et al., 2017). Some metal nanocomposites have also shown antidiabetic activities. Virk (2018) synthesized polydispersed gold-silver nanocomposites from *Trigonella foenum graecum* aqueous seed extract and evaluated their antidiabetic effects in diabetic rats. The results showed an ameliorative effect on serum alanine aminotransferase, creatinine, and blood urea levels with a significant profound antihyperglycemic effect (Virk, 2018). Gold nanoparticles functionalized with an active component of *Cassia auriculata*, namely propanoic acid 2-(3-acetoxy-4,4,14-trimethylandrost-8-en-17-yl) (PAT), were synthesized by Venkatachalam et al. (2013) to study their antidiabetic effect. The results revealed that the functionalized gold nanoparticles (size 12–41 nm) reduced plasma glucose, cholesterol, and triglyceride level in diabetic male albino rats induced by alloxan, as well as enhanced inhibitory activity of tyrosine phosphatase 1B enzyme (Venkatachalam et al., 2013). Rahim Pouran et al. (2020) utilized the seed extract of *Silybum marianum* for the synthesis of globular shaped, polycrystalline 35–42 nm zinc oxide-silver nanocomposites. The nanocomposite was intraperitoneally injected into alloxan-mediated type 1 diabetic Wistar male rats for 16 days. The results revealed that the biogenic nanocomposites possessed efficient antidiabetic activity by improving insulin and high-density lipoprotein cholesterol in plasma, reducing fasting plasma glucose, total cholesterol, and total triglycerides in the diabetic rat model (Rahim Pouran et al., 2020).

Metal oxide nanoparticles as nanomedicines for diabetes treatment

Various nanosized metal oxides have been reported to be beneficial as antidiabetic agents. Zinc oxide nanoparticles are extensively reported for their antidiabetic properties. Umrani & Paknikar (2014) evaluated the antidiabetic effect of ~10 nm spherical zinc oxide nanoparticles in streptozotocin-induced diabetic rats. The results showed that the oxide nanoparticles caused a 70% elevated insulin level in serum, a 29% decrease in blood glucose level, a 40% decrease in esterified fatty acids, and a 48% reduction in triglyceride level. The nanoparticles were proposed to be promising for both type 1 and 2 diabetes treatment (Umrani & Paknikar, 2014). Shwetha et al. (2020) synthesized 20–40 nm spherical ZnO nanoparticles from *Areca catechu* leaf extracts and demonstrated their antidiabetic activity. The results showed enhanced glucose uptake aided by yeast cells (Shwetha et al., 2020). Green zinc oxide nanoparticles synthesized from the leaf extract of *Hibiscus subdariffa* (Bala et al., 2015), synthetic 80–100 nm ZnO (Nazarizadeh & Asri-Rezaie, 2016), ZnO nanoparticles with dipeptidyl peptidase (DPP-4) inhibitor to restore the structure and function of beta cells (El-Gharbawy et al., 2016), and ZnO nanoparticles with pleiotropic antidiabetic effects (Asani et al., 2016) are additional reported works demonstrating the antidiabetic potential of zinc oxide nanoparticles. Javed et al. (2017) also demonstrated the antidiabetic effect of PEG and polyvinylpyrrolidone-doped nanosized copper oxide particles via inhibition of the α-amylase enzyme (Javed, Ahmed, et al., 2017; Javed, Usman, et al., 2016). Magnetic iron oxide nanoparticles also show enhanced antidiabetic activity. A hybrid nanogel formed by chitosan-coated cadmium telluride quantum dots and superparamagnetic iron oxide nanoparticles was evaluated for its antidiabetic properties by Shen et al. (2012). The study showed enhanced antidiabetic activity and the nanogel was proposed as a dietary supplement for diabetic patients (Shen et al., 2012). Athithan et al. (2020) synthesized spherical magnetite (Fe_3O_4) nanoparticles with 23 nm crystallite size using fruit extracts of *Annona muricata* and assessed their antidiabetic property. The study showed an enhanced ability to inhibit α-amylase activity (Athithan et al., 2020). In their previous work, the authors of this book have demonstrated the potential of magnesium oxide nanoparticles as a promising antidiabetic agent for reversing insulin resistance, a potential alternative to insulin therapy (Jeevanandam et al., 2016, 2017; Jeevanandam, Danquah, et al., 2015; Jeevanandam, Pal, et al., 2019; Jeevanandam, San Chan, et al., 2016). The authors showed that *Amaranthus tricolor*-synthesized MgO nanoparticles (~ 30 nm size) reversed insulin resistance in diabetic (3T3-L1 adipose) cells up to 30%, compared to chemically synthesized MgO (Jeevanandam, 2017). Silver-doped nanosized indium oxide particles as in vitro inhibitors of α-amylase and α-glucosidase (Naik et al., 2016), the impact of superparamagnetic quercetin-conjugated nanosized iron oxide particles on the memory of diabetes-induced rats (Ebrahimpour et al., 2018), and the effect of *Lawsonia inermis* herbal-mediated ceria oxide nanoparticles on fasting insulin levels (Kalakotla et al., 2019) are some additional studies on the antidiabetic properties of nanosized metal oxide particles. Several metallic nanoparticles have been employed recently for targeted and controlled delivery of glucagon-like peptide-1 (GLP-1). GLP-1 is a gastrointestinal peptide hormone, which is used as a novel therapeutic agent for the treatment of diabetes. It can stimulate the secretion of insulin, the expression of specific antidiabetic genes, and the proliferation of beta cells (Ma et al., 2017). Pérez-Ortiz et al. demonstrated gold nanoparticles to be efficient in the targeted delivery of GLP-1 in normoglycemic rats. The study showed that GLP-1 immobilized on the surface of the gold nanoparticles possessed an enhanced ability to reduce glycemic levels in rat models (Pérez-Ortiz et al., 2017).

Phytosynthesized metal and metal oxide nanoparticles as nanomedicines for diabetes treatment

A significant proportion of nanosized metal and metal oxide particles that possess antidiabetic activity are synthesized from plant extracts. Several studies have indicated that surface-coated metal nanoparticles with phytochemicals influence bioavailability and bioactivity in diabetic animal models (Ahmad, Senapati, et al., 2003; Ahmad, Griffete, et al., 2015; Ahmad, Srivastava, et al., 2021; Andra, et al., 2019; Paul, et al., 2021). The antidiabetic effects of phytochemicals and metal nanoparticles are vital in the formulation of effective metal-based nanomedicines for diabetes (Javed, Ahmed, et al., 2017; Javed, Usman, et al., 2016). Phytochemicals, such as flavonoids, phenols, terpenoids, tannins, and lignins are reported to possess antidiabetic properties. For instance, Subha et al. (2021) coated flavonoids extracted from *Nigella sativa* on the surface of spherical and triangular-shaped, 20—30 nm gold nanoparticles to evaluate their antidiabetic and antioxidant activity. The results show that the phytochemical-coated nanoparticles demonstrated enhanced α-amylase and α-glucosidase inhibition activity (Subha et al., 2021). In addition, some plant proteins, enzymes, and vitamins have been reported to possess antidiabetic activity (Kayarohanam & Kavimani, 2015; Tiwari & Rao, 2002). The formulation of nanomedicines for diabetes treatment requires careful selection of plants that generate phytochemicals with good antidiabetic activity. Table 5.1 lists various nanosized metal and metal oxide—based particles synthesized via plant extracts with specific physicochemical and antidiabetic properties.

Table 5.1 List of some phytosynthesized metal and metal oxide nanoparticles with antidiabetic properties.

Metal nanoparticles

Nanoparticles	Characteristics	Antidiabetic effects	References
Gold	Synthesis: Aqueous extract of *Cassia fistula* stem bark Size: 55—99 nm	Decrease in blood glucose level and effect on serum biochemistry and body weight	Daisy and Saipriya (2012)
Gold	Synthesis: Aqueous *Cassia auriculata* leaf extract Size: 15—25 nm, triangular and spherical shape	Antihyperglycemic activity	Ganesh Kumar et al. (2011)
Gold	Synthesis: Aqueous leaf extract of *Gymnema sylvestre* Size: 50 nm spherical shaped nanoparticles	Substantial decrease in the level of blood glucose in diabetic rats induced by alloxan	Karthick et al. (2014)
Gold	Synthesis: *Padina boergesenii* brown seaweed extract Size: 2—100 nm polydispersed nanoparticles with mixed morphologies	Inhibits α-glucosidase enzymatic activity	Senthilkumar et al. (2015)
Gold	Synthesis: Leaf extract of *Chamaecostus cuspidatus* Size: 50 nm spherical and monodispersed nanoparticles	Induces hypoglycemic activity by increasing insulin sensitivity and glucose uptake	Ponnanikajamideen et al. (2018)
Gold	Synthesis: Flavonoids from *Nigella sativa*	Inhibition of alpha-amylase and alpha-glucosidase activity	Subha et al. (2021)

(Continued)

Table 5.1 (Continued)

Metal nanoparticles

Nanoparticles	Characteristics	Antidiabetic effects	References
Silver	Size: 20–30 nm spherical and triangular shape Synthesis: Aqueous extract of *Tephrosia tinctoria*	Enhanced glucose uptake rate by inhibiting carbohydrate digestive enzymes	Rajaram et al. (2015)
Silver	Size: Less than 100 nm spherical nanoparticles Synthesis: Leaf extract of *Punica granatum*	Inhibits enzymes that leads to carbohydrate digestion such as α-glucosidases and α-amylase	Saratale et al. (2018)
Silver	Size: 35–60 nm size spherical nanoparticles Synthesis: Aqueous leaf extract of *Loniceria japonica*	Inhibits α-glucosidases and α-amylase	Balan et al. (2016)
Silver	Size: 53 nm spherical and hexagonal nanoparticles Synthesis: Methanol/water bark extract of *Eysenhardtia polystachya*	Sustained hyperglycemia, hyperlipidemic state, and reduced serum insulin level in zebra fish	Garcia Campoy et al. (2018)
Silver	Size: Monodispersed, spherical, 5–21 nm nanoparticles Synthesis: Aqueous leaf extract of *Ocimum basilicum*	Reduced postprandial hyperglycemia in diabetic patients by inhibiting α-glucosidases and α-amylase enzymes	Malapermal et al. (2017)
Silver	Size: Stable, 3–25 nm nanoparticles Synthesis: Seed extract of herb *Psoralea corylifolia* Size: 15–25 nm spherical, triangle, cuboidal, and tetrahedral nanoparticles	Inhibits tyrosine phosphatase 1B	Shanker et al. (2017)

(Continued)

Table 5.1 (Continued)

Metal nanoparticles

Nanoparticles	Characteristics	Antidiabetic effects	References
Silver	Synthesis: *Musa paradisiaca* stem extract Size: 30–60 nm sized particles with spherical and flat plate-like morphology	Normalizes glucose, galactose and insulin level	Anbazhagan et al. (2017)
Copper	Synthesis: Aqueous medicinal plant *Dioscorea bulbifera* tuber extract Size: 86 nm monodispersed nanoparticles	Inhibits α-amylase and α-glucosidase enzyme	Ghosh et al. (2015)
Selenium	Size: 10–80 nm spherical nanoparticles	Decreases hepatic, markers of renal function, total lipid, cholesterol, triglyceride, low-density lipoprotein cholesterol level, and glucose-6-phosphatase activity. Increases malic enzyme, hexokinase, glucose-6-phosphate dehydrogenase activities	Al-Quraishy et al. (2015)
Platinum	Synthesis: *Whitania somnifera* leaf extract Size: 12 nm spherical particles	Reduces plasma glucose levels	Li et al. (2017)
PAT functionalized gold	Synthesis: Based on anactive component of *Cassia auriculate* namely propanoic acid 2-(3-acetoxy-4,4,14-trimethylandrost-8-en-17-yl) (PAT) Size: 12–41 nm nanoparticles	Reduces plasma glucose, cholesterol and triglyceride level by inhibiting protein tyrosine phosphatase 1B enzyme	Venkatachalam et al. (2013)

(Continued)

Table 5.1 (Continued)

Metal nanoparticles

Nanoparticles	Characteristics	Antidiabetic effects	References
Zinc oxide-silver	Synthesis: *Silybum marianum* seed extract Size: 35–42 nm globular nanoparticles	Improves insulin and high-density lipoprotein cholesterol in plasma, reduces fasting plasma glucose, total cholesterol and total triglycerides in diabetic rat model	Rahim Pouran et al. (2020)

Metal oxide nanoparticles

Nanoparticles	Characteristics	Antidiabetic effects	References
Zinc oxide	Size: 10 nm sized spherical nanoparticles	Improved glucose tolerance, high insulin in the serum. Decreased blood glucose level, esterified fatty acids, and triglyceride levels	Umrani and Paknikar (2014)
Zinc oxide	Synthesis: *Hibiscus subdariffa* leaf extract Size: 12–46 nm spherical and 50–60 nm dumbbell nanoparticles	Inhibits α-amylase and α-glucosidase enzyme	Bala et al. (2015)
Zinc oxide	Synthesis: Sol-gel method Size: 20 nm nanoparticles.	Induce diabetic dysfunction by regulating glucose tolerance, fructosamine, insulin levels, pancreas histology and pancreatic superoxide dismutase activity with Vildagliptin (DPP4 inhibitor)	El-Gharbawy et al. (2016)

(Continued)

Table 5.1 (Continued)

Metal nanoparticles

Nanoparticles	Characteristics	Antidiabetic effects	References
Zinc oxide	Synthesis: *Andrographis paniculata* leaf extract Size: 96–115 and ∼57 nm spherical and hexagonal nanoparticles respectively	Inhibits α-amylase enzyme	Rajakumar et al. (2018)
Zinc oxide	Synthesis: Microwave assisted fruit extract of *Vaccinium arctostaphylos* Size: <20 nm spherical nanoparticles	Reduced cholesterol level	Bayrami et al. (2018)
Zinc oxide	Synthesis: *Silybum marianum* seed extract Size: <40 nm spherical nanoparticles	Regulates levels of insulin, total cholesterol, blood glucose, total triglyceride, and lipoprotein density	Mohammadi Arvanag et al. (2019)
Zinc oxide	Synthesis: *Ficus palmata* leaf extract Size: 34 nm spherical nanoparticles	Inhibits alpha-amylase and alpha-glucosidase activity	Sati et al. (2020)
Zinc oxide	Synthesis: *Areca catechu* leaf extract Size: 20–40 nm spherical nanoparticles	Elevates glucose uptake aided by yeast cells	Shwetha et al. (2020)
PEG and PVP doped copper oxide	Synthesis: Coprecipitation method Size: 25–90 nm spherical nanoparticles	Inhibition of α-amylase enzyme	Javed, Ahmed et al. (2017), Javed, Zia et al. (2017)
Iron oxide	Synthesis: Chitosan coated cadmium telluride quantum dots and superparamagnetic iron oxide hybrid nanogels Size: ∼160 nm core-shell nanogel	Improved insulin sensitivity	Shen et al. (2012)

(Continued)

Table 5.1 (Continued)

Metal nanoparticles

Nanoparticles	Characteristics	Antidiabetic effects	References
Magnetite (Fe_3O_4) nanoparticles	Synthesis: *Annona muricata* fruit extract Size: 23 nm spherical nanoparticles	Inhibits alpha–amylase activity	Athithan et al. (2020)
Magnesium oxide	Synthesis: Aqueous leaf extract of *Amaranthus tricolor* at pH 3 Size: 18—40 nm hexagonal nanoparticles	Reversed insulin resistance, expresses GLUT4 protein	Jeevanandam (2017)
Ceria oxide	Synthesis: *Lawsonia inermis* herbal extract Size: 50 nm spherical nanoparticles	Restored fasting insulin levels	Kalakotla et al. (2019)
Silver-doped indium oxide	Synthesis: Solution combustion method with pristine indium trioxide Size: 20—30 nm spherical nanoparticles	In vitro inhibitors of α-glucosidase and α-amylase	Naik et al. (2016)
Chlorine and fluorine-doped zinc oxide	Synthesis: Chemical solution method Size: 59 and 18 nm round and rod-shaped nanoparticles respectively	Protective changes in the retinal oxidation of diabetic rats	El-Rahman et al. (2016)

Mechanism of metal and metal oxide nanoparticles in diabetes treatment

The cellular mechanism of metal nanoparticles in exhibiting antidiabetic effects plays a crucial role in enhancing their efficacy in diabetes treatment (Mahalingam, 2020). From Table 5.1, a common antidiabetic mechanism is through the inhibition of carbohydrate digestive enzymes such as α-amylase and α-glucosidase (Balan et al., 2016). Certain metal nanoparticles, such as silver and gold, inhibit the protein tyrosine phosphatase 1B enzyme (Shanker et al., 2017; Venkatachalam et al., 2013). The enzyme inhibitory effects of these metal nanoparticles lead to the increase in glucose uptake, insulin sensitivity, glucose transporter (GLUT) 2 expression, triglyceride level, and

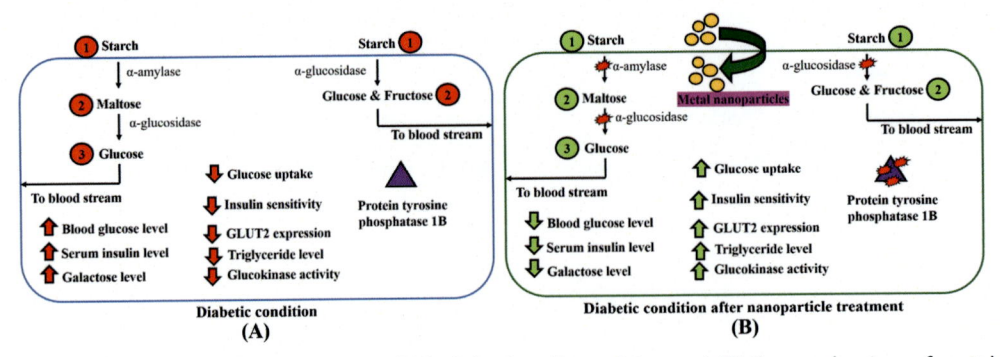

Figure 5.2 Schematic representation of (A) diabetic cell condition and (B) the mechanism of metal nanoparticles in the diabetes management.

glucokinase activity. These events further help in the regulation of blood glucose, serum insulin, and galactose level to help in diabetes management as shown in Fig. 5.2. Metal nanoparticles demonstrate to be more beneficial in managing the hyperglycemic conditions in diabetic patients than beta cell proliferation or enhanced insulin secretion (Huang & Cheng, 2020; Ibarra-Leal et al., 2019). Thus metal nanoparticles can be useful as antihyperglycemic nanomedicines in combination with conventional drugs.

Metal oxide nanoparticles exhibit a slightly different mechanism of antidiabetic activity, compared to metal nanoparticles. In nanosized metal particles, as they are formed by metal ions, they directly inhibit the enzymes to exhibit antidiabetic activity. Contrarily, metal oxide nanoparticles either directly bypass the cell membrane (Petri-Fink et al., 2008) or disintegrate into its ionic states and penetrate the cell (Verma & Stellacci, 2010). The difference between these two behaviors of metal oxide nanoparticles upon cellular interaction is highly significant in drafting their antidiabetic mechanism. Similar to metals, direct penetration of metal oxide nanoparticles into the cells exhibit antidiabetic activity via inhibition of α-amylase, α-glucosidase, and protein tyrosine phosphatase 1B enzymes to reduce hyperglycemic conditions in diabetic patients (Javed, Ahmed, et al., 2017; Javed, Usman, et al., 2016; Naik et al., 2016). The disintegration of nanosized metal oxide particles into ionic states generate metal ions, which acts as trace elements for the activation of unique enzymes such as protein kinase B, 5′ adenosine monophosphate–activated protein kinase (AMPK), poly (ADP-ribose) polymerase (PARS), and phosphoinositide 3-kinase (PI3K) enzymes (Asani et al., 2016; Jeevanandam, 2017). The activation of these enzymes leads to reversal of insulin resistance, elevated glucose uptake, insulin sensitivity, GLUT4 expression, etc., which eventually reduces blood glucose, serum insulin, and galactose level as shown in Fig. 5.3. Metal oxide nanoparticles have the potential to act as both enzyme inhibitors and activators depending on their stability and interaction with diabetic cells.

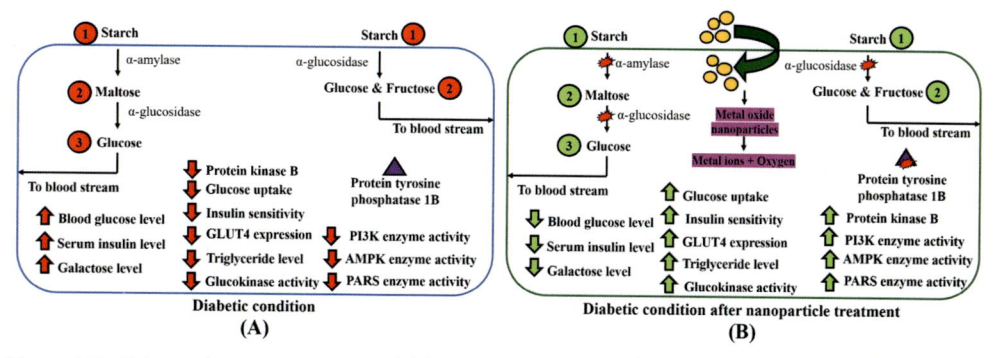

Figure 5.3 Schematic representation of (A) diabetic cell condition and (B) the mechanism of metal oxide nanoparticles in diabetes management.

Future perspective

Several research reports have demonstrated that nanosized metal and oxide-based metal particles are potential candidates as nanomedicines for the treatment of diabetes. However, there are challenges relating to the stability and biocompatibility of nanoparticles, proper understanding of the specific molecular mechanisms of metal and metal oxide nanoparticles in diabetic cells, and the lack of efficient treatment modules. These challenges hinder their widespread application as nanomedicines to treat diabetes. Advanced research on metal nanomedicines for diabetes treatment has focused on reducing hyperglycemic conditions, in combination with conventional insulin therapy and other medications. There are opportunities for researchers to focus on developing nanomedicines that can reverse and normalize diabetic conditions. The development of nanomedicines with metal oxide nanoparticles, which can act as trace metal supplements to reverse insulin resistance, is a growing area of research. Virus formulations of metal nanoparticles can be engineered to target diabetic cells. It uses nano-machines to synthesize metal nanoparticles in viral capsids and can be developed further to release the nanoparticles in a controlled fashion and terminate release after normalization of diabetic conditions (Choudhary, 2016; Shahgolzari et al., 2020). Viral formulations of metal nanoparticles can become a nano-molecular-robot to advance early-stage diabetes management.

Conclusion

This chapter presents a summary of nanosized metal and oxide-based metal particles in terms of their synthesis approaches, properties, and potentials as nanomedicines for diabetes treatment. Whilst various metal and metal oxide nanoparticles have been developed to date, only a few have been reported to possess antidiabetic activity. The chapter accounts for metal and metal oxide nanoparticles with antidiabetic activity, particularly those that are synthesized from plant extracts. The combined effect of phytochemicals and metal nanoparticles is highly important in enhancing antidiabetic activity. The antidiabetic mechanism of metal and metal oxide nanoparticles is also illustrated as an initial step to understanding the molecular basis of the formulation of metal-based nanomedicines for diabetes treatment.

References

Aghazadeh, M., Karimzadeh, I., Maragheh, M. G., & Ganjali, M. R. (2018). Enhancing the supercapacitive and superparamagnetic performances of iron oxide nanoparticles through yttrium cations electrochemical doping. *Materials Research, 21*(5).

Ahmad, A., Senapati, S., Khan, M. I., Kumar, R., Ramani, R., Srinivas, V., & Sastry, M. (2003). Intracellular synthesis of gold nanoparticles by a novel alkalotolerant actinomycete, *Rhodococcus* species. *Nanotechnology, 14*(7), 824.

Ahmad, R., Griffete, N., Lamouri, A., Felidj, N., Chehimi, M. M., & Mangeney, C. (2015). Nanocomposites of gold nanoparticles@ molecularly imprinted polymers: Chemistry, processing, and applications in sensors. *Chemistry of Materials, 27*(16), 5464–5478.

Ahmad, R., Srivastava, S., Ghosh, S., & Khare, S. K. (2021). Phytochemical delivery through nanocarriers: A review. *Colloids and Surfaces B: Biointerfaces, 197*, 111389. <https://doi.org/10.1016/j.colsurfb.2020.111389>

Ahmed, D. S., Mohammed, M. R., & Mohammed, M. K. A. (2020). Synthesis of multi-walled carbon nanotubes decorated with ZnO/Ag nanoparticles by co-precipitation method. *Nanoscience & Nanotechnology-Asia, 10*(2), 127–133.

Alagiri, M., Rameshkumar, P., & Pandikumar, A. (2017). Gold nanorod-based electrochemical sensing of small biomolecules: A review. *Microchimica Acta, 184*(9), 3069–3092.

Al-Quraishy, S., Dkhil, M. A., & Abdel Moneim, A. E. (2015). Anti-hyperglycemic activity of selenium nanoparticles in streptozotocin-induced diabetic rats. *International Journal of Nanomedicine, 10*, 6741–6756.

Alkaladi, A., Abdelazim, A. M., & Afifi, M. (2014). Antidiabetic activity of zinc oxide and silver nanoparticles on streptozotocin-induced diabetic rats. *International Journal of Molecular Sciences, 15*(2), 2015–2023.

Anbazhagan, P., Murugan, K., Jaganathan, A., et al. (2017). Mosquitocidal, antimalarial and antidiabetic potential of musa paradisiaca-synthesized silver nanoparticles: In vivo and in vitro approaches. *Journal of Cluster Science, 28*(1), 91–107.

Andra, S., Balu, S. K., Jeevanandham, J., Muthalagu, M., Vidyavathy, M., Chan, Y. S., & Danquah, M. K. (2019). Phytosynthesized metal oxide nanoparticles for pharmaceutical applications. *Naunyn-Schmiedeberg's Archives of Pharmacology, 392*(7), 755–771. <https://doi.org/10.1007/s00210-019-01666-7>

Anh, N. T. N., & Doong, R. (2018). One-step synthesis of size-tunable gold@ sulfur-doped graphene quantum dot nanocomposites for highly selective and sensitive detection of nanomolar 4-nitrophenol in aqueous solutions with complex matrix. *ACS Applied Nano Materials, 1*(5), 2153–2163.

Arbain, R., Othman, M., & Palaniandy, S. (2011). Preparation of iron oxide nanoparticles by mechanical milling. *Minerals Engineering, 24*(1), 1–9. <https://doi.org/10.1016/j.mineng.2010.08.025>

Ariga, K., Aono, M., Eid, K., Wang, H., & Wang, L. (2017). Chapter 6 – Nanoarchitectonic metals. In *Supra-materials nanoarchitectonics* (pp. 135–171). William Andrew Publishing. <https://doi.org/10.1016/B978-0-323-37829-1.00006-7>

Asani, S. C., Umrani, R. D., & Paknikar, K. M. (2016). In vitro studies on the pleotropic antidiabetic effects of zinc oxide nanoparticles. *Nanomedicine: Nanotechnology, Biology, and Medicine, 11*(13), 1671–1687.

Aslan, K., Leonenko, Z., Lakowicz, J. R., & Geddes, C. D. (2005). Fast and slow deposition of silver nanorods on planar surfaces: Application to metal-enhanced fluorescence. *The Journal of Physical Chemistry B, 109*(8), 3157–3162. <https://doi.org/10.1021/jp045186t>

Athithan, A. S. S., Jeyasundari, J., Renuga, D., & Jacob, Y. B. A. (2020). *Annona muricata* fruit mediated biosynthesis, physicochemical characterization of magnetite (Fe_3O_4) nanoparticles and assessment of its in vitro antidiabetic activity. *Rasayan Journal of Chemistry, 13*(3), 1759−1766.

Babushkina, I. V., Gladkova, E. V., Belova, S. V., & Norkin, I. A. (2017). Application of preparations containing copper nanoparticles for the treatment of experimental septic wounds. *Bulletin of Experimental Biology and Medicine, 164*(2), 162−164.

Bala, N., Saha, S., Chakraborty, M., Maiti, M., Das, S., Basu, R., & Nandy, P. (2015). Green synthesis of zinc oxide nanoparticles using *Hibiscus subdariffa* leaf extract: Effect of temperature on synthesis, antibacterial activity and anti-diabetic activity. *RSC Advances, 5*(7), 4993−5003.

Balan, K., Qing, W., Wang, Y., Liu, X., Palvannan, T., Wang, Y., Ma, F., & Zhang, Y. (2016). Antidiabetic activity of silver nanoparticles from green synthesis using *Lonicera japonica* leaf extract. *Rsc Advances, 6*(46), 40162−40168.

Bayrami, A., Parvinroo, S., Habibi-Yangjeh, A., et al. (2018). Bio-extract-mediated ZnO nanoparticles: microwave-assisted synthesis, characterization and antidiabetic activity evaluation. *Artificial Cells, Nanomedicine, and Biotechnology, 46*(4), 730−739.

Beyerlein, I. J., Mara, N. A., Bhattacharyya, D., Alexander, D. J., & Necker, C. T. (2011). Texture evolution via combined slip and deformation twinning in rolled silver−copper cast eutectic nanocomposite. *International Journal of Plasticity, 27*(1), 121−146.

Bhuyan, T., Mishra, K., Khanuja, M., Prasad, R., & Varma, A. (2015). Biosynthesis of zinc oxide nanoparticles from *Azadirachta indica* for antibacterial and photocatalytic applications. *Materials Science in Semiconductor Processing, 32*, 55−61. <https://doi.org/10.1016/j.mssp.2014.12.053>

Brennan, S. A., Ní Fhoghlú, C., Devitt, B. M., O'Mahony, F. J., Brabazon, D., & Walsh, A. (2015). Silver nanoparticles and their orthopaedic applications. *The Bone & Joint Journal, 97*(5), 582−589.

Caillard, A., Brault, P., Mathias, J., Charles, C., Boswell, R. W., & Sauvage, T. (2005). Deposition and diffusion of platinum nanoparticles in porous carbon assisted by plasma sputtering. *Surface and Coatings Technology, 200*(1), 391−394. <https://doi.org/10.1016/j.surfcoat.2005.01.033>

Carabineiro, S. (2017). Applications of gold nanoparticles in nanomedicine: Recent advances in vaccines. *Molecules (Basel, Switzerland), 22*(5), 857.

Carpenter, E. E., Kumbhar, A., Wiemann, J. A., Srikanth, H., Wiggins, J., Zhou, W., & O'Connor, C. J. (2000). Synthesis and magnetic properties of gold−iron−gold nanocomposites. *Materials Science and Engineering: A, 286*(1), 81−86.

Carter, C. B., & Norton, M. G. (2013). Sols, gels, and organic chemistry. In *Ceramic materials: Science and engineering* (pp. 400−411). New York: Springer. <https://doi.org/10.1007/978-0-387-46271-4_22>

Chaudhari, A. A., Nath, S., Kate, K., Dennis, V., Singh, S. R., Owen, D. R., Palazzo, C., Arnold, R. D., Miller, M. E., & Pillai, S. R. (2016). A novel covalent approach to bio-conjugate silver coated single walled carbon nanotubes with antimicrobial peptide. *Journal of Nanobiotechnology, 14*(1), 58.

Chaudhary, S., Kumar, S., Umar, A., Singh, J., Rawat, M., & Mehta, S. K. (2017). Europium-doped gadolinium oxide nanoparticles: A potential photoluminescencent probe for highly selective and sensitive detection of Fe^{3+} and Cr^{3+} ions. *Sensors and Actuators B: Chemical, 243*, 579−588. <https://doi.org/10.1016/j.snb.2016.12.002>

Chen, Y., Zhao, Y., Zou, X., & Sun, L. (2020). Porous silica nanocarriers with gold/carbon quantum dots for photo-chemotherapy and cellular imaging. *Journal of Drug Delivery Science and Technology, 61*, 102141.

Choudhary, M. I. (2016). *Frontiers in drug design & discovery: Volume 10*. Bentham Science Publishers. https://books.google.pt/books?id = 4pshEAAAQBAJ.

Crespo, P., Litrán, R., Rojas, T. C., Multigner, M., de la Fuente, J. M., Sánchez-López, J. C., García, M. A., Hernando, A., Penadés, S., & Fernández, A. (2004). Permanent magnetism, magnetic anisotropy, and hysteresis of thiol-capped gold nanoparticles. *Physical Review Letters, 93*(8), 087204. <https://doi.org/10.1103/PhysRevLett.93.087204>

Cuevas, R., Durán, N., Diez, M. C., Tortella, G. R., & Rubilar, O. (2015). Extracellular biosynthesis of copper and copper oxide nanoparticles by *Stereum hirsutum*, a native white-rot fungus from Chilean forests. *Journal of Nanomaterials, 16*(1), 57.

Daisy, P., & Saipriya, K. (2012). Biochemical analysis of Cassia fistula aqueous extract and phytochemically synthesized gold nanoparticles as hypoglycemic treatment for diabetes mellitus. *International Journal of Nanomedicine, 7*, 1189–1202.

Damonte, L. C., MendozaZélis, L. A., Marí Soucase, B., & Hernández Fenollosa, M. A. (2004). Nanoparticles of ZnO obtained by mechanical milling. *Powder Technology, 148*(1), 15–19. <https://doi.org/10.1016/j.powtec.2004.09.014>

Dang, T. M. D., Le, T. T. T., Fribourg-Blanc, E., & Dang, M. C. (2011). Synthesis and optical properties of copper nanoparticles prepared by a chemical reduction method. *Advances in Natural Sciences: Nanoscience and Nanotechnology, 2*(1), 015009.

Daniel, S. C. G. K., Joseph, P., & Sivakumar, M. (2018). Biosynthesized silver nanoparticle based hybrid materials. *Nanoscience & Nanotechnology-Asia, 8*(1), 4–12. <https://doi.org/10.2174/2210681207666170421114719>

Daraee, H., Eatemadi, A., Abbasi, E., FekriAval, S., Kouhi, M., & Akbarzadeh, A. (2016). Application of gold nanoparticles in biomedical and drug delivery. *Artificial Cells, Nanomedicine, and Biotechnology, 44*(1), 410–422.

Darr, J. A., Zhang, J., Makwana, N. M., & Weng, X. (2017). Continuous hydrothermal synthesis of inorganic nanoparticles: Applications and future directions. *Chemical Reviews, 117*(17), 11125–11238.

Dolgaev, S. I., Simakin, A. V., Voronov, V. V., Shafeev, G. A., & Bozon-Verduraz, F. (2002). Nanoparticles produced by laser ablation of solids in liquid environment. *Applied Surface Science, 186*(1), 546–551. <https://doi.org/10.1016/S0169-4332(01)00634-1>

Dong, H., Chen, Y. C., & Feldmann, C. (2015). Polyol synthesis of nanoparticles: Status and options regarding metals, oxides, chalcogenides, and non-metal elements. *Green Chemistry, 17*(8), 4107–4132. <https://doi.org/10.1039/C5GC00943J>

Dujardin, E., Peet, C., Stubbs, G., Culver, J. N., & Mann, S. (2015). Organization of metallic nanoparticles using tobacco mosaic virus templates. *Nano Letters, 3*(3), 413–417.

Durán, N., Marcato, P. D., Durán, M., Yadav, A., Gade, A., & Rai, M. (2011). Mechanistic aspects in the biogenic synthesis of extracellular metal nanoparticles by peptides, bacteria, fungi, and plants. *Applied Microbiology and Biotechnology, 90*(5), 1609–1624.

Dutta Choudhury, S., Badugu, R., Ray, K., & Lakowicz, J. R. (2012). Silver–gold nanocomposite substrates for metal-enhanced fluorescence: Ensemble and single-molecule spectroscopic studies. *The Journal of Physical Chemistry C, 116*(8), 5042–5048.

Ebrahimpour, S., Esmaeili, A., & Beheshti, S. (2018). Effect of quercetin-conjugated superparamagnetic iron oxide nanoparticles on diabetes-induced learning and memory impairment in rats. *International Journal of Nanomedicine, 13*, 6311.

Ederth, J., Johnsson, P., Niklasson, G. A., Hoel, A., Hultåker, A., Heszler, P., Granqvist, C. G., van Doorn, A. R., Jongerius, M. J., & Burgard, D. (2003). Electrical and optical properties of thin films consisting of tin-doped indium oxide nanoparticles. *Physical Review B, 68*(15), 155410. <https://doi.org/10.1103/PhysRevB.68.155410>

Ehsan, M. A., Khaledi, H., Pandikumar, A., Rameshkumar, P., Huang, N. M., Arifin, Z., & Mazhar, M. (2015). Nitrite ion sensing properties of $ZnTiO_3–TiO_2$ composite thin films deposited from a zinc–titanium molecular complex. *New Journal of Chemistry, 39*(9), 7442–7452.

El-Gharbawy, R. M., Emara, A. M., & Abu-Risha, S. E.-S. (2016). Zinc oxide nanoparticles and a standard antidiabetic drug restore the function and structure of beta cells in Type-2 diabetes. *Biomedicine & Pharmacotherapy, 84*, 810–820.

El-Rahman, S. N. A., Reda, S. M., & AlGhannam, S. M. (2016). Synthesis and characterization of nano-doped zinc oxide and its application as protective oxidative changes in the retina of diabetic rats. *Journal of Diabetes & Metabolism, 7*(691), 2.

Fedlheim, D. L., & Foss, C. A. (2001). *Metal nanoparticles: Synthesis, characterization, and applications.* CRC Press.

Feng, H., Yang, Y., You, Y., Li, G., Guo, J., Yu, T., Shen, Z., Wu, T., & Xing, B. (2009). Simple and rapid synthesis of ultrathin gold nanowires, their self-assembly and application in surface-enhanced Raman scattering. *Chemical Communications, 15*, 1984–1986. <https://doi.org/10.1039/B822507A>

Förster, H., Wolfrum, C., & Peukert, W. (2012). Experimental study of metal nanoparticle synthesis by an arc evaporation/condensation process. *Journal of Nanoparticle Research, 14*(7), 926. <https://doi.org/10.1007/s11051-012-0926-1>

Franci, G., Falanga, A., Galdiero, S., Palomba, L., Rai, M., Morelli, G., & Galdiero, M. (2015). Silver nanoparticles as potential antibacterial agents. *Molecules (Basel, Switzerland), 20*(5), 8856−8874.

Franke, M. E., Koplin, T. J., & Simon, U. (2006). Metal and metal oxide nanoparticles in chemiresistors: Does the nanoscale matter? *Small (Weinheim an der Bergstrasse, Germany), 2*(1), 36−50.

Ganesh Kumar, V., Dinesh Gokavarapu, S., Rajeswari, A., et al. (2011). Facile green synthesis of gold nanoparticles using leaf extract of antidiabetic potent Cassia auriculata. *Colloids and Surfaces B: Biointerfaces, 87*(1), 159−163.

Garcia Campoy, A. H., Perez Gutierrez, R. M., Manriquez-Alvirde, G., et al. (2018). Protection of silver nanoparticles using Eysenhardtia polystachya in peroxide-induced pancreatic β-Cell damage and their antidiabetic properties in zebrafish. *International Journal of Nanomedicine, 13*, 2601−2612.

Garfinkel, D. A., Pakeltis, G., Tang, N., Ivanov, I. N., Fowlkes, J. D., Gilbert, D. A., & Rack, P. D. (2020). Optical and magnetic properties of Ag−Ni bimetallic nanoparticles assembled via pulsed laser-induced dewetting. *ACS Omega, 5*(30), 19285−19292.

GhavamiNejad, A., Aguilar, L. E., Ambade, R. B., Lee, S.-H., Park, C. H., & Kim, C. S. (2015). Immobilization of silver nanoparticles on electropolymerized polydopamine films for metal implant applications. *Colloids and Interface Science Communications, 6*, 5−8.

GhavamiNejad, A., Park, C. H., & Kim, C. S. (2016). In situ synthesis of antimicrobial silver nanoparticles within antifouling zwitterionic hydrogels by catecholic redox chemistry for wound healing application. *Biomacromolecules, 17*(3), 1213−1223.

Ghorbani, H. R., Mehr, F. P., Pazoki, H., & Rahmani, B. M. (2015). Synthesis of ZnO nanoparticles by precipitation method. *Oriental Journal of Chemistry, 31*(2), 1219−1221.

Ghosh, S., More, P., Nitnavare, R., Jagtap, S., Chippalkatti, R., Derle, A., ... Chopade, B. A. (2015). Antidiabetic and antioxidant properties of copper nanoparticles synthesized by medicinal plant *Dioscorea bulbifera. Journal of Nanomedicine & Nanotechnology* (S6), 1.

Gold, K., Slay, B., Knackstedt, M., & Gaharwar, A. K. (2018). Antimicrobial activity of metal and metal-oxide based nanoparticles. *Advanced Therapeutics, 1*(3), 1700033. <https://doi.org/10.1002/adtp.201700033>

Govorov, A. O., & Richardson, H. H. (2007). Generating heat with metal nanoparticles. *Nano Today, 2*(1), 30−38.

Graczyk, H., Bryan, L. C., Lewinski, N., Suarez, G., Coullerez, G., Bowen, P., & Riediker, M. (2015). Physicochemical characterization of nebulized superparamagnetic iron oxide nanoparticles (SPIONs). *Journal of Aerosol Medicine and Pulmonary Drug Delivery, 28*(1), 43−51. <https://doi.org/10.1089/jamp.2013.1117>

Grumezescu, A. M., Mashitah, M. D., Chan, Y. S., & Jason, J. (2016). Chapter 8 − Antimicrobial properties of nanobiomaterials and the mechanism. In *Nanobiomaterials in antimicrobial therapy* (pp. 261−312). William Andrew Publishing. <https://doi.org/10.1016/B978-0-323-42864-4.00008-7>

Guo, B.-L., Han, P., Guo, L.-C., Cao, Y.-Q., Li, A.-D., Kong, J.-Z., Zhai, H.-F., & Wu, D. (2015). The antibacterial activity of Ta-doped ZnO nanoparticles. *Nanoscale Research Letters, 10*(1), 336. <https://doi.org/10.1186/s11671-015-1047-4>

Guzman, J., Carrettin, S., Fierro-Gonzalez, J. C., Hao, Y., Gates, B. C., & Corma, A. (2005). CO oxidation catalyzed by supported gold: Cooperation between gold and nanocrystalline rare-earth supports forms reactive surface superoxide and peroxide species. *Angewandte Chemie International Edition, 117*(30), 4856−4859.

Habibi, M. H., & Karimi, B. (2015). Doctor blade sol−gel deposition of a nano-composite copper-zinc oxide on borosilicate glass for advanced oxidative degradation of textille dye in water environment. *Iranian Journal of Environmental Technology, 1*, 31−38.

Hassanjani-Roshan, A., Vaezi, M. R., Shokuhfar, A., & Rajabali, Z. (2011). Synthesis of iron oxide nanoparticles via sonochemical method and their characterization. *Particuology, 9*(1), 95−99. <https://doi.org/10.1016/j.partic.2010.05.013>

Hayashi, H., & Hakuta, Y. (2010). Hydrothermal synthesis of metal oxide nanoparticles in supercritical water. *Materials, 3*(7), 3794−3817.

Hayat, M. A. (2012). *Colloidal gold: Principles, methods, and applications.* Elsevier.

He, L., Weniger, F., Neumann, H., & Beller, M. (2016). Synthesis, characterization, and application of metal nanoparticles supported on nitrogen-doped carbon: Catalysis beyond electrochemistry. *Angewandte Chemie International Edition*, *55*(41), 12582−12594.

Huang, S.-Y., & Cheng, C.-C. (2020). Spontaneous self-assembly of single-chain amphiphilic polymeric nanoparticles in water. *Nanomaterials*, *10*(10). <https://doi.org/10.3390/nano10102006>

Huang, X.-W., Liang, H., Li, Z., Zhou, J., Chen, X., Bai, S.-M., & Yang, H.-H. (2017). Monodisperse phase transfer and surface bioengineering of metal nanoparticles via a silk fibroin protein corona. *Nanoscale*, *9*(8), 2695−2700.

Huber, D. L. (2005). Synthesis, properties, and applications of iron nanoparticles. *Small (Weinheim an der Bergstrasse, Germany)*, *1*(5), 482−501.

Ibarra-Leal, J. J., Yocupicio, L., Apolinar-Iribe, A., Díaz-Reval, I., Parra-Delgado, H., Limón-Miranda, S., Sánchez-Pastor, E., & Virgen-Ortiz, A. (2019). In vivo zinc oxide nanoparticles induces acute hyperglycemic response a dose-dependent and route of administration in healthy and diabetic rats. *Beilstein Archives*. <https://doi.org/10.3762/bxiv.2019.75.v1>

Iravani, S., Korbekandi, H., Mirmohammadi, S. V., & Zolfaghari, B. (2014). Synthesis of silver nanoparticles: Chemical, physical and biological methods. *Research in Pharmaceutical Sciences*, *9*(6), 385−406. <https://www.ncbi.nlm.nih.gov/pubmed/26339255>

Javed, R., Ahmed, M., Haq, I., ul, Nisa, S., & Zia, M. (2017). PVP and PEG doped CuO nanoparticles are more biologically active: Antibacterial, antioxidant, antidiabetic and cytotoxic perspective. *Materials Science and Engineering: C*, *79*, 108−115. <https://doi.org/10.1016/j.msec.2017.05.006>

Javed, R., Usman, M., Tabassum, S., & Zia, M. (2016). Effect of capping agents: Structural, optical and biological properties of ZnO nanoparticles. *Applied Surface Science*, *386*, 319−326.

Javed, R., Zia, M., Yücesan, B., & Gürel, E. (2017). Abiotic stress of ZnO-PEG, ZnO-PVP, CuO-PEG and CuO-PVP nanoparticles enhance growth, sweetener compounds and antioxidant activities in shoots of Stevia rebaudiana Bertoni. *IET Nanobiotechnology*, *11*(7), 898−902.

Jeevanandam, J. (2017). Enhanced synthesis and delivery of magnesium oxide nanoparticles for reverse insulin resistance in type 2 diabetes mellitus.

Jeevanandam, J., Chan, Y. S., & Danquah, M. K. (2016). Biosynthesis of metal and metal oxide nanoparticles. *ChemBioEng Reviews*, *3*(2), 55−67.

Jeevanandam, J., Chan, Y. S., & Danquah, M. K. (2017). Calcination-dependent morphology transformation of sol-gel-synthesized MgO nanoparticles. *ChemistrySelect*, *2*(32), 10393−10404.

Jeevanandam, J., Danquah, M. K., Debnath, S., Meka, V. S., & Chan, Y. S. (2015). Opportunities for nano-formulations in type 2 diabetes mellitus treatments. *Current Pharmaceutical Biotechnology*, *16*(10), 853−870.

Jeevanandam, J., Kulabhusan, P. K., Sabbih, G., Akram, M., & Danquah, M. K. (2020). Phytosynthesized nanoparticles as a potential cancer therapeutic agent. *3 Biotech*, *10*(12), 1−26.

Jeevanandam, J., Pal, K., & Danquah, M. K. (2019). Virus-like nanoparticles as a novel delivery tool in gene therapy. *Biochimie*, *157*, 38−47.

Jeevanandam, J., San Chan, Y., & Danquah, M. K. (2016). Nano-formulations of drugs: Recent developments, impact and challenges. *Biochimie*, *128*, 99−112.

Ji, K., Deng, J., Zang, H., Han, J., Arandiyan, H., & Dai, H. (2015). Fabrication and high photocatalytic performance of noble metal nanoparticles supported on 3DOM InVO$_4$−BiVO$_4$ for the visible-light-driven degradation of rhodamine B and methylene blue. *Applied Catalysis B: Environmental*, *165*, 285−295. <https://doi.org/10.1016/j.apcatb.2014.10.005>

Jiang, B., Wu, Q., Deng, N., Chen, Y., Zhang, L., Liang, Z., & Zhang, Y. (2016). Hydrophilic GO/Fe$_3$O$_4$/Au/PEG nanocomposites for highly selective enrichment of glycopeptides. *Nanoscale*, *8*(9), 4894−4897.

Jin, R., & Nobusada, K. (2014). Doping and alloying in atomically precise gold nanoparticles. *Nano Research*, *7*(3), 285−300.

Kalakotla, S., Jayarambabu, N., Mohan, G. K., Mydin, R. B. S. M. N., & Gupta, V. R. (2019). A novel pharmacological approach of herbal mediated cerium oxide and silver nanoparticles with improved biomedical activity in comparison with *Lawsonia inermis*. *Colloids and Surfaces B: Biointerfaces*, *174*, 199−206. <https://doi.org/10.1016/j.colsurfb.2018.11.014>

Karthick, V., Kumar, V. G., Dhas, T. S., et al. (2014). Effect of biologically synthesized gold nanoparticles on alloxan-induced diabetic rats—An in vivo approach. *Colloids and Surfaces B: Biointerfaces, 122,* 505−511.

Kashin, A. S., & Ananikov, V. P. (2011). A SEM study of nanosized metal films and metal nanoparticles obtained by magnetron sputtering. *Russian Chemical Bulletin, 60*(12), 2602−2607. <https://doi.org/10.1007/s11172-011-0399-x>

Kathiraven, T., Sundaramanickam, A., Shanmugam, N., & Balasubramanian, T. (2015). Green synthesis of silver nanoparticles using marine algae *Caulerpa racemosa* and their antibacterial activity against some human pathogens. *Applied Nanoscience, 5*(4), 499−504.

Katwal, R., Kaur, H., Sharma, G., Naushad, M., & Pathania, D. (2015). Electrochemical synthesized copper oxide nanoparticles for enhanced photocatalytic and antimicrobial activity. *Journal of Industrial and Engineering Chemistry, 31,* 173−184.

Kaviyarasu, K., Geetha, N., Kanimozhi, K., MariaMagdalane, C., Sivaranjani, S., Ayeshamariam, A., Kennedy, J., & Maaza, M. (2017). In vitro cytotoxicity effect and antibacterial performance of human lung epithelial cells A549 activity of zinc oxide doped TiO_2 nanocrystals: Investigation of biomedical application by chemical method. *Materials Science and Engineering: C, 74,* 325−333. <https://doi.org/10.1016/j.msec.2016.12.024>

Kayarohanam, S., & Kavimani, S. (2015). Current trends of plants having antidiabetic activity: A review. *Journal of Bioanalysis & Biomedicine, 7*(2), 55.

Kelly, K. L., Coronado, E., Zhao, L. L., & Schatz, G. C. (2003). The optical properties of metal nanoparticles: The influence of size, shape, and dielectric environment. *The Journal of Physical Chemistry B, 107*(3), 668−677. <https://doi.org/10.1021/jp026731y>

Kim, H. Y., Lee, H. M., & Henkelman, G. (2012). CO oxidation mechanism on CeO_2-supported Au nanoparticles. *Journal of the American Chemical Society, 134*(3), 1560−1570. <https://doi.org/10.1021/ja207510v>

Kokate, M., Garadkar, K., & Gole, A. (2016). Zinc-oxide-silica-silver nanocomposite: Unique one-pot synthesis and enhanced catalytic and anti bacterial performance. *Journal of Colloid and Interface Science, 483,* 249−260. <https://doi.org/10.1016/j.jcis.2016.08.039>

Kreibig, U. (1974). Electronic properties of small silver particles: The optical constants and their temperature dependence. *Journal of Physics F: Metal Physics, 4*(7), 999. <http://stacks.iop.org/0305-4608/4/i=7/a=007>

Kumar, C. G., & Poornachandra, Y. (2015). Biodirected synthesis of Miconazole-conjugated bacterial silver nanoparticles and their application as antifungal agents and drug delivery vehicles. *Colloids and Surfaces B: Biointerfaces, 125,* 110−119.

Lee, J., Lee, P., Lee, H., Lee, D., Lee, S. S., & Ko, S. H. (2012). Very long Ag nanowire synthesis and its application in a highly transparent, conductive and flexible metal electrode touch panel. *Nanoscale, 4*(20), 6408−6414. <https://doi.org/10.1039/C2NR31254A>

Lee, N., & Hyeon, T. (2012). Designed synthesis of uniformly sized iron oxide nanoparticles for efficient magnetic resonance imaging contrast agents. *Chemical Society Reviews, 41*(7), 2575−2589.

Leung, K. C.-F., Xuan, S., Zhu, X., Wang, D., Chak, C.-P., Lee, S.-F., Ho, W. K. W., & Chung, B. C. T. (2012). Gold and iron oxide hybrid nanocomposite materials. *Chemical Society Reviews, 41*(5), 1911−1928.

Li, G. R., Xu, H., Lu, X. F., Feng, J. X., Tong, Y. X., & Su, C. Y. (2013). Electrochemical synthesis of nanostructured materials for electrochemical energy conversion and storage. *Nanoscale, 5*(10), 4056−4069. <https://doi.org/10.1039/c3nr00607g>

Li, L., Yan, J., Wang, T., Zhao, Z.-J., Zhang, J., Gong, J., & Guan, N. (2015). Sub-10 nm rutile titanium dioxide nanoparticles for efficient visible-light-driven photocatalytic hydrogen production. *Nature Communications, 6,* 5881. <https://doi.org/10.1038/ncomms6881>

Li, M.-G., Shang, Y.-J., Gao, Y.-C., Wang, G.-F., & Fang, B. (2005). Preparation of novel mercury-doped silver nanoparticles film glassy carbon electrode and its application for electrochemical biosensor. *Analytical Biochemistry, 341*(1), 52−57.

Li, X., Wei, J., Aifantis, K. E., Fan, Y., Feng, Q., Cui, F.-Z., & Watari, F. (2016). Current investigations into magnetic nanoparticles for biomedical applications. *Journal of Biomedical Materials Research. Part A, 104*(5), 1285−1296. <https://doi.org/10.1002/jbm.a.35654>

Li, Y., Zhang, J., Gu, J., & Chen, S. (2017). Biosynthesis of polyphenol-stabilised nanoparticles and assessment of anti-diabetic activity. *Journal of Photochemistry and Photobiology B: Biology, 169*, 96−100. <https://doi.org/10.1016/j.jphotobiol.2017.02.017>

Lin, Y., Cooper, C., Wang, M., Adams, J. J., Genzer, J., & Dickey, M. D. (2015). Handwritten, soft circuit boards and antennas using liquid metal nanoparticles. *Small (Weinheim an der Bergstrasse, Germany), 11*(48), 6397−6403.

Ling, D., Lee, N., & Hyeon, T. (2015). Chemical synthesis and assembly of uniformly sized iron oxide nanoparticles for medical applications. *Accounts of Chemical Research, 48*(5), 1276−1285.

Linic, S., Aslam, U., Boerigter, C., & Morabito, M. (2015). Photochemical transformations on plasmonic metal nanoparticles. *Nature Materials, 14*(6), 567.

Liu, Y., Szeifert, J. M., Feckl, J. M., Mandlmeier, B., Rathousky, J., Hayden, O., Fattakhova-Rohlfing, D., & Bein, T. (2010). Niobium-doped titania nanoparticles: synthesis and assembly into mesoporous films and electrical conductivity. *ACS Nano, 4*(9), 5373−5381. <https://doi.org/10.1021/nn100785j>

Ma, H., Li, X., Yan, T., Li, Y., Zhang, Y., Wu, D., Wei, Q., & Du, B. (2016). Electrochemiluminescent immunosensing of prostate-specific antigen based on silver nanoparticles-doped Pb (II) metal-organic framework. *Biosensors and Bioelectronics, 79*, 379−385.

Ma, P. C., Tang, B. Z., & Kim, J.-K. (2008). Effect of CNT decoration with silver nanoparticles on electrical conductivity of CNT-polymer composites. *Carbon, 46*(11), 1497−1505. <https://doi.org/10.1016/j.carbon.2008.06.048>

Ma, S., We, L., Yang, H., Deng, S., & Jevnikar, A. M. (2017). Emerging technologies to achieve oral delivery of GLP-1 and GLP-1 analogs for treatment of type 2 diabetes mellitus (T2DM). *Canadian Journal of Biotechnology, 1*(1), 1.

Magdalane, C. M., Kaviyarasu, K., Vijaya, J. J., Siddhardha, B., & Jeyaraj, B. (2016). Photocatalytic activity of binary metal oxide nanocomposites of CeO_2/CdO nanospheres: Investigation of optical and antimicrobial activity. *Journal of Photochemistry and Photobiology B: Biology, 163*, 77−86. <https://doi.org/10.1016/j.jphotobiol.2016.08.013>

Mahalingam, D. A. G. (2020). Green synthesized metal nanoparticles, characterization and its antidiabetic activities—A review. *Research Journal of Pharmacy and Technology, 13*(1), 468−474.

Malapermal, V., Botha, I., Krishna, S. B. N., et al. (2017). Enhancing antidiabetic and antimicrobial performance of *Ocimum basilicum*, and *Ocimum sanctum* (L.) using silver nanoparticles. *Saudi Journal of Biological Sciences, 24*(6), 1294−1305.

Maria Magdalane, C., Kaviyarasu, K., Raja, A., Arularasu, M. V., Mola, G. T., Isaev, A. B., Al-Dhabi, N. A., Arasu, M. V., Jeyaraj, B., Kennedy, J., & Maaza, M. (2018). Photocatalytic decomposition effect of erbium doped cerium oxide nanostructures driven by visible light irradiation: Investigation of cytotoxicity, antibacterial growth inhibition using catalyst. *Journal of Photochemistry and Photobiology B: Biology, 185*, 275−282. <https://doi.org/10.1016/j.jphotobiol.2018.06.011>

Maria, K. H., & Mieno, T. (2015). Synthesis of single-walled carbon nanotubes by low-frequency bipolar pulsed arc discharge method. *Vacuum, 113*, 11−18. <https://doi.org/10.1016/j.vacuum.2014.11.025>

Matijević, E., Andreescu, S., Ornatska, M., Erlichman, J. S., Estevez, A., & Leiter, J. C. (2012). Biomedical applications of metal oxide nanoparticles. In *Fine particles in medicine and pharmacy* (pp. 57−100). Springer US. <https://doi.org/10.1007/978-1-4614-0379-1_3>

Mazumder, B., Ray, S., Pal, P., & Pathak, Y. (2019). *Nanotechnology: Therapeutic, nutraceutical, and cosmetic advances*. CRC Press.

Meyer, D. E., Sikdar, S. K., Hutson, N. D., & Bhattacharyya, D. (2007). Examination of sulfur-functionalized, copper-doped iron nanoparticles for vapor-phase mercury capture in entrained-flow and fixed-bed systems. *Energy & Fuels, 21*(5), 2688−2697.

Milanese, M., Colangelo, G., Creti, A., Lomascolo, M., Iacobazzi, F., & De Risi, A. (2016). Optical absorption measurements of oxide nanoparticles for application as nanofluid in direct absorption solar power systems − Part I: Water-based nanofluids behavior. *Solar Energy Materials and Solar Cells, 147*, 315−320.

Min, Y.-L., Wan, Y., & Yu, S.-H. (2009). Au@ Y_2O_3: Eu^{3+} rare earth oxide hollow sub-microspheres with encapsulated gold nanoparticles and their optical properties. *Solid State Sciences, 11*(1), 96−101.

Mishra, P. K., Mishra, H., Ekielski, A., Talegaonkar, S., & Vaidya, B. (2017). Zinc oxide nanoparticles: A promising nanomaterial for biomedical applications. *Drug Discovery Today*, *22*(12), 1825–1834.

Mohammadi Arvanag, F., Bayrami, A., Habibi-Yangjeh, A., et al. (2019). A comprehensive study on antidiabetic and antibacterial activities of ZnO nanoparticles biosynthesized using Silybum marianum L seed extract. *Materials Science and Engineering: C*, *97*, 397–405.

Mohanpuria, P., Rana, N. K., & Yadav, S. K. (2008). Biosynthesis of nanoparticles: Technological concepts and future applications. *Journal of Nanoparticle Research*, *10*(3), 507–517.

Mohapatra, S., Mishra, Y. K., Avasthi, D. K., Kabiraj, D., Ghatak, J., & Varma, S. (2008). Synthesis of gold-silicon core-shell nanoparticles with tunable localized surface plasmon resonance. *Applied Physics Letters*, *92*(10), 103105.

Munir, T., Mahmood, A., Kashif, M., Sohail, A., Shifa, M. S., Sharif, M., & Manzoor, S. (2020). Impact of Ni dopant on optical and magnetic properties of ZnO nanoparticles for biomedical applications. *Journal of Ovonic Research*, *16*(3), 165–171.

Muñoz, J. E., Cervantes, J., Esparza, R., & Rosas, G. (2007). Iron nanoparticles produced by high-energy ball milling. *Journal of Nanoparticle Research*, *9*(5), 945–950. <https://doi.org/10.1007/s11051-007-9226-6>

Naik, M. Z., Meena, S. N., Ghadi, S. C., Naik, M. M., & Salker, A. V. (2016). Evaluation of silver-doped indium oxide nanoparticles as in vitro α-amylase and α-glucosidase inhibitors. *Medicinal Chemistry Research*, *25*(3), 381–389.

Naito, M., Yokoyama, T., Hosokawa, K., & Nogi, K. (2008). Chapter 6 — Evaluation methods for properties of nanostructured body. In *Nanoparticle technology handbook* (3rd ed., pp. 301–363). Elsevier. <https://doi.org/10.1016/B978-0-444-64110-6.00006-8>

Naranjo, L. P., de Araújo, C. B., Malta, O. L., Cruz, P. A. S., & Kassab, L. R. P. (2005). Enhancement of Pr^{3+} luminescence in $PbO-GeO_2$ glasses containing silver nanoparticles. *Applied Physics Letters*, *87*(24), 241914.

Nazarizadeh, A., & Asri-Rezaie, S. (2016). Comparative study of antidiabetic activity and oxidative stress induced by zinc oxide nanoparticles and zinc sulfate in diabetic rats. *AAPS PharmSciTech*, *17*(4), 834–843.

Neațu, S., Maciá-Agulló, J. A., Concepción, P., & Garcia, H. (2014). Gold–copper nanoalloys supported on TiO_2 as photocatalysts for CO_2 reduction by water. *Journal of the American Chemical Society*, *136*(45), 15969–15976.

Nethi, S. K., Bollu, V. S., & Patra, C. R. (2020). Rare earth-based nanoparticles: Biomedical applications, pharmacological and toxicological significance. In: *Nanoparticles and their biomedical applications* (pp. 1–43). Springer.

Ojha, D. P., Joshi, M. K., & Kim, H. J. (2017). Photo-Fenton degradation of organic pollutants using a zinc oxide decorated iron oxide/reduced graphene oxide nanocomposite. *Ceramics International*, *43*(1, Part B), 1290–1297. <https://doi.org/10.1016/j.ceramint.2016.10.079>

Oves, M., Arshad, M., Khan, M. S., Ahmed, A. S., Azam, A., & Ismail, I. M. I. (2015). Anti-microbial activity of cobalt doped zinc oxide nanoparticles: Targeting water borne bacteria. *Journal of Saudi Chemical Society*, *19*(5), 581–588. https://doi.org/10.1016/j.jscs.2015.05.003.

Palza, H. (2015). Antimicrobial polymers with metal nanoparticles. *International Journal of Molecular Sciences*, *16*(1), 2099–2116.

Pathania, D., Gupta, D., Kothiyal, N. C., Eldesoky, G. E., & Naushad, M. (2016). Preparation of a novel chitosan-g-poly (acrylamide)/Zn nanocomposite hydrogel and its applications for controlled drug delivery of ofloxacin. *International Journal of Biological Macromolecules*, *84*, 340–348.

Paucar Álvarez, C., Sarmiento, J. S. C., Freitas, S. C., & García, C. (2017). *Solvothermal synthesis of magnesium oxide-substituted hydroxyapatite nanoparticles as antibacterial nanomaterial for biomedical applications* (Vol. 381, pp. 8–14). Trans Tech Publications.

Paul, R. K., Kesharwani, P., & Raza, K. (2021). Recent update on nano-phytopharmaceuticals in the management of diabetes. *Journal of Biomaterials Science, Polymer Edition*, *32*, 2046–2068. <https://doi.org/10.1080/09205063.2021.1952381>

Pavlatou, E. A., & Gorgoulis, V. G. (2021). Nanomedicine: Photo-activated nanostructured titanium dioxide, as a promising anticancer agent. *Pharmacology & Therapeutics*, *222*, 107795.

Pazos-Pérez, N., Baranov, D., Irsen, S., Hilgendorff, M., Liz-Marzán, L. M., & Giersig, M. (2008). Synthesis of flexible, ultrathin gold nanowires in organic media. *Langmuir: The ACS Journal of Surfaces and Colloids, 24*(17), 9855–9860. <https://doi.org/10.1021/la801675d>

Peng, X.-H., Qian, X., Mao, H., Wang, A. Y., Chen, Z. G., Nie, S., & Shin, D. M. (2008). Targeted magnetic iron oxide nanoparticles for tumor imaging and therapy. *International Journal of Nanomedicine, 3*(3), 311–321. <https://www.ncbi.nlm.nih.gov/pubmed/18990940>

Pérez-Ortiz, M., Zapata-Urzúa, C., Acosta, G. A., Álvarez-Lueje, A., Albericio, F., & Kogan, M. J. (2017). Gold nanoparticles as an efficient drug delivery system for GLP-1 peptides. *Colloids and Surfaces B: Biointerfaces, 158*, 25–32.

Petri-Fink, A., Steitz, B., Finka, A., Salaklang, J., & Hofmann, H. (2008). Effect of cell media on polymer coated superparamagnetic iron oxide nanoparticles (SPIONs): Colloidal stability, cytotoxicity, and cellular uptake studies. *European Journal of Pharmaceutics and Biopharmaceutics, 68*(1), 129–137.

Piasecki, P., Piasecki, A., Pan, Z., Mu, R., & Morgan, S. H. (2010). Formation of Ag nanoparticles and enhancement of Tb^{3+} luminescence in Tb and Ag co-doped lithium-lanthanum-aluminosilicate glass. *Journal of Nanophotonics, 4*(1), 043522.

Ponnanikajamideen, M., Rajeshkumar, S., Vanaja, M., et al. (2018). In vivo type 2 diabetes and wound-healing effects of antioxidant gold nanoparticles synthesized using the insulin plant chamaecostus cuspidatus in albino rats. *Canadian Journal of Diabetes.*

Priyadarshini, E., Rawat, K., Prasad, T., & Bohidar, H. B. (2018). Antifungal efficacy of Au@ carbon dots nanoconjugates against opportunistic fungal pathogen, *Candida albicans. Colloids and Surfaces B: Biointerfaces, 163*, 355–361.

Qi, H., & Hegmann, T. (2006). Formation of periodic stripe patterns in nematic liquid crystals doped with functionalized gold nanoparticles. *Journal of Materials Chemistry, 16*(43), 4197–4205.

Qiang, Y., Antony, J., Sharma, A., Nutting, J., Sikes, D., & Meyer, D. (2006). Iron/iron oxide core-shell nanoclusters for biomedical applications. *Journal of Nanoparticle Research, 8*(3), 489–496. <https://doi.org/10.1007/s11051-005-9011-3>

Rafique, M., Shaikh, A. J., Rasheed, R., Tahir, M. B., Bakhat, H. F., Rafique, M. S., & Rabbani, F. (2017). A review on synthesis, characterization and applications of copper nanoparticles using green method. *Nano, 12*(4), 1750043.

Rahim Pouran, S., Bayrami, A., Mohammadi Arvanag, F., Habibi-Yangjeh, A., Darvishi Cheshmeh Soltani, R., Singh, R., AbdulRaman, A. A., Chae, K. H., Khataee, A., & Kang, H. K. (2020). Biogenic integrated ZnO/Ag nanocomposite: Surface analysis and in vivo practices for the management of type 1 diabetes complications. *Colloids and Surfaces B: Biointerfaces, 189*, 110878. <https://doi.org/10.1016/j.colsurfb.2020.110878>

Rai, M., Kon, K., Chudobova, D., Cihalova, K., Kopel, P., Melichar, L., Ruttkay-Nedecky, B., Vaculovicova, M., Adam, V., & Kizek, R. (2015). Chapter 13 — Complexes of metal-based nanoparticles with chitosan suppressing the risk of *Staphylococcus aureus* and *Escherichia coli* infections. In *Nanotechnology in diagnosis, treatment and prophylaxis of infectious diseases* (pp. 217–232). Academic Press. <https://doi.org/10.1016/B978-0-12-801317-5.00013-X>

Rajakumar, G., Thiruvengadam, M., Mydhili, G., et al. (2018). Green approach for synthesis of zinc oxide nanoparticles from Andrographis paniculata leaf extract and evaluation of their antioxidant, anti-diabetic, and anti-inflammatory activities. *Bioprocess and Biosystems Engineering, 41*(1), 21–30.

Rajaram, K., Aiswarya, D. C., & Sureshkumar, P. (2015). Green synthesis of silver nanoparticle using Tephrosia tinctoria and its antidiabetic activity. *Materials Letters, 138*, 251–254.

Rajeshkumar, S., & Sandhiya, D. (2020). *Biomedical applications of zinc oxide nanoparticles synthesized using eco-friendly method. Nanoparticles and their biomedical applications* (pp. 65–93). Springer.

Rajeshwari, A., Kavitha, S., Alex, S. A., Kumar, D., Mukherjee, A., Chandrasekaran, N., & Mukherjee, A. (2015). Cytotoxicity of aluminum oxide nanoparticles on *Allium cepa* root tip—Effects of oxidative stress generation and biouptake. *Environmental Science and Pollution Research, 22*(14), 11057–11066.

Rak, M. J., Saadé, N. K., Friščić, T., & Moores, A. (2014). Mechanosynthesis of ultra-small monodisperse amine-stabilized gold nanoparticles with controllable size. *Green Chemistry, 16*(1), 86–89. <https://doi.org/10.1039/C3GC41827H>

Ren, L., Li, Y., Hou, J., Wang, T., Yang, Y., & Zhao, X. (2020). Fabrication and cavity-size-dependent photocatalytic property of TiO_2 hollow nanoparticles with tunable cavity size. *Materials Research Bulletin, 126*, 110744.

Richter, K., Birkner, A., & Mudring, A.-V. (2010). Stabilizer-free metal nanoparticles and metal−metal oxide nanocomposites with long-term stability prepared by physical vapor deposition into ionic liquids. *Angewandte Chemie International Edition, 49*(13), 2431−2435. <https://doi.org/10.1002/anie.200901562>

Rubilar, O., Rai, M., Tortella, G., Diez, M. C., Seabra, A. B., & Durán, N. (2013). Biogenic nanoparticles: Copper, copper oxides, copper sulphides, complex copper nanostructures and their applications. *Biotechnology Letters, 35*(9), 1365−1375.

Sabziparvar, N., Saeedi, Y., Nouri, M., Najafi Bozorgi, A. S., Alizadeh, E., Attar, F., Akhtari, K., Mousavi, S. E., & Falahati, M. (2018). Investigating the interaction of silicon dioxide nanoparticles with human hemoglobin and lymphocyte cells by biophysical, computational, and cellular studies. *The Journal of Physical Chemistry B, 122*(15), 4278−4288.

Sahoo, S. K., Misra, R., & Parveen, S. (2012). Nanoparticles: A boon to drug delivery, therapeutics, diagnostics and imaging. In *Nanomedicine in cancer* (pp. 73−124). Pan Stanford.

Samat, N. A., & Nor, R. M. (2013). Sol−gel synthesis of zinc oxide nanoparticles using *Citrus aurantifolia* extracts. *Ceramics International, 39*, S545−S548.

Sánchez-López, E., Gomes, D., Esteruelas, G., Bonilla, L., Lopez-Machado, A., LauraGalindo, R., Cano, A., Espina, M., Ettcheto, M., & Camins, A. (2020). Metal-based nanoparticles as antimicrobial agents: An overview. *Nanomaterials, 10*(2), 292.

Santra, C., Rahman, S., Bojja, S., James, O. O., Sen, D., Maity, S., Mohanty, A. K., Mazumder, S., & Chowdhury, B. (2013). Barium, calcium and magnesium doped mesoporous ceria supported gold nanoparticle for benzyl alcohol oxidation using molecular O_2. *Catalysis Science & Technology, 3*(2), 360−370.

Sarafraz, M. M., & Hormozi, F. (2016). Comparatively experimental study on the boiling thermal performance of metal oxide and multi-walled carbon nanotube nanofluids. *Powder Technology, 287*, 412−430.

Saratale, R. G., Saratale, G. D., Shin, H. S., Jacob, J. M., Pugazhendhi, A., Bhaisare, M., & Kumar, G. (2018). New insights on the green synthesis of metallic nanoparticles using plant and waste biomaterials: Current knowledge, their agricultural and environmental applications. *Environmental Science and Pollution Research, 25*(11), 10164−10183.

Saratale, R. G., Shin, H. S., Kumar, G., et al. (2018). Exploiting antidiabetic activity of silver nanoparticles synthesized using Punica granatum leaves and anticancer potential against human liver cancer cells (HepG2). *Artificial Cells, Nanomedicine, and Biotechnology, 46*(1), 211−222.

Sasikala, R., Rani, S. K., Karthikeyan, K., & Easwaramoorthy, D. (2016). Synthesis and antibacterial studies of lanthanum, cerium and erbium loaded copper oxide nanoparticles. *DJ Journal of Engineering Chemistry and Fuel, 1*(4), 43−51.

Sati, S. C., Kour, G., Bartwal, A. S., & Sati, M. D. (2020). Biosynthesis of metal nanoparticles from leaves of *Ficus palmata* and evaluation of their anti-inflammatory and anti-diabetic activities. *Biochemistry, 59*(33), 3019−3025. <https://doi.org/10.1021/acs.biochem.0c00388>

Schmid, G. (2010). *Nanoparticles*. Wiley VCH.

Schmid, G., Kreyling, W. G., & Simon, U. (2017). Toxic effects and biodistribution of ultrasmall gold nanoparticles. *Archives of Toxicology, 91*(9), 3011−3037.

Senapati, S., Syed, A., Moeez, S., Kumar, A., & Ahmad, A. (2012). Intracellular synthesis of gold nanoparticles using alga *Tetraselmis kochinensis*. *Materials Letters, 79*, 116−118.

Senthilkumar, P., Priya, L., Kumar, R. S., et al. (2015). Potent α-glucosidase inhibitory activity of green synthesized gold nanoparticles from the brown seaweed Padina boergesenii. *International Journal of Recent Advances in Multidisciplinary Research, 2*(11), 917−923.

Shahgolzari, M., Pazhouhandeh, M., Milani, M., Yari Khosroushahi, A., & Fiering, S. (2020). Plant viral nanoparticles for packaging and in vivo delivery of bioactive cargos. *WIREs Nanomedicine and Nanobiotechnology, 12*(5), e1629. <https://doi.org/10.1002/wnan.1629>

Shanker, K., Mohan, G. K., Hussain, M. A., Jayarambabu, N., & Pravallika, P. L. (2017). Green biosynthesis, characterization, in vitro antidiabetic activity, and investigational acute toxicity studies of some herbal-mediated silver nanoparticles on animal models. *Pharmacognosy Magazine, 13*(49), 188−192. <https://doi.org/10.4103/0973-1296.197642>

Sharma, N., Jandaik, S., Kumar, S., Chitkara, M., & Sandhu, I. S. (2016). Synthesis, characterisation and antimicrobial activity of manganese- and iron-doped zinc oxide nanoparticles. *Journal of Experimental Nanoscience*, *11*(1), 54—71. <https://doi.org/10.1080/17458080.2015.1025302>

Shen, J.-M., Xu, L., Lu, Y., Cao, H.-M., Xu, Z.-G., Chen, T., & Zhang, H.-X. (2012). Chitosan-based luminescent/magnetic hybrid nanogels for insulin delivery, cell imaging, and antidiabetic research of dietary supplements. *International Journal of Pharmaceutics*, *427*(2), 400—409.

Sherin, S., Sheeja, S., Devi, R. S., Balachandran, S., Soumya, R. S., & Abraham, A. (2017). In vitro and in vivo pharmacokinetics and toxicity evaluation of curcumin incorporated titanium dioxide nanoparticles for biomedical applications. *Chemico-Biological Interactions*, *275*, 35—46.

Shi, D., Sadat, M. E., Dunn, A. W., & Mast, D. B. (2015). Photo-fluorescent and magnetic properties of iron oxide nanoparticles for biomedical applications. *Nanoscale*, *7*(18), 8209—8232. <https://doi.org/10.1039/C5NR01538C>

Shwetha, U. R., Latha, M. S., Rajith Kumar, C. R., Kiran, M. S., & Betageri, V. S. (2020). Facile synthesis of zinc oxide nanoparticles using novel *Areca catechu* leaves extract and their in vitro antidiabetic and anticancer studies. *Journal of Inorganic and Organometallic Polymers and Materials*, *30*, 4876—4883.

Siaw, Y. M., Jeevanandam, J., Hii, Y. S., & San Chan, Y. (2020). Photo-irradiation coupled biosynthesis of magnesium oxide nanoparticles for antibacterial application. *Naunyn-Schmiedeberg's Archives of Pharmacology*, *393*, 2253—2264.

Singh, N., Mondal, D., Sharma, M., Devkar, R. V., Dubey, S., & Prasad, K. (2015). Sustainable processing and synthesis of nontoxic and antibacterial magnetic nanocomposite from spider silk in neoteric solvents. *ACS Sustainable Chemistry & Engineering*, *3*(10), 2575—2581. <https://doi.org/10.1021/acssuschemeng.5b00810>

Singh, S. K., Giri, N. K., Rai, D. K., & Rai, S. B. (2010). Enhanced upconversion emission in Er^{3+}-doped tellurite glass containing silver nanoparticles. *Solid State Sciences*, *12*(8), 1480—1483.

Song, M., Liu, T., Shi, C., Zhang, X., & Chen, X. (2016). Bioconjugated manganese dioxide nanoparticles enhance chemotherapy response by priming tumor-associated macrophages toward M1-like phenotype and attenuating tumor hypoxia. *ACS Nano*, *10*(1), 633—647.

Sosa, I. O., Noguez, C., & Barrera, R. G. (2003). Optical properties of metal nanoparticles with arbitrary shapes. *The Journal of Physical Chemistry B*, *107*(26), 6269—6275. <https://doi.org/10.1021/jp0274076>

Stepanov, A. L. (2016). Nonlinear optical properties of metal nanoparticles in silicate glass. In *Glass nanocomposites* (pp. 165—179). Elsevier.

Subha, V., Arya, P. N., Kousika, Y., Sivaramakrishnan, R., Pugazhendhi, A., Rishivarathan, I., Jose, S., & Ilangovan, R. (2021). *Nigella sativa* flavonoids surface coated gold NPs (Au-NPs) enhancing antioxidant and anti-diabetic activity. *Process Biochemistry*. Available from https://www.sciencedirect.com/science/article/pii/S1359511321000106.

Swihart, M. T. (2003). Vapor-phase synthesis of nanoparticles. *Current Opinion in Colloid & Interface Science*, *8*(1), 127—133. <https://doi.org/10.1016/S1359-0294(03)00007-4>

Tamayo, L., Azócar, M., Kogan, M., Riveros, A., & Páez, M. (2016). Copper-polymer nanocomposites: An excellent and cost-effective biocide for use on antibacterial surfaces. *Materials Science and Engineering: C*, *69*, 1391—1409.

Tang, Z., Wu, J., Yu, X., Hong, R., Zu, X., Lin, X., Luo, H., Lin, W., & Yi, G. (2021). Fabrication of Au nanoparticle arrays on flexible substrate for tunable localized surface plasmon resonance. *ACS Applied Materials & Interfaces*, *13*, 9281—9288.

Teng, M.-H., Lin, H.-Y., Chiu, C.-C., & Huang, Y.-C. (2017). Using liquid organic compounds to improve the encapsulation efficiency in the synthesis of graphite encapsulated metal nanoparticles by an arc-discharge method. *Diamond and Related Materials*, *80*, 133—139. <https://doi.org/10.1016/j.diamond.2017.10.011>

Tharchanaa, S. B., Priyanka, K., Preethi, K., & Shanmugavelayutham, G. (2021). Facile synthesis of Cu and CuO nanoparticles from copper scrap using plasma arc discharge method and evaluation of antibacterial activity. *Materials Technology*, *36*(2), 97—104.

Thareja, R. K., & Shukla, S. (2007). Synthesis and characterization of zinc oxide nanoparticles by laser ablation of zinc in liquid. *Applied Surface Science*, *253*(22), 8889—8895. <https://doi.org/10.1016/j.apsusc.2007.04.088>

Tiwari, A. K., & Rao, J. M. (2002). Diabetes mellitus and multiple therapeutic approaches of phytochemicals: Present status and future prospects. *Current Science, 83,* 30−38.

Tiwari, P. M., Vig, K., Dennis, V. A., & Singh, S. R. (2011). Functionalized gold nanoparticles and their biomedical applications. *Nanomaterials, 1*(1), 31−63.

Tjong, S. C. (2013). Recent progress in the development and properties of novel metal matrix nanocomposites reinforced with carbon nanotubes and graphene nanosheets. *Materials Science and Engineering: R: Reports, 74*(10), 281−350.

Tommalieh, M. J., Ibrahium, H. A., Awwad, N. S., & Menazea, A. A. (2020). Gold nanoparticles doped polyvinyl alcohol/chitosan blend via laser ablation for electrical conductivity enhancement. *Journal of Molecular Structure, 1221,* 128814.

Umrani, R. D., & Paknikar, K. M. (2014). Zinc oxide nanoparticles show antidiabetic activity in streptozotocin-induced Type 1 and 2 diabetic rats. *Nanomedicine, 9*(1), 89−104.

Vanaja, M., Rajeshkumar, S., Paulkumar, K., Gnanajobitha, G., Chitra, K., Malarkodi, C., & Annadurai, G. (2015). Fungal assisted intracellular and enzyme based synthesis of silver nanoparticles and its bactericidal efficiency. *International Research Journal of Pharmaceutical and Biosciences, 2,* 8−19.

Varnavski, O., Ispasoiu, R. G., Balogh, L., Tomalia, D., & Goodson Iii, T. (2001). Ultrafast time-resolved photoluminescence from novel metal−dendrimer nanocomposites. *The Journal of Chemical Physics, 114*(5), 1962−1965.

Venkatachalam, M., Govindaraju, K., Sadiq, A. M., Tamilselvan, S., Kumar, V. G., & Singaravelu, G. (2013). Functionalization of gold nanoparticles as antidiabetic nanomaterial. *Spectrochimica Acta Part A: Molecular and Biomolecular Spectroscopy, 116,* 331−338.

Verma, A., & Stellacci, F. (2010). Effect of surface properties on nanoparticle−cell interactions. *Small (Weinheim an der Bergstrasse, Germany), 6*(1), 12−21.

Vincent, M., Hartemann, P., & Engels-Deutsch, M. (2016). Antimicrobial applications of copper. *International Journal of Hygiene and Environmental Health, 219*(7), 585−591.

Virk, P. (2018). Antidiabetic activity of green gold-silver nanocomposite with Trigonella foenum graecum L. seeds extract on streptozotocin-induced diabetic rats. *Pakistan Journal of Zoology, 50*(2).

Wakayama, H., & Yonekura, H. (2017). Block copolymer-based nanocomposites with exotic self-assembled structures induced by a magnetic field. *Macromolecular Research, 25*(3), 201−205.

Wang, X., Yang, Y., Dong, J., Bei, F., & Ai, S. (2014). Lanthanum-functionalized gold nanoparticles for coordination−bonding recognition and colorimetric detection of methyl parathion with high sensitivity. *Sensors and Actuators B: Chemical, 204,* 119−124.

Wei, L., Lu, J., Xu, H., Patel, A., Chen, Z.-S., & Chen, G. (2015). Silver nanoparticles: Synthesis, properties, and therapeutic applications. *Drug Discovery Today, 20*(5), 595−601.

Wu, M.-C., Hiltunen, J., Sápi, A., Avila, A., Larsson, W., Liao, H.-C., Huuhtanen, M., Tóth, G., Shchukarev, A., & Laufer, N. (2011). Nitrogen-doped anatase nanofibers decorated with noble metal nanoparticles for photocatalytic production of hydrogen. *ACS Nano, 5*(6), 5025−5030.

Wu, W., Wu, Z., Yu, T., Jiang, C., & Kim, W.-S. (2015). Recent progress on magnetic iron oxide nanoparticles: Synthesis, surface functional strategies and biomedical applications. *Science and Technology of Advanced Materials, 16*(2), 023501.

Xiao, J., Zhu, Y., Huddleston, S., Li, P., Xiao, B., Farha, O. K., & Ameer, G. A. (2018). Copper metal−organic framework nanoparticles stabilized with folic acid improve wound healing in diabetes. *ACS Nano, 12*(2), 1023−1032.

Xu, J., Yang, H., Fu, W., Du, K., Sui, Y., Chen, J., Zeng, Y., Li, M., & Zou, G. (2007). Preparation and magnetic properties of magnetite nanoparticles by sol−gel method. *Journal of Magnetism and Magnetic Materials, 309*(2), 307−311.

Yadav, N. C., Chhillar, A. K., & Rana, J. S. (2020). Detection of pathogenic bacteria with special emphasis to biosensors integrated with gold nanoparticles. *Sensors International, 1,* 100028.

Yan, Y., Warren, S. C., Fuller, P., & Grzybowski, B. A. (2016). Chemoelectronic circuits based on metal nanoparticles. *Nature Nanotechnology, 11*(7), 603−608.

Yang, D.-S., Bhattacharjya, D., Inamdar, S., Park, J., & Yu, J.-S. (2012). Phosphorus-doped ordered mesoporous carbons with different lengths as efficient metal-free electrocatalysts for oxygen reduction reaction in alkaline media. *Journal of the American Chemical Society, 134*(39), 16127−16130.

Yang, G. W. (2007). Laser ablation in liquids: Applications in the synthesis of nanocrystals. *Progress in Materials Science, 52*(4), 648–698. <https://doi.org/10.1016/j.pmatsci.2006.10.016>

Yousefi, R., Jamali-Sheini, F., Cheraghizade, M., Khosravi-Gandomani, S., Sáaedi, A., Huang, N. M., Basirun, W. J., & Azarang, M. (2015). Enhanced visible-light photocatalytic activity of strontium-doped zinc oxide nanoparticles. *Materials Science in Semiconductor Processing, 32*, 152–159. <https://doi.org/10.1016/j.mssp.2015.01.013>

Youssef, A. M., El-Sayed, S. M., El-Sayed, H. S., Salama, H. H., & Dufresne, A. (2016). Enhancement of Egyptian soft white cheese shelf life using a novel chitosan/carboxymethyl cellulose/zinc oxide bionanocomposite film. *Carbohydrate Polymers, 151*, 9–19. <https://doi.org/10.1016/j.carbpol.2016.05.023>

Yu, H., Chen, M., Rice, P. M., Wang, S. X., White, R. L., & Sun, S. (2005). Dumbbell-like bifunctional $Au-Fe_3O_4$ nanoparticles. *Nano Letters, 5*(2), 379–382. <https://doi.org/10.1021/nl047955q>

Zaleska-Medynska, A., Marchelek, M., Diak, M., & Grabowska, E. (2016). Noble metal-based bimetallic nanoparticles: The effect of the structure on the optical, catalytic and photocatalytic properties. *Advances in Colloid and Interface Science, 229*, 80–107. <https://doi.org/10.1016/j.cis.2015.12.008>

Zhang, H., Li, Y., Ivanov, I. A., Qu, Y., Huang, Y., & Duan, X. (2010). Plasmonic modulation of the upconversion fluorescence in $NaYF_4$:Yb/Tm hexaplate nanocrystals using gold nanoparticles or nanoshells. *Angewandte Chemie International Edition, 49*(16), 2865–2868.

Zhang, M., Zhao, Y., Yan, L., Peltier, R., Hui, W., Yao, X., Cui, Y., Chen, X., Sun, H., & Wang, Z. (2016). Interfacial engineering of bimetallic Ag/Pt nanoparticles on reduced graphene oxide matrix for enhanced antimicrobial activity. *ACS Applied Materials & Interfaces, 8*(13), 8834–8840.

Zhao, M.-Q., Zhang, Q., Zhang, W., Huang, J.-Q., Zhang, Y., Su, D. S., & Wei, F. (2010). Embedded high density metal nanoparticles with extraordinary thermal stability derived from guest–host mediated layered double hydroxides. *Journal of the American Chemical Society, 132*(42), 14739–14741.

Zhou, Y., Yu, S. H., Cui, X. P., Wang, C. Y., & Chen, Z. Y. (1999). Formation of silver nanowires by a novel solid–liquid phase arc discharge method. *Chemistry of Materials, 11*(3), 545–546. <https://doi.org/10.1021/cm981122h>

Zhu, W., Wu, Z., Foo, G. S., Gao, X., Zhou, M., Liu, B., Veith, G. M., Wu, P., Browning, K. L., Lee, H. N., Li, H., Dai, S., & Zhu, H. (2017). Taming interfacial electronic properties of platinum nanoparticles on vacancy-abundant boron nitride nanosheets for enhanced catalysis. *Nature Communications, 8*, 15291. <https://doi.org/10.1038/ncomms15291>

Zmojda, J., Kochanowicz, M., Miluski, P., Jadach, R., Pisarski, W. A., Pisarska, J., Ferrari, M., Righini, G., & Dorosz, D. (2017). *Photoluminescence of antimony-germanate-silicate glass doped with europium ions and silver nanoparticles* (pp. 1–4). IEEE.

CHAPTER 6

Biosynthesized nanoparticles for diabetes treatment

Introduction

Physical and chemical approaches are considered conventional methods for the synthesis of nanoparticles (Ling et al., 2015; Naveed Ul Haq et al., 2017). The simplicity of the synthesis process and high nanoparticles yields are major advantages of physical and chemical approaches (Parashar et al., 2009). However, the high process cost (energy demand) associated with physical methods and the use of toxic chemicals in chemical synthesis constitute major drawbacks (Gudikandula & Charya Maringanti, 2016). Biosynthesis approaches involving biological organisms are an alternative to chemical and physical approaches. Biosynthesis approaches have the potential to offer inexpensive synthesis costs, the yield of nontoxic nanoparticles with high biocompatibility and bioactivity, and large-scale nanoparticles production (Parveen et al., 2016). Microbial- and plant-mediated approaches are the two main types of nanoparticles biosynthesis that are widely used (Prasad et al., 2016; Singh, Kim, Wang, et al., 2016; Singh, Kim, Zhang, et al., 2016). Plant-mediated nanoparticles synthesis approaches possess advantages such as the availability of a variety of plant species, the presence of various phytochemicals that can facilitate the generation of nanoparticles with desired size and morphological properties, and less toxicity compared to chemical methods (Duan et al., 2015). Even though plant-mediated approaches are beneficial, the use of pesticides for plant growth can affect the phytochemical content (Zhao et al., 2006). Also, challenges to plant use, such as biopiracy, can restrict the application of known plants for large-scale nanoparticles production (Barhoum et al., 2020; Bueno & Ritoré, 2019; Kędziora et al., 2013). Bacteria (Singh et al., 2015), fungi (Kitching et al., 2015), and algae (Patel et al., 2015) are the main microbes used for the synthesis of nanoparticles. They are subclassified into intra- and extracellular synthesis (Hasan, 2015; Garg et al., 2020; Singh et al., 2015). Microbially synthesized nanoparticles are widely employed as drug carriers (Kumar & Poornachandra, 2015), in medical imaging (Raveendran et al., 2013), biosensing (Luo et al., 2014), nutraceuticals (Semyonov et al., 2014), antimicrobials (Gajbhiye et al., 2009), and nanomedicines (Singh, Kim, Wang, et al., 2016; Singh, Kim, Zhang, et al., 2016). In addition, phytochemicals extracted from unique parts of plants are explored for use in various pharmaceutical and biomedical applications (Habibullah et al., 2021; Rónavári et al., 2021). The potential to generate less toxic nanoparticles with high bioavailability, biocompatibility, and bioactivity makes biosynthesized nanoparticles more suitable for the treatment of diseases such as diabetes (Abdelghany et al., 2018; Yahyaei et al., 2018). The biosynthesized nanoparticles have been used in the diagnosis of diabetes via glucose biosensors (Dönmez, 2020; Lv et al., 2018), targeted insulin delivery as an effective insulin carrier (Khalil et al., 2017; Mahalingam, 2020), and as nanomedicines with antidiabetic activity (Javed et al., 2021; Jeevanandam, Chan, & Danquah, 2020; Jeevanandam, Kulabhusan, et al., 2020; Popli et al., 2018). The ability to convert complex precursors into nanoparticles, the capability to fine-tune reaction parameters to obtain desired nanoparticle properties, and the potential to scale up production, and the ability to enhance antidiabetic activity with less toxicity via the use of enzymes or proteins as capping agents are the main advantages of using biosynthesized nanoparticles in diabetes treatment (Jeevanandam, Danquah, & Pan, 2021; Jeevanandam, Pan, et al., 2021; Velusamy et al., 2016). This chapter discusses various nanoparticles synthesized from microbes and plants, as well as their ability for use in developing glucose sensors, for controlled insulin therapy, and formulating antidiabetic nanomedicines.

Synthesis of nanoparticles via microbes

The conversion and reduction of complex precursors into nanosized particles is crucial in nanoparticles biosynthesis. This step occurs intracellularly within the organism or extracellularly via extracted metabolites secreted by the organism. Nanoparticles can be synthesized by using organisms such as bacteria, fungi, algae, viruses, and archaic bacteria as shown in Fig. 6.1.

Bacterial nanoparticles synthesis

Bacteria are prokaryotes of nanometers to micrometers in size and can grow in almost any environment (Glick & Glick, 2015). They can either be harmful or useful to a living organism (Newman & Reynolds, 2005). Bacteria are classified based on Gram staining and oxygen utilization for growth. Gram-positive and Gram-negative bacterial classification is based on the presence of peptidoglycan in the cell wall (Díaz De Rienzo et al., 2015). Aerobic and anaerobic bacterial classifications are based on the requirement of oxygen for growth (Unden et al., 2016).

Nanoparticles are synthesized by bacteria either intracellularly or extracellularly. Ahmad, Mukherjee, et al. (2003) and Ahmad, Senapati, et al. (2003) demonstrated that gold nanoparticles can be synthesized intracellularly using a novel alkalotolerant bacterium, *Rhodococcus* sp. The gold nanoparticles are 5–15 nm in size, monodispersed, and spherical, and present in the cell wall and cytoplasmic membrane of the bacterium (Ahmad, Mukherjee, et al., 2003; Ahmad, Senapati, et al., 2003). Gold nanoparticles have also been synthesized intracellularly using a probiotic bacterium, *Lactobacillus kimchicus* DCY51[T],

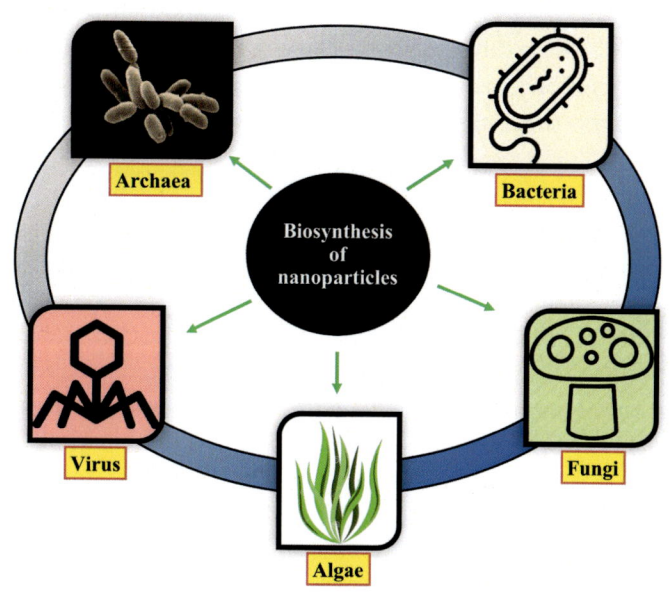

Figure 6.1 Organisms used for biosynthesis of nanoparticles.

isolated from Korean kimchi, which is a traditional food prepared by fermenting vegetables with probiotic lactic acid bacteria. The results showed 5—30 nm spherical and stable gold nanoparticles with antioxidant properties (Markus et al., 2016). Intracellular synthesized nanoparticles can bind to the bacterial cellular components, making it difficult to separate and obtain pure nanoparticles (Kalathil et al., 2011). Bacterial synthesis of nanoparticles via metabolites that are secreted extracellularly is common in synthesizing pure and stable nanoparticles. Bacterial enzymes such as sulfite reductase (Gholami-Shabani et al., 2015), keratinase (Lateef et al., 2015), and proteases (Bharde et al., 2007) are reported as reducing and stabilizing agents for nanoparticle synthesis. Shivaji et al. (2011) reported that cell-free culture supernatants containing extracellular metabolites of five psychrophilic bacterial species, namely, *Pseudomonas proteolytica*, *Pseudomonas antarctica*, *Pseudomonas meridiana*, *Arthrobacter kerguelensis*, *Arthrobacter gangotriensis*, and two mesophilic bacteria, *Bacillus indicus* and *Bacillus cecembensis*, can be used for the synthesis of silver nanoparticles. The results showed silver nanoparticles with sizes in the range of 6—13 nm and stable for about 8 months in dark (Shivaji et al., 2011). Silver nanoparticles have also been synthesized using extracellular polymeric substances obtained from electroactive bacteria (Li et al., 2016). Spherical 20—40 nm cadmium sulfate nanoparticles have been synthesized from extracellular polymeric substances extracted from the marine bacteria *Pseudomonas aeruginosa* JP-11 (Raj et al., 2016). To date, existing knowledge on specific bacterial metabolites and/or their synergistic effects and mechanism responsible for extracellular nanoparticles synthesis is limited. A major advantage of using bacteria for nanoparticles synthesis is their ability to precipitate bulk metals into nanosized structures and facilitate large-scale production with reduced toxicity. However, bacterial biosynthesis of nanoparticles is time-consuming and requires significant purification steps as well as rigorous approaches to control size, morphology, and crystallinity (Jeevanandam, Chan, & Danquah, 2016).

Bacteria-mediated approaches are used to synthesize various nanoparticles from metals to complex composites. Metal nanoparticles such as gold (Narayanan & Sakthivel, 2010), silver (Siddiqi et al., 2018), copper (Saif Hasan et al., 2008), platinum (Konishi et al., 2007), selenium, and tellurium (Zonaro et al., 2015) have been synthesized using specific bacterial strains. Iron, mercury, lead, manganese, palladium, nickel, cobalt, lithium, rhodium, and ruthenium are some metals reported to be synthesized as nanosized particles via intra- and extracellular bacteria-mediated synthesis approaches. Metal sulfates of silver, iron, and cadmium have also been synthesized using bacterial metabolites as a reducing agent (Park et al., 2016). Similarly, iron (Wu et al., 2008), copper (Singh et al., 2010), titanium (Jha et al., 2009), zinc (Jayaseelan et al., 2012), manganese (Salunke et al., 2015), and silicon (Yamamoto & Kuroda, 2016) in its oxidized forms have been synthesized as nanoparticles using bacterial metabolite extracts. Other than metal oxides, bimetallic and multimetallic nanoparticles such as cadmium−zinc, cadmium−selenide, cadmium−telluride, zinc−selenide, cadmium−cesium, praseodymium−gadolinium, strontium−gadolinium, iron−silver, iron−cobalt, iron−manganese, cadmium−selenium−zinc, iron−cobalt−nickel, iron−cobalt−manganese, cadmium−selenium−zinc−tellurium, and silver−cadmium−selenium−zinc have also been synthesized using recombinant strains of *Escherichia coli* (Park et al., 2016). Polymeric nanoparticles have been synthesized by bacterial extracts (Wu et al., 2014) and in some cases, bacterial polymers are coated with nanoparticles for desired applications (Gao et al., 2015). Research interest is focusing on the use of bacteria-mediated nanoparticles synthesis approaches to fabricate complex nanosized composites with enhanced properties (Khalid et al., 2017; Ullah et al., 2016). It is notable that *E. coli* (Gurunathan et al., 2009) and *Bacillus* species (Momin et al., 2019) are the highly used bacteria for the synthesis of various nanoparticles.

Fungi-mediated nanoparticle synthesis

Fungi are eukaryotic and filamentous organisms with spores (Webster & Weber, 2007). Unicellular and filamentous fungi, such as yeast and molds, are major classifications of fungi based on their structure (Gadd, 2007). Various types of fungi are used for the synthesis of nanoparticles. Similar to bacteria, fungi-mediated nanoparticles synthesis occurs either intracellularly or extracellularly (Ahmad et al., 2005). Fungi possess the ability to convert complex organic materials into simpler compounds through which they convert complex precursors into nanosized particles (Sanghi & Verma, 2009). Mukherjee, Ahmad, Mandal, Senapati, Sainkar, Khan, Parishcha, et al. (2001)

and Mukherjee, Ahmad, Mandal, Senapati, Sainkar, Khan, Ramani, et al. (2001) demonstrated that *Verticillium* fungal species can be used for the intracellular synthesis of silver nanoparticles. They showed an intracellular reduction of silver ions to generate 25 ± 12 nm silver nanoparticles attached to the fungal cell wall (Mukherjee, Ahmad, Mandal, Senapati, Sainkar, Khan, Parishcha, et al., 2001; Mukherjee, Ahmad, Mandal, Senapati, Sainkar, Khan, Ramani, et al., 2001). Salvadori et al. (2015) showed that the dead biomass of *Hypocrea lixii* can be used for biosorption-mediated synthesis of spherical nickel oxide nanoparticles (Salvadori et al., 2015). The extraction and downstream purification of nanoparticles attached to the internal cellular or cell wall components of fungi can be challenging.

Extracellular fungi-mediated nanoparticles synthesis approaches are used for the fabrication of various nanoparticles. Fungi contain a large amount of proteins and enzymes that can be extracted and utilized as reducing and stabilizing agents for nanoparticles synthesis (Jeevanandam, Chan, & Danquah, 2016). Several studies have reported the extracellular synthesis of silver and gold nanoparticles from the fungal extracts of *Aspergillus fumigatus* (Bhainsa & D'Souza, 2006) and *Fusarium oxysporum* (Ahmad, Mukherjee, et al., 2003; Ahmad, Senapati, et al., 2003; Mukherjee et al., 2002). Balakumaran et al. (2015) demonstrated that endophytic fungi can be beneficial for the synthesis of small size silver nanoparticles. They found that *Guignardia mangiferae* (Bios PTK 4) and other endophytic fungi are beneficial for synthesizing 5–30 nm spherical silver nanoparticles (Balakumaran et al., 2015). Cuevas et al. (2015) showed that a white-rot fungus named *Stereum hirsutum* can be used to synthesize copper and copper oxide nanoparticles (Cuevas et al., 2015). The fungal extracts of *Saccharomyces cerevisiae* (yeast) have been used by Roy et al. (2015) for the synthesis of silver nanoparticles (Roy et al., 2015). Tryptophan, tyrosine (Bhainsa & D'Souza, 2006), sulfate reductase (Sastry et al., 2003), hydrolyzing proteins (Bharde et al., 2006), membrane-associated cytochrome, and redox proteins (Kitching et al., 2015) are the fungal proteins and enzymes that are reported to be responsible for the formation of nanoparticles. However, there is still a limited understanding of the mechanism of specific fungal proteins and/or combinations that facilitates the formation of unique nanoparticles with defined characteristics. Simple extraction of extracellular enzymes, cultivation of fungi for effective metal distribution as a catalyst, ability to synthesize nanoparticles with desired size, morphology, and surface charge are some additional advantages of using fungi to synthesize nanoparticles. However, fungi-mediated nanoparticles synthesis approaches are challenged by long biosynthesis duration and high process cost resulting from downstream process (Jeevanandam, Chan, & Danquah, 2016).

Metals, metal oxides, metal-organic frameworks, polymers, and composites are the types of nanoparticles that can be synthesized using fungi. Silver (Vigneshwaran et al., 2007), gold (Mukherjee, Ahmad, Mandal, Senapati, Sainkar, Khan, Parishcha, et al.,

2001; Mukherjee, Ahmad, Mandal, Senapati, Sainkar, Khan, Ramani, et al., 2001), selenium (Zare et al., 2013), platinum (Syed & Ahmad, 2012), copper (Cuevas et al., 2015), and cadmium selenide (Ahmad et al., 2002) are examples of metal nanoparticles that are synthesized by fungi. These metal nanoparticles are synthesized by various fungal species with more efficient scale-up methods (Rai et al., 2016). Metal oxide nanoparticles of copper (Honary et al., 2012; Saravanakumar et al., 2018), iron (Bharde et al., 2006), zirconia (Bansal et al., 2004), and cerium (Khan & Ahmad, 2013) have also been synthesized using various fungal species. Various metal oxide nanoparticles, such as antimony oxide, have been reported to be synthesized using fungal extracts (Shah et al., 2015). Polymers extracted from fungi and fungal proteins and enzymes can be useful for the synthesis of novel polymeric nanoparticles (Fakhrullin et al., 2009; Saravanakumar et al., 2018). In addition, core-shell nanoparticles (Khatami et al., 2018), doped nanoparticles (Zhao et al., 2018), bimetallic nanoparticles (Romero-Núñez et al., 2019), and nanocomposites (Gottardo et al., 2019) have been synthesized from fungal extracts. Research is exploring the use of fungal proteins as a template for nanoparticles synthesis (Zhang et al., 2019).

Algae-mediated nanoparticles synthesis

Algae are mostly photosynthetic organisms. Depending on the nature of their growth environment, algae can be classified as halophytic, in highly concentrated saltwater bodies (Ward et al., 2015); symbiotic, in association with other organisms (Birkeland et al., 2015); cryophytic, in cold snowy and icy regions (Knack et al., 2015); thermophytic, in hot media or surfaces (Sahoo et al., 2015); and lithophytes, in moist stones and rock surfaces (Bahuguna et al., 2016). Algae are also classified as macro- and microalgae depending on their multi- and unicellular structures (Barbarino & Lourenço, 2005), and aquatic and terrestrial depending on their habitat (Cai et al., 2018). Microalgae are extensively used for the synthesis of nanoparticles.

Nanoparticles can be synthesized either intra- or extracellular using microalgae. Senapati et al. (2012) demonstrated the intracellular synthesis of gold nanoparticles using the alga *Tetraselmis kochinensis*. They synthesized that 5—35 nm gold nanoparticles are formed in the cell wall and cytoplasmic membrane of the algae (Senapati et al., 2012). Gold nanoparticles were also synthesized by Feurtet-Mazel et al. (2016) from the freshwater diatom, *Eolimna minima* developed as river biofilms. The diatoms converted complex gold precursors to simple nanosized gold intracellularly. The gold nanoparticles protect the diatoms from high salt concentrations via detoxification (Feurtet-Mazel et al., 2016). The ability to take up heavy metals and the presence of carboxyl groups and intracellular phosphates that can convert complex organic matter into simpler nanosized particles are some of the advantages of using intracellular microalgal nanoparticle synthesis approaches. However, as with other organisms, downstream processes of purifying the nanoparticles remain a challenge (Govindaraju et al., 2008).

Extracellular nanoparticles synthesis based on algal extracts has been used for the fabrication of a wide range of nanoparticles. Proteins (Navarro et al., 2008), polysaccharides (El-Rafie et al., 2013), enzymes (Kalabegishvili et al., 2012), and pigments (Singh et al., 2013) present in microalgae have been used as reducing and stabilizing agents to synthesize nanoparticles. Microalgal pigments play a unique role in binding with the precursor complex, reduce them into nanosized particles, and stabilize them by preventing agglomeration. Phycopigments such as fucoxanthin (Singh et al., 2013), soluble fucoidan (Agarwal et al., 2017), chlorophyll *a* (Matorin et al., 2013), carotenoids, and xanthophylls (Bunghez et al., 2011) are reported as useful pigments for the synthesis of nanoparticles. Singaravelu et al. (2007) reported the generation of monodispersed gold nanoparticles extracellularly using a marine alga, *Sargassum wightii* Greville. They obtained 8–12 nm gold nanoparticles with high stability (Singaravelu et al., 2007). The same alga has been used for the synthesis of silver nanoparticles by Govindaraju et al. (2009). They produced 8–27 nm silver nanoparticles that possessed antibacterial activity against *Bombyx mori* (Govindaraju et al., 2009). Shankar et al. (2016) provided a list of microalgae with biocomponents that can be extracted for the synthesis of gold and silver nanoparticles (Shankar et al., 2016). *Chlorella pyrenoidosa* has been used by Aziz et al. (2015) for the synthesis of 100 nm silver nanoparticles with antibacterial and photocatalytic properties. Difficulties in the cultivation of high-density microalgae, multistep approaches required for extracting biocomponents from microalgae, and long nanoparticles synthesis duration remain as challenges to large-scale nanoparticles production from microalgae (LewisOscar et al., 2016).

Algae have been used for the synthesis of a wide variety of nanoparticles such as metals, metal oxides, carbon-based nanoparticles, polymers, magnetic and nanocomposites. Metal nanoparticles such as gold (Arockiya Aarthi Rajathi et al., 2012), silver (Merin et al., 2010), palladium (Momeni & Nabipour, 2015), and platinum (Brayner et al., 2007) have been synthesized using algal extracts. Microalgal extracts are also used to synthesize metal oxide nanoparticles of zinc (Azizi et al., 2014), cerium (Taylor et al., 2016), aluminum (Koopi & Buazar, 2018), copper (Abboud et al., 2014), and iron (Mahdavi et al., 2013). Morphologies such as nanorods (Bhutiya et al., 2018) and nanofibers (Wang et al., 2017) have also been synthesized using microalgal extracts. Carbon-based nanoparticles, such as carbon nanotubes (CNTs) (Rhiem et al., 2015), carbon dots (Ramanan et al., 2016), and graphene (Sharma et al., 2018), have been synthesized from microalgal extracts. Polymeric nanoparticles (Ge et al., 2015), nanocomposites (Aphale et al., 2015), core-shell nanostructures (Srivastava et al., 2012), doped nanoparticles (Saha et al., 2018), and bimetallic nanostructures (Tuo et al., 2017) have also been synthesized using biocomponents extracted from microalgae.

Other microbes used for nanoparticles synthesis

Viruses are unique organisms that can also be employed for the synthesis of nanoparticles. Viruses are nanosized organisms with a genetic material covered by a protein coat called the capsid. Douglas et al. (2002) proposed that the protein capsid of cowpea chlorotic mottle virus can be modified into cages via protein engineering and used as a constrained template for the synthesis of nanomaterials (Douglas et al., 2002). Tobacco mosaic viruses have been used as a template for the synthesis of platinum nanotubes with high surface area. The platinum nanotubes showed 65% high catalytic mass activity compared to normal platinum nanoparticles (Górzny et al., 2010). Cobalt—platinum and iron—platinum bimetallic nanowires have been synthesized by Tsukamoto et al. (2007) using the tobacco mosaic virus as a biotemplate (Tsukamoto et al., 2007). Novel virus-like organosilica nanoparticles (Jiang et al., 2019), virus-derived designer shells assembled in vitro around inorganic nanoparticles (Wege et al., 2018), and virus spike and membrane lytic mimicking nanoparticles for superior binding and endosomal escape (Liu et al., 2018) are some virus-inspired nanostructures or nanoparticles.

Archaea are a subgroup of bacteria that possess a cell wall with no peptidoglycan (Kates et al., 1993). Archaea have also been employed in the synthesis of nanoparticles. Ashiuchi et al. (2013) extracted poly-glutamate from archaea to form nanosized polymers and demonstrated their ability as an antimicrobial thermoplastic material (Ashiuchi et al., 2013). Similarly, halophilic archaea species such as *Halococcus salifodinae* BK_3 have been used for the synthesis of silver nanoparticles (Srivastava et al., 2013). Selenium nanoparticles (Tugarova & Kamnev, 2017), quantum dots (Bruna et al., 2019), and cruxrhodopsin nanoparticles (Taran et al., 2017) are reported to be synthesized from archaic bacteria. Photosynthetic bacteria, also known as cyanobacteria, have also been used for the synthesis of nanoparticles. Filamentous cyanobacteria for the synthesis of gold and palladium nanoparticles (Lengke et al., 2006, 2007) and marine cyanobacteria for the synthesis of cadmium sulfate and silver nanoparticles (Ali et al., 2011; MubarakAli et al., 2012) have been reported. Chemosynthetic bacteria are non-photosynthetic and depend on nutrients from the environment for their growth (Russell et al., 2018). These microbes are commonly used for the intracellular synthesis of nanoparticles (Mashjoor et al., 2019). Various metal and metal oxide nanoparticles can be synthesized using chemosynthetic bacteria (Timko et al., 2008; Ren et al., 2012). The synthesis of nanoparticles using these microbes is in the early research stages, hence more technological efforts are required to advance this approach.

Synthesis of nanoparticles via plant extract and natural products

Various nanoparticles, such as metals (Solgi & Taghizadeh, 2020), metal oxides (Varghese et al., 2020), carbon (Zaib et al., 2021), polymer (Jeevanandam, Danquah,

& Pan, 2021; Jeevanandam, Pan, et al., 2021), and composites (Huang et al., 2020), have been synthesized using phytochemicals extracted from plants (Jeevanandam, Chan, & Danquah, 2020; Jeevanandam, Kulabhusan, et al., 2020) or via intracellular synthesis (Jeevanandam, Danquah, & Pan, 2021; Jeevanandam, Pan, et al., 2021; Saim et al., 2021). Neem leaf extracts have been used to develop zinc oxide—nickel oxide nanocomposites via a hydrothermal approach (Hessien et al., 2021). Bhavyasree and Xavier (2020) synthesized copper oxide—carbon nanoparticles using the aqueous leaf extract of *Adhatoda vasica* Nees. The nanocomposites had a nanoflakes morphology with 7—11 nm of thickness and exhibited antimicrobial activity against *E. coli*, *Klebsiella pneumoniae*, *P. aeruginosa*, *Staphylococcus aureus*, *Candida albicans*, and *Aspergillus niger* (Bhavyasree & Xavier, 2020). Noohpisheh et al. (2020) prepared novel silver—zinc oxide nanocomposites by using the aqueous leaf extract of *Trigonella foenum-graecum*. The nanocomposites were 75 nm in size with −37.5 mV zeta potential. The zinc oxide nanoparticles had rod and plate morphologies, and the silver nanoparticles had a spherical morphology (Noohpisheh et al., 2020). Yulizar et al. (2021) synthesized vanadium pentoxide—iron oxide nanocomposites using the phytochemicals of *Foeniculum vulgare* stem powder extracted using a 1:1 volume ratio of *n*-hexane and methanol. The generated nanoparticles were semirectangular in morphology with 90 nm in size (Yulizar et al., 2021). Moharrami and Motamedi (2020) also synthesized magnetite-functionalized cellulose nanocrystals using agricultural wastes, such as sugar-beet pulp. The nanocrystals were used for the synthesis of starch-based hydrogel nanocomposites via enzymatic hydrolysis as a potential nanoadsorbent (Moharrami & Motamedi, 2020). In addition, Khatami et al. (2018) have reported the synthesis of several core-shell nanoparticles using natural plant products.

Microbially synthesized nanoparticles as glucose sensors

Microbially synthesized nanoparticles can be used to synthesize glucose sensors, to enhance glucose detection ability, sensitivity, stability, and in situ monitoring efficacy, and to minimize toxic reactions in the environment after disposal (Scognamiglio, 2013). The following sections discuss some of these developed technologies focusing on bacteria and fungi.

Bacteria

Bacterial metabolites are coated over nanoparticles as they act as reducing and stabilizing agents for nanosized particles formation, and these metabolites help in binding with glucose for detection. In certain cases, proteins and enzymes of bacteria are coated over presynthesized nanoparticles to enhance their glucose-binding efficiency. Scognamiglio et al. (2007) developed a novel nonconsuming analyte fluorescence biosensor by using a glucose-binding protein extracted from a mutant bacterial strain. The protein was coupled with a donor—acceptor fluorophore to enable glucose detection via fluorescence resonance energy transfer—based biosensing (Scognamiglio et al., 2007). Wang et al. (2010) reported the fabrication of a bioenzymatic glucose biosensor with the help of a nanocomposite synthesized using bacterial cellulose nanofibers and gold nanoparticles. Horseradish peroxidase and glucose oxidase enzymes were immobilized over the nanocomposites with a modified glassy carbon electrode. The nanocomposite showed a sensitive amperometric response to glucose via electron mediator. The biosensor showed promise for the determination of glucose in human blood cells to diagnose diabetes (Wang et al., 2010). Yoon et al. (2011) fabricated a novel optical glucose biosensor by coupling a bacterial glucose-binding protein with near-infrared fluorescent single-walled CNTs. The optical biosensor demonstrated efficient glucose detection ability with high bacterial protein selectivity (Yoon et al., 2011). Yao et al. (2018) synthesized bacterial cellulose nanofibers loaded with quantum dots that are reported to be useful as a color-tunable luminescent macrofibers to detect glucose in a complex mixture. The cellulose nanofibers were proposed to be significant in the fabrication of highly sensitive fiber-based biosensors to detect and monitor glucose levels in diabetic patients (Yao et al., 2018). Vaidyanathan et al. (2016) synthesized a unique bacterial-induced, hollow cylindrical nickel oxide nanoparticles and demonstrated its efficacy for enzyme-less detection of glucose. The results demonstrated the potential of the nanomaterial as an electrochemical amperometric glucose sensor with high sensitivity and low detection limit (Vaidyanathan et al., 2016).

Fungi

Fungal extracts have also been used to synthesize nanoparticles and to coat nanoparticles to enhance their glucose detection ability. Wang et al. (2008) demonstrated the fabrication of a high-performance amperometric glucose biosensor using fungi-extracted chitosan. The study reported the synthesis of copper nanoparticle—chitosan—CNT nanocomposites with a modified glassy carbon electrode in which glucose oxidase was immobilized. The biosensor showed excellent glucose sensitivity, a rapid response time of fewer than 4 s, and good selectivity (Wang et al., 2008). Pyranose oxidase, a periplasmic homotetrameric flavoprotein present in wood degrading fungi, was immobilized onto a gold nanoparticle—polyaniline—silver chloride—gelatin nanocomposite matrix by Ozdemir et al. (2010) for the detection of glucose. The result revealed better glucose detection in complex biological samples than spectrophotometric Trinder reaction—based commercial glucose enzyme assay kit (Ozdemir et al., 2010). Luo et al. (2004) have also used fungal-extracted chitosan—glucose oxidase—gold nanoparticles to fabricate a novel biocomposite sensor for the detection of glucose (Luo et al., 2004). Solairaj et al. (2017) synthesized copper nanoparticles that were immobilized on fungi-derived chitin nanostructures with the potential for use as electrochemical glucose sensors (Solairaj et al., 2017). Politi et al. (2016) synthesized 10 ± 5 nm gold nanoparticles with high stability using a small amphiphilic fungal protein known as hydrophobin Vmh2. The fungal protein coated on gold nanoparticles showed promise for glucose detection in diabetes management (Politi et al., 2016).

Phytosynthesized nanoparticles as glucose sensors

Similar to microbially synthesized nanoparticles, phytosynthesized nanostructures have been used to develop glucose sensors. Dayakar et al. (2018) fabricated biogenic silver nanoparticles using *Ocimum tenuiflorum* for nonenzymatic glucose biosensing. The biosensor possessed a high sensitivity of 895.8 $\mu A/mM/cm^2$, a linear range of 1−8.9 mM, less than 4 s response time, and a low detection limit of 0.0048 μM (Dayakar et al., 2018). Similarly, Maarebia et al. (2019) developed silver nanoparticles using *Myrmecodia pendans* extract with and without polyvinyl alcohol to develop nanoparticles-based biosensors with enhanced ability to detect 71.71 mg/dL glucose concentration in blood (Maarebia et al., 2019). Dayakar et al. (2017) demonstrated the synthesis of novel zinc oxide nanoparticles with *O. tenuiflorum* leaf extract for use as a glucose sensor via a nonenzymatic approach. The nanoparticles were spherical, 10−20 nm in size, and showed 631.30 $\mu A/mM/cm^2$ reproducible sensitivity, 1−8.6 mM linear dynamic range, 0.042 μM detection limit, and <4 s response time (Dayakar et al., 2017). Wahab et al. (2019) produced novel gold nanoparticles using aqueous leaf extract of *Muntingia calabura* for the detection of blood glucose levels. The nanoparticles were 78.2 nm and showed a 1−4 mM measurement range and 0.1500 mM concentration of minimum detection limit (Wahab et al., 2019). Even though phytosynthesized nanoparticles can be used to fabricate glucose sensors, challenges relating to their stability and large-scale synthesis persist.

Biosynthesized nanoparticles for controlled insulin therapy

Microbially synthesized nanoparticles are used in insulin therapy as potential drug carriers for controlled insulin release in addition to their biocompatibility, bioavailability, and less toxicity. Bacteria and fungi synthesized nanoparticles are widely used as insulin carriers and delivery tools. Bacterial magnetic nanoparticles as carriers of recombinant DNA (Tang et al., 2012), poly(lactic-*co*-glycolic acid) (PLGA) nanoparticles coated with bacterial invasion for gastrointestinal drug delivery (Dawson & Halbert, 2000), and engineered bacteria as microbots for the delivery of drug-carrying nanoparticles (Akin et al., 2007) are some applications of bacterial nanoparticles as drug carriers. Kumar and Poornachandra (2015) synthesized novel miconazole-conjugated bacterial silver nanoparticles and demonstrated their efficacy as a drug delivery vehicle (Kumar & Poornachandra, 2015). Bacterial cellulose-coated nanoparticles are part of nanosized drug carriers (Tsai et al., 2018). Magnetosomes are magnetic nanoparticles synthesized by bacteria and are gaining attention as a novel magnetic drug delivery system (Dieudonné et al., 2019). Also, nanoscale bacteria-enabled autonomous drug delivery systems called NanoBEADS (Suh et al., 2019) and metal-organic framework nanoparticles that mimic bacteria are reported to possess potential for drug delivery (Guo et al., 2019).

Chitosan extracted from bacteria and chitin from fungi are commonly used biopolymers as nanodrug carriers for targeted insulin delivery (Dhanasekaran et al., 2018; Lin et al., 2007). Lectin extracted from fungi was used to modify solid lipid nanoparticles to develop a carrier for oral administration of insulin. The nanoparticles possessed 60% of insulin entrapment efficiency and demonstrated good pharmacological bioavailability (Zhang et al., 2006). Viral nanoparticles have also been utilized as targeted delivery systems for the controlled delivery of insulin. Wu et al. (2018) showed that biomimetic virus-like and charge-reversible nanoparticles can be used to overcome mucus and epithelial barriers in oral delivery of insulin (Wu et al., 2018). Chen et al. (2018) reported that noninfectious Hepatitis E viral capsid can be used to deliver insulin subcutaneously (Chen et al., 2018). Gold nanoparticles synthesized via leaf powder extract of the insulin plant *Chamaecostus cuspidatus* were used as an antioxidant agent to achieve 50% of free radical inhibition. The phytochemical-coated nanoparticles were 50 nm in size, monodispersed, and spherical with the ability to improve insulin level and control blood glucose and glycogen levels after 21 days of treatment in a diabetic mice model (Ponnanikajamideen et al., 2019). Vinotha et al. (2019) synthesized biogenic zinc oxide nanoparticles via insulin-rich *Costus igneus* leaf extracts. The nanoparticles were hexagonal, 26.55 nm in size, and possessed the ability to inhibit 74% of alpha-amylase, 82% of alpha-glucosidase enzymes, and 75% of antioxidant ability at 100 μg/mL concentration (Vinotha et al., 2019).

Biosynthesized nanoparticles with antidiabetic effect

Nanoparticles synthesized from microbial polymers such as chitosan, cellulose, and chitin can be useful in reducing hyperglycemic diabetic conditions with effective antidiabetic properties. Vitta and Thiruvengadam (2012) reported some positive aspects of multifunctional bacterial cellulose and nanoparticles-embedded composites as promising antidiabetic agents to reduce hyperglycemic conditions (Vitta & Thiruvengadam, 2012). Bacakova et al. (2019) reported that celluloses from the bacteria *Gluconacetobacter* and algae *Cladophora* can be used to incorporate antidiabetic nanoparticles with cotton fibers for wound healing in diabetes (Bacakova et al., 2019). It has been proposed by Fu et al. (2013) that bacterial cellulose nanomaterials are highly beneficial in repairing skin damage among diabetic patients due to their superior biocompatibility (Fu et al., 2013). Bacterial cellulose nanofibers can be blended with cotton fibers to develop smart wound dresses for the treatment of chronic diabetic wounds (Gianino et al., 2018). Fungal chitosan and algal chitin polymers can also be used as antidiabetic agents for the treatment of diabetes complications. Ausar et al. (2003) formulated bread with chitosan from yeast and demonstrated improved levels of high- and low-density lipids in diabetic subjects (Ausar et al., 2003). El-Batal et al. (2020) synthesized gum Arabic polymer-stabilized bimetallic silver–gold nanoparticles via gamma-ray irradiation for the treatment of diabetic foot patients. The study showed that the nanoparticles were spherical, 18.58 nm in size, and possessed the ability to inhibit the bacteria *Enterobacter cloacae*, *E. coli*, *Acinetobacter baumannii*, *P. aeruginosa*, Gram-positive *Enterococcus faecalis*, *Enterococcus faecium*, *Bacillus subtilis*, *S. aureus*, and the fungal strains such as *Candida tropicalis* and *C. albicans* that can cause diabetic foot ulcer (El-Batal et al., 2020). Silver nanoparticles prepared using *Allium cepa* possessed enhanced antidiabetic activity via the inhibition of alpha-amylase and alpha-glucosidase (Jini & Sharmila, 2020). Also, zinc oxide nanoparticles prepared via phytochemicals extracted from *Urtica dioica* (Bayrami et al., 2020) and *Areca catechu* (Shwetha et al., 2020), silver nanoparticles via *Eryngium thyrsoideum* Boiss (Mahmoudi et al., 2021), and *Silybum marianum* (Shah et al., 2020), and gold nanoparticles via *Fritillaria cirrhosa* (Guo et al., 2020) have been reported to exhibit good antidiabetic activity. Table 6.1 presents a summary of different nanoparticles synthesized by microbes and plants, or coated with biomolecules for diagnosis, monitoring, and treatment of diabetes.

Future perspectives

Biomolecules generated from microbes and plants have found research applications in the development of nanomedicines for diabetes treatment. However, the utilization of nanoparticles synthesized by microbes and plants is in the early research stages. Biomolecules and nanoparticles synthesized from microbes and plants have also been used in developing glucose sensors to monitor glucose levels in diabetic subjects. However, only microbial and plant polymers coated with nanoparticles or synthesized as nanomaterials have mostly been explored in insulin delivery and antidiabetic applications. Among microbes, viruses as nanosized particles show promising ability to carry insulin to target sites in a controlled release fashion. More research efforts can focus on the engineering of viral nanoparticles to develop nanorobots using bioaffinity probes such as aptamers to target diabetic cells for insulin release. Nanorobots represent a future opportunity that can leverage structure–function relationships associated with diabetes cells and nanotechnology to develop targeted delivery platforms based on specific antigenic signals for diabetes treatments. While nanoparticles with antidiabetic properties as nanomedicines can be synthesized using microbes, there are concerns relating to potential interactions between nanoparticles and the human genome that may lead to mutation (Manna & Bandyopadhyay, 2017; Song et al., 2019; Zhang et al., 2020). Utilization of multilevel drug screening and nanoparticles purification

Table 6.1 Nanoparticles synthesized via microbes for diabetes monitoring and treatment.

Nanoparticles	Application in diabetes	Treatment	Reference
Bacterial cellulose–gold nanocomposites	Bacterial cellulose nanofibers loaded with quantum dots	High-sensitive fiber-based glucose biosensors	Wang et al. (2010), Yoon et al. (2011, 2018)
Bacterial glucose-binding protein coated fluorescent single-walled carbon nanotubes	Bioenzymatic glucose biosensor Optical glucose biosensor		
Bacterial-induced, hollow cylindrical nickel oxide nanoparticles	Enzyme-less glucose sensors	High-performance amperometric glucose biosensor	Vaidyanathan et al. (2016)
Copper nanoparticles-fungal chitosan-carbon nanotube			Wang et al. (2008)
Fungal pyranose oxidase immobilized gold nanoparticles—polyaniline—silver chloride—gelatin nanocomposite	Spectrophotometric glucose biosensor	Biocomposite glucose sensor	Ozdemir et al. (2010)
Fungal-extracted chitosan—glucose oxidase—gold nanoparticles			Luo et al. (2004)
Fungal glucose oxidase-coated nanoparticles	Rapid glucose sensor		Dubey et al. (2017)
Copper nanoparticles immobilized in fungal-derived chitosan nanostructures	Electrochemical glucose sensor		Solairaj et al. (2017)
Fungal hyphophobin Vmh2-synthesized gold nanoparticles	Glucose monitoring biosensor		Politi et al. (2016)
Algal extracts mediated gold and silver nanoparticles M13 virus phages as manganese dioxide nanowires template	Optical glucose biosensor	Electrochemical glucose biosensor at neutral pH	Rassas et al. (2019), Han et al. (2016)
Silver nanoparticles using *Ocimum tenuiflorum*	Nonenzymatic glucose biosensor		Dayakar et al. (2018)
Silver nanoparticles using *Myrmecodia pendans*	Blood glucose biosensor		Maarebia et al. (2019)

Zinc oxide nanoparticles from *Ocimum tenuiflorum* leaf extract	Nonenzymatic glucose biosensor		Dayakar et al. (2017)
Gold nanoparticles via aqueous *Muntingia calabura* leaf extract	Nanosize chitosan and chitin extracted from bacteria and algae		
	Blood glucose biosensor		Wahab et al. (2019)
	Targeted insulin delivery		Lin et al. (2007), Dhanasekaran et al. (2018)
Fungal lectin modified solid lipid nanoparticles		Drug carrier for oral administration of insulin	Zhang et al. (2006)
Biomimetic virus-like and charge-reversible nanoparticles	Oral delivery of insulin		Wu et al. (2018)
Noninfectious Hepatitis E viral nanosized capsid	Subcutaneous delivery of insulin		Chen et al. (2018)
Multifunctional bacterial cellulose and nanoparticles-embedded composites			
Reduced hyperglycemic condition as an antidiabetic agent			Vitta and Thiruvengadam (2012)
Cellulose from bacteria	Gluconacetobacter and algae *Cladophora* coated antidiabetic nanoparticles		Bacakova et al. (2019)
	Bacterial cellulose nanomaterials		Fu et al. (2013)
	Wound healing in diabetes		
	Skin damage repairs in diabetes		

(*Continued*)

Table 6.1 (Continued)

Nanoparticles	Application in diabetes	Treatment	Reference
Bacterial cellulose nanofibers	Gum Arabic polymer-stabilized bimetallic silver−gold nanoparticles		Coradi et al. (2017)
Wound healing of diabetic foot disease		Antidiabetic and antimicrobial activity for diabetic foot ulcer treatment	El-Batal et al. (2020)
Zinc oxide nanoparticles using *Urtica dioica* and *Areca catechu*		Gold nanoparticles using *Fritillaria cirrhosa*	
Antidiabetic activity			Bayrami et al. (2020), Shwetha et al. (2020)
Antidiabetic activity			Guo et al. (2020)

strategies along with extensive preclinical and clinical trials may be useful in addressing these concerns to promote the application of microbially synthesized nanoparticles for diabetes treatments.

Conclusion

This chapter provides an overview of the synthesis of nanoparticles from microbes and plants and their applications in glucose monitoring and diabetes treatment. While a wide variety of nanoparticles synthesized from microbes exist, only a few have been explored for their insulin delivery efficacy and antidiabetic activity. However, a significant number of biosynthesized nanoparticles have been explored for the fabrication of glucose biosensors to monitor sugar levels in diabetic subjects. With more research advancements, biosynthesized nanoparticles can find applications in the development of nanorobots for enhanced diabetes treatment.

References

Abboud, Y., Saffaj, T., Chagraoui, A., El Bouari, A., Brouzi, K., Tanane, O., & Ihssane, B. (2014). Biosynthesis, characterization and antimicrobial activity of copper oxide nanoparticles (CONPs) produced using brown alga extract (*Bifurcaria bifurcata*). *Applied Nanoscience, 4*(5), 571−576. Available from https://doi.org/10.1007/s13204-013-0233-x.

Abdelghany, T. M., Al-Rajhi, A. M. H., Al Abboud, M. A., Alawlaqi, M. M., Magdah, A. G., Helmy, E. A. M., & Mabrouk, A. S. (2018). Recent advances in green synthesis of silver nanoparticles and their applications: About future directions. A review. *BioNanoScience, 8*(1), 5−16.

Agarwal, H., Venkat Kumar, S., & Rajeshkumar, S. (2017). A review on green synthesis of zinc oxide nanoparticles − An eco-friendly approach. *Resource-Efficient Technologies, 3*(4), 406−413. Available from https://doi.org/10.1016/j.reffit.2017.03.002.

Ahmad, A., Mukherjee, P., Mandal, D., Senapati, S., Khan, M. I., Kumar, R., & Sastry, M. (2002). Enzyme mediated extracellular synthesis of CdS nanoparticles by the fungus, *Fusarium oxysporum*. *Journal of the American Chemical Society, 124*(41), 12108−12109. Available from https://doi.org/10.1021/ja0272960.

Ahmad, A., Mukherjee, P., Senapati, S., Mandal, D., Khan, M. I., Kumar, R., & Sastry, M. (2003). Extracellular biosynthesis of silver nanoparticles using the fungus *Fusarium oxysporum*. *Colloids and Surfaces B: Biointerfaces, 28*(4), 313−318. Available from https://doi.org/10.1016/S0927-7765(02)00174-1.

Ahmad, A., Senapati, S., Khan, M. I., Kumar, R., Ramani, R., Srinivas, V., & Sastry, M. (2003). Intracellular synthesis of gold nanoparticles by a novel alkalotolerant actinomycete, *Rhodococcus* species. *Nanotechnology, 14*(7), 824−828. Available from https://doi.org/10.1088/0957-4484/14/7/323.

Ahmad, A., Senapati, S., Khan, M. I., Kumar, R., & Sastry, M. (2005). Extra-/intracellular biosynthesis of gold nanoparticles by an alkalotolerant fungus, *Trichothecium* sp. *Journal of Biomedical Nanotechnology, 1*(1), 47−53.

Akin, D., Sturgis, J., Ragheb, K., Sherman, D., Burkholder, K., Robinson, J. P., Bhunia, A. K., Mohammed, S., & Bashir, R. (2007). Bacteria-mediated delivery of nanoparticles and cargo into cells. *Nature Nanotechnology, 2*, 441. Available from https://doi.org/10.1038/nnano.2007.149.

Ali, D. M., Sasikala, M., Gunasekaran, M., & Thajuddin, N. (2011). Biosynthesis and characterization of silver nanoparticles using marine cyanobacterium, *Oscillatoria willei* NTDM01. *Digest Journal of Nanomaterials and Biostructures, 6*(2), 385−390.

Aphale, A., Chattopadhyay, A., Mahakalakar, K., & Patra, P. (2015). Synthesis and electrochemical analysis of algae cellulose-polypyrrole-graphene nanocomposite for supercapacitor electrode. *Journal of Nanoscience and Nanotechnology*, *15*(8), 6225–6229. Available from https://doi.org/10.1166/jnn.2015.10280.

Arockiya Aarthi Rajathi, F., Parthiban, C., Ganesh Kumar, V., & Anantharaman, P. (2012). Biosynthesis of antibacterial gold nanoparticles using brown alga, *Stoechospermum marginatum* (kützing). *Spectrochimica Acta Part A: Molecular and Biomolecular Spectroscopy*, *99*, 166–173. Available from https://doi.org/10.1016/j.saa.2012.08.081.

Ashiuchi, M., Fukushima, K., Oya, H., Hiraoki, T., Shibatani, S., Oka, N., Nishimura, H., Hakuba, H., Nakamori, M., & Kitagawa, M. (2013). Development of antimicrobial thermoplastic material from archaeal poly-γ-l-glutamate and its nanofabrication. *ACS Applied Materials & Interfaces*, *5*(5), 1619–1624. Available from https://doi.org/10.1021/am3032025.

Ausar, S. F., Morcillo, M., León, A. E., Ribotta, P. D., Masih, R., Vilaro Mainero, M., Amigone, J. L., Rubin, G., Lescano, C., Castagna, L. F., Beltramo, D. M., Diaz, G., & Bianco, I. D. (2003). Improvement of HDL- and LDL-cholesterol levels in diabetic subjects by feeding bread containing chitosan. *Journal of Medicinal Food*, *6*(4), 397–399. Available from https://doi.org/10.1089/109662003772519985.

Aziz, N., Faraz, M., Pandey, R., Shakir, M., Fatma, T., Varma, A., Barman, I., & Prasad, R. (2015). Facile algae-derived route to biogenic silver nanoparticles: synthesis, antibacterial, and photocatalytic properties. *Langmuir: The ACS Journal of Surfaces and Colloids*, *31*(42), 11605–11612. Available from https://doi.org/10.1021/acs.langmuir.5b03081.

Azizi, S., Ahmad, M. B., Namvar, F., & Mohamad, R. (2014). Green biosynthesis and characterization of zinc oxide nanoparticles using brown marine macroalga *Sargassum muticum* aqueous extract. *Materials Letters*, *116*, 275–277. Available from https://doi.org/10.1016/j.matlet.2013.11.038.

Bacakova, L., Pajorova, J., Bacakova, M., Skogberg, A., Kallio, P., Kolarova, K., & Svorcik, V. (2019). Versatile application of nanocellulose: from industry to skin tissue engineering and wound healing. *Nanomaterials*, *9*(2), 164.

Bahuguna, Y. M., Gairola, S., Uniyal, P. L., & Bhatt, A. B. (2016). Moss flora of Kedarnath Wildlife Sanctuary (KWLS), Garhwal Himalaya, India. *Proceedings of the National Academy of Sciences, India Section B: Biological Sciences*, *86*(4), 931–943. Available from https://doi.org/10.1007/s40011-015-0531-z.

Balakumaran, M. D., Ramachandran, R., & Kalaichelvan, P. T. (2015). Exploitation of endophytic fungus, *Guignardia mangiferae* for extracellular synthesis of silver nanoparticles and their in vitro biological activities. *Microbiological Research*, *178*, 9–17. Available from https://doi.org/10.1016/j.micres.2015.05.009.

Bansal, V., Rautaray, D., Ahmad, A., & Sastry, M. (2004). Biosynthesis of zirconia nanoparticles using the fungus *Fusarium oxysporum*. *Journal of Materials Chemistry*, *14*(22), 3303–3305. Available from https://doi.org/10.1039/B407904C.

Barbarino, E., & Lourenço, S. O. (2005). An evaluation of methods for extraction and quantification of protein from marine macro- and microalgae. *Journal of Applied Phycology*, *17*(5), 447–460. Available from https://doi.org/10.1007/s10811-005-1641-4.

Barhoum, A., Jeevanandam, J., Rastogi, A., Samyn, P., Boluk, Y., Dufresne, A., Danquah, M. K., & Bechelany, M. (2020). Plant celluloses, hemicelluloses, lignins, and volatile oils for the synthesis of nanoparticles and nanostructured materials. *Nanoscale*, *12*(45), 22845–22890.

Bayrami, A., Haghgooie, S., Pouran, S. R., Arvanag, F. M., & Habibi-Yangjeh, A. (2020). Synergistic antidiabetic activity of ZnO nanoparticles encompassed by *Urtica dioica* extract. *Advanced Powder Technology*, *31*(5), 2110–2118.

Bhainsa, K. C., & D'Souza, S. F. (2006). Extracellular biosynthesis of silver nanoparticles using the fungus *Aspergillus fumigatus*. *Colloids and Surfaces B: Biointerfaces*, *47*(2), 160–164. Available from https://doi.org/10.1016/j.colsurfb.2005.11.026.

Bharde, A., Kulkarni, A., Rao, M., Prabhune, A., & Sastry, M. (2007). Bacterial enzyme mediated biosynthesis of gold nanoparticles. *Journal of Nanoscience and Nanotechnology*, 7(12), 4369–4377. Available from https://doi.org/10.1166/jnn.2007.891.

Bharde, A., Rautaray, D., Bansal, V., Ahmad, A., Sarkar, I., Yusuf, S. M., Sanyal, M., & Sastry, M. (2006). Extracellular biosynthesis of magnetite using fungi. *Small (Weinheim an der Bergstrasse, Germany)*, *2*(1), 135−141. Available from https://doi.org/10.1002/smll.200500180.

Bhavyasree, P. G., & Xavier, T. S. (2020). Green synthesis of copper oxide/carbon nanocomposites using the leaf extract of *Adhatoda vasica* Nees, their characterization and antimicrobial activity. *Heliyon*, *6*(2), e03323. Available from https://doi.org/10.1016/j.heliyon.2020.e03323.

Bhutiya, P. L., Mahajan, M. S., Abdul Rasheed, M., Pandey, M., Zaheer Hasan, S., & Misra, N. (2018). Zinc oxide nanorod clusters deposited seaweed cellulose sheet for antimicrobial activity. *International Journal of Biological Macromolecules*, *112*, 1264−1271. Available from https://doi.org/10.1016/j.ijbiomac.2018.02.108.

Birkeland, C., Muller-Parker, G., D'Elia, C. F., & Cook, C. B. (2015). Interactions between corals and their symbiotic algae. *Coral reefs in the anthropocene* (pp. 99−116). Springer Netherlands. Available from https://doi.org/10.1007/978-94-017-7249-5_5.

Brayner, R., Barberousse, H., Hemadi, M., Djedjat, C., Yéprémian, C., Coradin, T., Livage, J., Fiévet, F., & Couté, A. (2007). Cyanobacteria as bioreactors for the synthesis of Au, Ag, Pd, and Pt nanoparticles via an enzyme-mediated route. *Journal of Nanoscience and Nanotechnology*, *7*(8), 2696−2708. Available from https://doi.org/10.1166/jnn.2007.600.

Bruna, N., Collao, B., Tello, A., Caravantes, P., Díaz-Silva, N., Monrás, J. P., Órdenes-Aenishanslins, N., Flores, M., Espinoza-Gonzalez, R., Bravo, D., & Pérez-Donoso, J. M. (2019). Synthesis of salt-stable fluorescent nanoparticles (quantum dots) by polyextremophile halophilic bacteria. *Scientific Reports*, *9*(1), 1953. Available from https://doi.org/10.1038/s41598-018-38330-8.

Bueno, J., & Ritoré, S. (2019). Bioprospecting model for a new Colombia drug discovery initiative in the pharmaceutical industry. *Analysis of science, technology, and innovation in emerging economies* (pp. 37−63). Springer.

Bunghez, I. R., Fierascu, R. C., Dumitriu, I., Ghiurea, M., & Ion, R. M. (2011). Biosynthesis of silver and gold nanoparticles via pigments extracted from *Spinacia oleracea*. *Environmental Engineering & Management Journal (EEMJ)*, *10*(2).

Cai, L., Zhou, G., Tong, H., Tian, R. M., Zhang, W., Ding, W., Liu, S., Huang, H., & Qian, P. Y. (2018). Season structures prokaryotic partners but not algal symbionts in subtropical hard corals. *Applied Microbiology and Biotechnology*, *102*(11), 4963−4973. Available from https://doi.org/10.1007/s00253-018-8909-5.

Chen, C. C., Baikoghli, M. A., & Cheng, R. H. (2018). Tissue targeted nanocapsids for oral insulin delivery via drink. *Pharmaceutical Patent Analyst*, *7*(3), 121−127. Available from https://doi.org/10.4155/ppa-2017-0041.

Coradi, C. S., Flumignan, C. D. Q., Laks, R., Flumignan, R. L. G., Alvarenga, B. H., Figueiredo, G. Z. C., Navarro, T. P., Dardik, A., Junqueira, D., & Cisneros, L. (2017). Vascular diseases for the non-specialist: an evidence-based guide. 35−45. Available from https://doi.org/10.1007/978-3-319-46059-8.

Cuevas, R., Duran, N., Diez, M. C., Tortella, G. R., & Rubilar, O. (2015). Extracellular biosynthesis of copper and copper oxide nanoparticles by *Stereum hirsutum*, a native white-rot fungus from chilean forests. *Journal of Nanomaterials*, *2015*, 7. Available from https://doi.org/10.1155/2015/789089.

Dawson, G. F., & Halbert, G. W. (2000). The in vitro cell association of invasin coated polylactide-co-glycolide nanoparticles. *Pharmaceutical Research*, *17*(11), 1420−1425.

Dayakar, T., Venkateswara Rao, K., Bikshalu, K., Rajendar, V., & Park, S.-H. (2017). Novel synthesis and structural analysis of zinc oxide nanoparticles for the non enzymatic glucose biosensor. *Materials Science and Engineering: C*, *75*, 1472−1479. Available from https://doi.org/10.1016/j.msec.2017.02.032.

Dayakar, T., Venkateswara Rao, K., Park, J., Sadasivuni, K. K., Ramachandra Rao, K., & Jaya rambabu, N. (2018). Non-enzymatic biosensing of glucose based on silver nanoparticles synthesized from *Ocimum tenuiflorum* leaf extract and silver nitrate. *Materials Chemistry and Physics*, *216*, 502−507. Available from https://doi.org/10.1016/j.matchemphys.2018.05.046.

Dhanasekaran, S., Rameshthangam, P., Venkatesan, S., Singh, S. K., & Vijayan, S. R. (2018). In vitro and in silico studies of chitin and chitosan based nanocarriers for curcumin and insulin delivery. *Journal of Polymers and the Environment*, *26*(10), 4095−4113.

Dieudonné, A., Pignol, D., & Prévéral, S. (2019). Magnetosomes: biogenic iron nanoparticles produced by environmental bacteria. *Applied Microbiology and Biotechnology*. Available from https://doi.org/10.1007/s00253-019-09728-9.

Douglas, T., Strable, E., Willits, D., Aitouchen, A., Libera, M., & Young, M. (2002). Protein engineering of a viral cage for constrained nanomaterials synthesis. *Advanced Materials*, *14*(6), 415−418. Available from https://doi.org/10.1002/1521-4095(20020318)14:6 < 415::AID-ADMA415 > 3.0.CO;2-W.

Duan, H., Wang, D., & Li, Y. (2015). Green chemistry for nanoparticle synthesis. *Chemical Society Reviews*, *44*(16), 5778−5792.

Dubey, M. K., Zehra, A., Aamir, M., Meena, M., Ahirwal, L., Singh, S., & Bajpai, V. K. (2017). Improvement strategies, cost effective production, and potential applications of fungal glucose oxidase (GOD): current updates. *Frontiers in microbiology*, *8*, 1032.

Díaz De Rienzo, M. A., Stevenson, P., Marchant, R., & Banat, I. M. (2015). Antibacterial properties of biosurfactants against selected Gram-positive and -negative bacteria. *FEMS Microbiology Letters*, *363* (2). Available from https://doi.org/10.1093/femsle/fnv224.

Dönmez, S. (2020). Green synthesis of zinc oxide nanoparticles using zingiber officinale root extract and their applications in glucose biosensor. *El-Cezeri Journal of Science and Engineering*, *7*(3), 1191−1200.

El-Batal, A. I., Abd Elkodous, M., El-Sayyad, G. S., Al-Hazmi, N. E., Gobara, M., & Baraka, A. (2020). Gum Arabic polymer-stabilized and Gamma rays-assisted synthesis of bimetallic silver-gold nanoparticles: Powerful antimicrobial and antibiofilm activities against pathogenic microbes isolated from diabetic foot patients. *International Journal of Biological Macromolecules*, *165*, 169−186. Available from https://doi.org/10.1016/j.ijbiomac.2020.09.160.

El-Rafie, H. M., El-Rafie, M. H., & Zahran, M. K. (2013). Green synthesis of silver nanoparticles using polysaccharides extracted from marine macro algae. *Carbohydrate Polymers*, *96*(2), 403−410. Available from https://doi.org/10.1016/j.carbpol.2013.03.071.

Fakhrullin, R. F., Zamaleeva, A. I., Morozov, M. V., Tazetdinova, D. I., Alimova, F. K., Hilmutdinov, A. K., Zhdanov, R. I., Kahraman, M., & Culha, M. (2009). Living fungi cells encapsulated in polyelectrolyte shells doped with metal nanoparticles. *Langmuir: the ACS Journal of Surfaces and Colloids*, *25* (8), 4628−4634. Available from https://doi.org/10.1021/la803871z.

Feurtet-Mazel, A., Mornet, S., Charron, L., Mesmer-Dudons, N., Maury-Brachet, R., & Baudrimont, M. (2016). Biosynthesis of gold nanoparticles by the living freshwater diatom *Eolimna minima*, a species developed in river biofilms. *Environmental Science and Pollution Research*, *23*(5), 4334−4339. Available from https://doi.org/10.1007/s11356-015-4139-x.

Fu, L., Zhang, J., & Yang, G. (2013). Present status and applications of bacterial cellulose-based materials for skin tissue repair. *Carbohydrate Polymers*, *92*(2), 1432−1442. Available from https://doi.org/10.1016/j.carbpol.2012.10.071.

Gadd, G. M. (2007). Geomycology: biogeochemical transformations of rocks, minerals, metals and radionuclides by fungi, bioweathering and bioremediation. *Mycological Research*, *111*(1), 3−49. Available from https://doi.org/10.1016/j.mycres.2006.12.001.

Gajbhiye, M., Kesharwani, J., Ingle, A., Gade, A., & Rai, M. (2009). Fungus-mediated synthesis of silver nanoparticles and their activity against pathogenic fungi in combination with fluconazole. *Nanomedicine: Nanotechnology, Biology and Medicine*, *5*(4), 382−386.

Gao, W., Fang, R. H., Thamphiwatana, S., Luk, B. T., Li, J., Angsantikul, P., Zhang, Q., Hu, C.-M. J., & Zhang, L. (2015). Modulating antibacterial immunity via bacterial membrane-coated nanoparticles. *Nano Letters*, *15*(2), 1403−1409. Available from https://doi.org/10.1021/nl504798g.

Garg, D., Sarkar, A., Chand, P., Bansal, P., Gola, D., Sharma, S., Khantwal, S., Mehrotra, R., Chauhan, N., & Bharti, R. K. (2020). Synthesis of silver nanoparticles utilizing various biological systems: mechanisms and applications—a review. *Progress in Biomaterials*, *9*, 81−95.

Ge, S., Agbakpe, M., Zhang, W., & Kuang, L. (2015). Heteroaggregation between PEI-coated magnetic nanoparticles and algae: Effect of particle size on algal harvesting efficiency. *ACS Applied Materials & Interfaces*, *7*(11), 6102−6108. Available from https://doi.org/10.1021/acsami.5b00572.

Gholami-Shabani, M., Shams-Ghahfarokhi, M., Gholami-Shabani, Z., Akbarzadeh, A., Riazi, G., Ajdari, S., Amani, A., & Razzaghi-Abyaneh, M. (2015). Enzymatic synthesis of gold nanoparticles using sulfite reductase purified from *Escherichia coli*: A green eco-friendly approach. *Process Biochemistry, 50*(7), 1076−1085. Available from https://doi.org/10.1016/j.procbio.2015.04.004.

Gianino, E., Miller, C., & Gilmore, J. (2018). Smart wound dressings for diabetic chronic wounds. *Bioengineering, 5*(3), 51.

Glick, B. R., & Glick, B. R. (2015). Introduction to plant growth-promoting bacteria. *Beneficial plant-bacterial interactions* (pp. 1−28). Springer International Publishing. Available from https://doi.org/10.1007/978-3-319-13921-0_1.

Górzny, M. Ł., Walton, A. S., & Evans, S. D. (2010). Synthesis of high-surface-area platinum nanotubes using a viral template. *Advanced Functional Materials, 20*(8), 1295−1300. Available from https://doi.org/10.1002/adfm.200902196.

Gottardo, B., Lemes, T. H., Byzynski, G., Paziani, M. H., von-Zeska-Kress, M. R., de Almeida, M. T. G., & Volanti, D. P. (2019). One-pot synthesis and antifungal activity of nontoxic silver-loaded hydroxyapatite nanocomposites against *Candida* species. *ACS Applied Nano Materials*. Available from https://doi.org/10.1021/acsanm.9b00091.

Govindaraju, K., Basha, S. K., Kumar, V. G., & Singaravelu, G. (2008). Silver, gold and bimetallic nano-particles production using single-cell protein (*Spirulina platensis*) Geitler. *Journal of Materials Science, 43*(15), 5115−5122. Available from https://doi.org/10.1007/s10853-008-2745-4.

Govindaraju, K., Kiruthiga, V., Kumar, V. G., & Singaravelu, G. (2009). Extracellular synthesis of silver nanoparticles by a marine alga, *Sargassum wightii* Grevilli and their antibacterial effects. *Journal of Nanoscience and Nanotechnology, 9*(9), 5497−5501. Available from https://doi.org/10.1166/jnn.2009.1199.

Gudikandula, K., & Charya Maringanti, S. (2016). Synthesis of silver nanoparticles by chemical and biological methods and their antimicrobial properties. *Journal of Experimental Nanoscience, 11*(9), 714−721.

Guo, A., Durymanov, M., Permyakova, A., Sene, S., Serre, C., & Reineke, J. (2019). Metal organic framework (MOF) particles as potential bacteria-mimicking delivery systems for infectious diseases: Characterization and cellular internalization in alveolar macrophages. *Pharmaceutical Research, 36*(4), 53. Available from https://doi.org/10.1007/s11095-019-2589-4.

Guo, Y., Jiang, N., Zhang, L., & Yin, M. (2020). Green synthesis of gold nanoparticles from *Fritillaria cirrhosa* and its anti-diabetic activity on Streptozotocin induced rats. *Arabian Journal of Chemistry, 13*(4), 5096−5106.

Gurunathan, S., Kalishwaralal, K., Vaidyanathan, R., Venkataraman, D., Pandian, S. R. K., Muniyandi, J., Hariharan, N., & Eom, S. H. (2009). Biosynthesis, purification and characterization of silver nanoparticles using *Escherichia coli*. *Colloids and Surfaces B: Biointerfaces, 74*(1), 328−335. Available from https://doi.org/10.1016/j.colsurfb.2009.07.048.

Habibullah, G., Viktorova, J., & Ruml, T. (2021). Current strategies for noble metal nanoparticle synthesis. *Nanoscale Research Letters, 16*(1), 1−12.

Han, L., Shao, C., Liang, B., & Liu, A. (2016). Genetically engineered phage-templated MnO2 nanowires: synthesis and their application in electrochemical glucose biosensor operated at neutral pH condition. *ACS applied materials & interfaces, 8*(22), 13768−13776.

Hasan, S. (2015). A review on nanoparticles: their synthesis and types. *Research Journal of Recent Sciences*, ISSN, 2277, 2502.

Hessien, M., Da'na, E., & Taha, A. (2021). Phytoextract assisted hydrothermal synthesis of ZnO−NiO nanocomposites using neem leaves extract. *Ceramics International, 47*(1), 811−816. Available from https://doi.org/10.1016/j.ceramint.2020.08.192.

Honary, S., Barabadi, H., Gharaei-Fathabad, E., & Naghibi, F. (2012). Green synthesis of copper oxide nanoparticles using *Penicillium aurantiogriseum*, *Penicillium citrinum* and *Penicillium waksmanii*. *Digest Journal of Nanomaterials and Biostructures, 7*(3), 999−1005.

Huang, X., Chang, L., Lu, Y., Li, Z., Kang, Z., Zhang, X., Liu, M., & Yang, D.-P. (2020). Plant-mediated synthesis of dual-functional Eggshell/Ag nanocomposites towards catalysis and antibacterial applications. *Materials Science and Engineering: C, 113*, 111015.

Javed, B., Ikram, M., Farooq, F., Sultana, T., & Raja, N. I. (2021). Biogenesis of silver nanoparticles to treat cancer, diabetes, and microbial infections: A mechanistic overview. *Applied Microbiology and Biotechnology, 105*, 2261−2275.

Jayaseelan, C., Rahuman, A. A., Kirthi, A. V., Marimuthu, S., Santhoshkumar, T., Bagavan, A., Gaurav, K., Karthik, L., & Rao, K. V. B. (2012). Novel microbial route to synthesize ZnO nanoparticles using *Aeromonas hydrophila* and their activity against pathogenic bacteria and fungi. *Spectrochimica Acta Part A: Molecular and Biomolecular Spectroscopy, 90,* 78−84. Available from https://doi.org/10.1016/j.saa.2012.01.006.

Jeevanandam, J., Chan, Y. S., & Danquah, M. K. (2016). Biosynthesis of metal and metal oxide nanoparticles. *ChemBioEng Reviews, 3*(2), 55−67. Available from https://doi.org/10.1002/cben.201500018.

Jeevanandam, J., Chan, Y. S., & Danquah, M. K. (2020). Cytotoxicity and insulin resistance reversal ability of biofunctional phytosynthesized MgO nanoparticles. *3 Biotech, 10,* 489.

Jeevanandam, J., Danquah, M. K., & Pan, S. (2021). Plant-derived nanobiomaterials as a potential next generation dental implant surface modifier. *Frontiers in Materials, 8,* 130.

Jeevanandam, J., Kulabhusan, P. K., Sabbih, G., Akram, M., & Danquah, M. K. (2020). Phytosynthesized nanoparticles as a potential cancer therapeutic agent. *3 Biotech, 10*(12), 1−26.

Jeevanandam, J., Pan, S., Harini, A., & Danquah, M. K. (2021). Challenges in the risk assessment of nanomaterial toxicity towards microbes. *Interfaces Between Nanomaterials and Microbes.*

Jha, A. K., Prasad, K., & Kulkarni, A. R. (2009). Synthesis of TiO_2 nanoparticles using microorganisms. *Colloids and Surfaces B: Biointerfaces, 71*(2), 226−229. Available from https://doi.org/10.1016/j.colsurfb.2009.02.007.

Jiang, Y., Liu, H., Wang, L., Zhou, L., Huang, Z., Ma, L., He, Y., Shi, L., & Gao, J. (2019). Virus-like organosilica nanoparticles for lipase immobilization: Characterization and biocatalytic applications. *Biochemical Engineering Journal, 144,* 125−134. Available from https://doi.org/10.1016/j.bej.2019.01.022.

Jini, D., & Sharmila, S. (2020). Green synthesis of silver nanoparticles from *Allium cepa* and its in vitro antidiabetic activity. *Materials Today: Proceedings, 22,* 432−438. Available from https://doi.org/10.1016/j.matpr.2019.07.672.

Kalabegishvili, T., Kirkesali, E., & Rcheulishvili, A. (2012). *Synthesis of gold nanoparticles by blue-green algae Spirulina platensis,* 14. Available from http://inis.iaea.org/search/search.aspx?orig_q = RN:43108872.

Kalathil, S., Lee, J., & Cho, M. H. (2011). Electrochemically active biofilm-mediated synthesis of silver nanoparticles in water. *Green Chemistry, 13*(6), 1482−1485. Available from https://doi.org/10.1039/C1GC15309A.

Kates, M., Kushner, D. J., Matheson, A. T., & Woese, C. R. (1993). Introduction the archaea: Their history and significance. *New comprehensive biochemistry* (26, pp. vii−xxix). Elsevier. Available from https://doi.org/10.1016/S0167-7306(08)60248-3.

Kędziora, A., Gorzelańczyk, K., & Bugla-Płoskońska, G. (2013). Positive and negative aspects of silver nanoparticles usage. *Biology International, 53,* 67−76.

Khalid, A., Khan, R., Ul-Islam, M., Khan, T., & Wahid, F. (2017). Bacterial cellulose-zinc oxide nanocomposites as a novel dressing system for burn wounds. *Carbohydrate Polymers, 164,* 214−221. Available from https://doi.org/10.1016/j.carbpol.2017.01.061.

Khalil, R. I., Burns, T. A., Radecka, I., Kowalczuk, M., Khalaf, T., Adamus, G., Johnston, B., & Khechara, P. M. (2017). Bacterial-derived polymer poly-y-glutamic acid (y-PGA)-based micro/nanoparticles as a delivery system for antimicrobials and other biomedical applications. *International Journal of Molecular Sciences, 18*(2). Available from https://doi.org/10.3390/ijms18020313.

Khan, S. A., & Ahmad, A. (2013). Fungus mediated synthesis of biomedically important cerium oxide nanoparticles. *Materials Research Bulletin, 48*(10), 4134−4138. Available from https://doi.org/10.1016/j.materresbull.2013.06.038.

Khatami, M., Alijani, H. Q., Nejad, M. S., & Varma, R. S. (2018). Core@ shell nanoparticles: Greener synthesis using natural plant products. *Applied Sciences, 8*(3), 411.

Kitching, M., Ramani, M., & Marsili, E. (2015). Fungal biosynthesis of gold nanoparticles: Mechanism and scale up. *Microbial Biotechnology, 8*(6), 904−917. Available from https://doi.org/10.1111/1751-7915.12151.

Knack, J. J., Wilcox, L. W., Delaux, P. M., Ané, J. M., Piotrowski, M. J., Cook, M. E., Graham, J. M., & Graham, L. E. (2015). Microbiomes of streptophyte algae and bryophytes suggest that a functional suite of microbiota fostered plant colonization of land. *International Journal of Plant Sciences, 176*(5), 405−420. Available from https://doi.org/10.1086/681161.

Konishi, Y., Ohno, K., Saitoh, N., Nomura, T., Nagamine, S., Hishida, H., Takahashi, Y., & Uruga, T. (2007). Bioreductive deposition of platinum nanoparticles on the bacterium *Shewanella algae. Journal of Biotechnology, 128*(3), 648−653. Available from https://doi.org/10.1016/j.jbiotec.2006.11.014.

Koopi, H., & Buazar, F. (2018). A novel one-pot biosynthesis of pure alpha aluminum oxide nanoparticles using the macroalgae *Sargassum ilicifolium*: A green marine approach. *Ceramics International, 44*(8), 8940−8945. Available from https://doi.org/10.1016/j.ceramint.2018.02.091.

Kumar, C. G., & Poornachandra, Y. (2015). Biodirected synthesis of miconazole-conjugated bacterial silver nanoparticles and their application as antifungal agents and drug delivery vehicles. *Colloids and Surfaces B: Biointerfaces, 125*, 110−119.

Lateef, A., Adelere, I. A., Gueguim-Kana, E. B., Asafa, T. B., & Beukes, L. S. (2015). Green synthesis of silver nanoparticles using keratinase obtained from a strain of *Bacillus safensis* LAU 13. *International Nano Letters, 5*(1), 29−35. Available from https://doi.org/10.1007/s40089-014-0133-4.

Lengke, M. F., Fleet, M. E., & Southam, G. (2006). Morphology of gold nanoparticles synthesized by filamentous cyanobacteria from gold(I) − thiosulfate and gold(III) − chloride complexes. *Langmuir: The ACS Journal of Surfaces and Colloids, 22*(6), 2780−2787. Available from https://doi.org/10.1021/la052652c.

Lengke, M. F., Fleet, M. E., & Southam, G. (2007). Synthesis of palladium nanoparticles by reaction of filamentous cyanobacterial biomass with a palladium(II) chloride complex. *Langmuire ACS Journal of Surfaces and Colloids, 23*(17), 8982−8987. Available from https://doi.org/10.1021/la7012446.

LewisOscar, F., Vismaya, S., Arunkumar, M., Thajuddin, N., Dhanasekaran, D., & Nithya, C. (2016). Algal nanoparticles: Synthesis and biotechnological potentials. *Algae-organisms for imminent biotechnology*. InTech.

Li, S.-W., Zhang, X., & Sheng, G.-P. (2016). Silver nanoparticles formation by extracellular polymeric substances (EPS) from electroactive bacteria. *Environmental Science and Pollution Research, 23*(9), 8627−8633. Available from https://doi.org/10.1007/s11356-016-6105-7.

Lin, Y.-H., Mi, F.-L., Chen, C.-T., Chang, W.-C., Peng, S.-F., Liang, H.-F., & Sung, H.-W. (2007). Preparation and characterization of nanoparticles shelled with chitosan for oral insulin delivery. *Biomacromolecules, 8*(1), 146−152. Available from https://doi.org/10.1021/bm0607776.

Ling, D., Lee, N., & Hyeon, T. (2015). Chemical synthesis and assembly of uniformly sized iron oxide nanoparticles for medical applications. *Accounts of Chemical Research, 48*(5), 1276−1285.

Liu, S., Yang, J., Jia, H., Zhou, H., Chen, J., & Guo, T. (2018). Virus spike and membrane-lytic mimicking nanoparticles for high cell binding and superior endosomal escape. *ACS Applied Materials & Interfaces, 10*(28), 23630−23637. Available from https://doi.org/10.1021/acsami.8b06934.

Luo, P., Liu, Y., Xia, Y., Xu, H., & Xie, G. (2014). Aptamer biosensor for sensitive detection of toxin A of *Clostridium difficile* using gold nanoparticles synthesized by *Bacillus stearothermophilus*. *Biosensors and Bioelectronics, 54*, 217−221.

Luo, X.-L., Xu, J.-J., Du, Y., & Chen, H.-Y. (2004). A glucose biosensor based on chitosan−glucose oxidase−gold nanoparticles biocomposite formed by one-step electrodeposition. *Analytical Biochemistry, 334*(2), 284−289. Available from https://doi.org/10.1016/j.ab.2004.07.005.

Lv, P., Zhou, H., Mensah, A., Feng, Q., Wang, D., Hu, X., Cai, Y., Amerigo Lucia, L., Li, D., & Wei, Q. (2018). A highly flexible self-powered biosensor for glucose detection by epitaxial deposition of gold nanoparticles on conductive bacterial cellulose. *Chemical Engineering Journal, 351*, 177−188. Available from https://doi.org/10.1016/j.cej.2018.06.098.

Maarebia, R. Z., Wahab, A. W., & Taba, P. (2019). Synthesis and characterization of silver nanoparticles using water extract of sarang semut (*Myrmecodia pendans*) for blood glucose sensors. *Jurnal Akta Kimia Indonesia (Indonesia Chimica Acta), 12*(1), 29−46.

Mahalingam, D. A. G. (2020). Green synthesized metal nanoparticles, characterization and its antidiabetic activities—a review. *Research Journal of Pharmacy and Technology, 13*(1), 468−474.

Mahdavi, M., Namvar, F., Ahmad, M., & Mohamad, R. (2013). Green biosynthesis and characterization of magnetic iron oxide (Fe_3O_4) nanoparticles using seaweed (*Sargassum muticum*) aqueous extract. *Molecules (Basel, Switzerland), 18*(5), 5954−5964.

Mahmoudi, F., Mahmoudi, F., Gollo, K. H., & Amini, M. M. (2021). Biosynthesis of novel silver nanoparticles using *Eryngium thyrsoideum* Boiss extract and comparison of their antidiabetic activity with chemical synthesized silver nanoparticles in diabetic rats. *Biological Trace Element Research, 199*(5), 1967−1978.

Manna, I., & Bandyopadhyay, M. (2017). Engineered nickel oxide nanoparticles affect genome stability in *Allium cepa* (L.). *Plant Physiology and Biochemistry, 121*, 206−215.

Markus, J., Mathiyalagan, R., Kim, Y.-J., Abbai, R., Singh, P., Ahn, S., Perez, Z. E. J., Hurh, J., & Yang, D. C. (2016). Intracellular synthesis of gold nanoparticles with antioxidant activity by probiotic *Lactobacillus kimchicus* DCY51T isolated from Korean kimchi. *Enzyme and Microbial Technology, 95,* 85−93. Available from https://doi.org/10.1016/j.enzmictec.2016.08.018.

Mashjoor, S., Yousefzadi, M., Zolgharnein, H., Kamrani, E., & Alishahi, M. (2019). Phyco-linked vs chemogenic magnetite nanoparticles: Route selectivity in nano-synthesis, antibacterial and acute zoo-planktonic responses. *Materials Science and Engineering: C.* Available from https://doi.org/10.1016/j.msec.2019.01.049.

Matorin, D. N., Todorenko, D. A., Seifullina, N. K., Zayadan, B. K., & Rubin, A. B. (2013). Effect of silver nanoparticles on the parameters of chlorophyll fluorescence and P700 reaction in the green alga *Chlamydomonas reinhardtii. Microbiology (Reading, England), 82*(6), 809−814. Available from https://doi.org/10.1134/S002626171401010X.

Merin, D. D., Prakash, S., & Bhimba, B. V. (2010). Antibacterial screening of silver nanoparticles synthesized by marine micro algae. *Asian Pacific Journal of Tropical Medicine, 3*(10), 797−799. Available from https://doi.org/10.1016/S1995-7645(10)60191-5.

Moharrami, P., & Motamedi, E. (2020). Application of cellulose nanocrystals prepared from agricultural wastes for synthesis of starch-based hydrogel nanocomposites: Efficient and selective nanoadsorbent for removal of cationic dyes from water. *Bioresource Technology, 313,* 123661. Available from https://doi.org/10.1016/j.biortech.2020.123661.

Momeni, S., & Nabipour, I. (2015). A simple green synthesis of palladium nanoparticles with *Sargassum* alga and their electrocatalytic activities towards hydrogen peroxide. *Applied Biochemistry and Biotechnology, 176*(7), 1937−1949. Available from https://doi.org/10.1007/s12010-015-1690-3.

Momin, B., Rahman, S., Jha, N., & Annapure, U. S. (2019). Valorization of mutant *Bacillus licheniformis* M09 supernatant for green synthesis of silver nanoparticles: Photocatalytic dye degradation, antibacterial activity, and cytotoxicity. *Bioprocess and Biosystems Engineering, 42*(4), 541−553. Available from https://doi.org/10.1007/s00449-018-2057-2.

MubarakAli, D., Gopinath, V., Rameshbabu, N., & Thajuddin, N. (2012). Synthesis and characterization of CdS nanoparticles using C-phycoerythrin from the marine cyanobacteria. *Materials Letters, 74,* 8−11. Available from https://doi.org/10.1016/j.matlet.2012.01.026.

Mukherjee, P., Ahmad, A., Mandal, D., Senapati, S., Sainkar, S. R., Khan, M. I., Parishcha, R., Ajaykumar, P. V., Alam, M., Kumar, R., & Sastry, M. (2001). Fungus-mediated synthesis of silver nanoparticles and their immobilization in the mycelial matrix: A novel biological approach to nanoparticle synthesis. *Nano Letters, 1*(10), 515−519. Available from https://doi.org/10.1021/nl0155274.

Mukherjee, P., Ahmad, A., Mandal, D., Senapati, S., Sainkar, S. R., Khan, M. I., Ramani, R., Parischa, R., Ajaykumar, P. V., Alam, M., Sastry, M., & Kumar, R. (2001). Bioreduction of AuCl4 − ions by the fungus, *Verticillium* sp. and surface trapping of the gold nanoparticles formed. *Angewandte Chemie International Edition, 40*(19), 3585−3588. Available from https://doi.org/10.1002/1521-3773(20011001)40:19 < 3585::AID-ANIE3585 > 3.0.CO;2-K.

Mukherjee, P., Senapati, S., Mandal, D., Ahmad, A., Khan, M. I., Kumar, R., & Sastry, M. (2002). Extracellular synthesis of gold nanoparticles by the fungus *Fusarium oxysporum. Chembiochem: A European Journal of Chemical Biology, 3*(5), 461−463. Available from https://doi.org/10.1002/1439-7633(20020503)3:5 < 461::AID-CBIC461 > 3.0.CO;2-X.

Narayanan, K. B., & Sakthivel, N. (2010). Biological synthesis of metal nanoparticles by microbes. *Advances in Colloid and Interface Science, 156*(1), 1−13. Available from https://doi.org/10.1016/j.cis.2010.02.001.

Navarro, E., Piccapietra, F., Wagner, B., Marconi, F., Kaegi, R., Odzak, N., Sigg, L., & Behra, R. (2008). Toxicity of silver nanoparticles to *Chlamydomonas reinhardtii. Environmental Science & Technology, 42*(23), 8959−8964. Available from https://doi.org/10.1021/es801785m.

Naveed Ul Haq, A., Nadhman, A., Ullah, I., Mustafa, G., Yasinzai, M., & Khan, I. (2017). Synthesis approaches of zinc oxide nanoparticles: The dilemma of ecotoxicity. *Journal of Nanomaterials, 2017.*

Newman, L. A., & Reynolds, C. M. (2005). Bacteria and phytoremediation: New uses for endophytic bacteria in plants. *Trends in Biotechnology, 23*(1), 6−8. Available from https://doi.org/10.1016/j.tibtech.2004.11.010.

Noohpisheh, Z., Amiri, H., Farhadi, S., & Mohammadi-gholami, A. (2020). Green synthesis of Ag-ZnO nanocomposites using *Trigonella foenum-graecum* leaf extract and their antibacterial, antifungal, antioxidant and photocatalytic properties. *Spectrochimica Acta Part A: Molecular and Biomolecular Spectroscopy, 240*, 118595. Available from https://doi.org/10.1016/j.saa.2020.118595.

Ozdemir, C., Yeni, F., Odaci, D., & Timur, S. (2010). Electrochemical glucose biosensing by pyranose oxidase immobilized in gold nanoparticle-polyaniline/AgCl/gelatin nanocomposite matrix. *Food Chemistry, 119*(1), 380–385. Available from https://doi.org/10.1016/j.foodchem.2009.05.087.

Parashar, U. K., Saxena, P. S., & Srivastava, A. (2009). Bioinspired synthesis of silver nanoparticles. *Digest Journal of Nanomaterials & Biostructures (DJNB), 4*(1).

Park, T. J., Lee, K. G., & Lee, S. Y. (2016). Advances in microbial biosynthesis of metal nanoparticles. *Applied Microbiology and Biotechnology, 100*(2), 521–534. Available from https://doi.org/10.1007/s00253-015-6904-7.

Parveen, K., Banse, V., & Ledwani, L. (2016). (1st ed., p. 020048) *Green synthesis of nanoparticles: Their advantages and disadvantages*, (Vol. 1724, p. 020048). AIP Publishing.

Patel, V., Berthold, D., Puranik, P., & Gantar, M. (2015). Screening of cyanobacteria and microalgae for their ability to synthesize silver nanoparticles with antibacterial activity. *Biotechnology Reports, 5*, 112–119. Available from https://doi.org/10.1016/j.btre.2014.12.001.

Politi, J., De Stefano, L., Rea, I., Gravagnuolo, A. M., Giardina, P., Methivier, C., Casale, S., & Spadavecchia, J. (2016). One-pot synthesis of a gold nanoparticle–Vmh2 hydrophobin nanobiocomplex for glucose monitoring. *Nanotechnology, 27*(19), 195701. Available from https://doi.org/10.1088/0957-4484/27/19/195701.

Ponnanikajamideen, M., Rajeshkumar, S., Vanaja, M., & Annadurai, G. (2019). In vivo type 2 diabetes and wound-healing effects of antioxidant gold nanoparticles synthesized using the insulin plant *Chamaecostus cuspidatus* in albino rats. *Canadian Journal of Diabetes, 43*(2), 82–89.e6. Available from https://doi.org/10.1016/j.jcjd.2018.05.006.

Popli, D., Anil, V., Subramanyam, A. B., Namratha, M. N., Ranjitha, V. R., Rao, S. N., Rai, R. V., & Govindappa, M. (2018). Endophyte fungi, *Cladosporium* species-mediated synthesis of silver nanoparticles possessing in vitro antioxidant, anti-diabetic and anti-Alzheimer activity. *Artificial Cells, Nanomedicine, and Biotechnology, 46*(sup1), 676–683. Available from https://doi.org/10.1080/21691401.2018.1434188.

Prasad, R., Pandey, R., & Barman, I. (2016). Engineering tailored nanoparticles with microbes: Quo vadis? *Wiley Interdisciplinary Reviews: Nanomedicine and Nanobiotechnology, 8*(2), 316–330.

Rai, M., Ingle, A., Gaikwad, S., Gupta, I., Yadav, A., Gade, A., & Duran, N. (2016). Fungi: Myconanofactory, mycoremediation and medicine. *Fungi: Applications and management strategies* (pp. 201–219). CRC Press.

Raj, R., Dalei, K., Chakraborty, J., & Das, S. (2016). Extracellular polymeric substances of a marine bacterium mediated synthesis of CdS nanoparticles for removal of cadmium from aqueous solution. *Journal of Colloid and Interface Science, 462*, 166–175. Available from https://doi.org/10.1016/j.jcis.2015.10.004.

Ramanan, V., Thiyagarajan, S. K., Raji, K., Suresh, R., Sekar, R., & Ramamurthy, P. (2016). Outright green synthesis of fluorescent carbon dots from eutrophic algal blooms for in vitro imaging. *ACS Sustainable Chemistry & Engineering, 4*(9), 4724–4731. Available from https://doi.org/10.1021/acssuschemeng.6b00935.

Raveendran, S., Poulose, A. C., Yoshida, Y., Maekawa, T., & Kumar, D. S. (2013). Bacterial exopolysaccharide based nanoparticles for sustained drug delivery, cancer chemotherapy and bioimaging. *Carbohydrate Polymers, 91*(1), 22–32.

Ren, F., He, X., Wang, K., & Yin, J. (2012). Biosynthesis of gold nanoparticles using catclaw buttercup (Radix Ranunculi Ternati) and evaluation of its colloidal stability. *Journal of Biomedical Nanotechnology, 8*(4), 586–593. Available from https://doi.org/10.1166/jbn.2012.1417.

Rhiem, S., Riding, M. J., Baumgartner, W., Martin, F. L., Semple, K. T., Jones, K. C., Schäffer, A., & Maes, H. M. (2015). Interactions of multiwalled carbon nanotubes with algal cells: Quantification of association, visualization of uptake, and measurement of alterations in the composition of cells. *Environmental Pollution, 196*, 431–439. Available from https://doi.org/10.1016/j.envpol.2014.11.011.

Romero-Núñez, A., González, G., Romero-Ibarra, J. E., Vega-González, A., Cruz-Jiménez, G., Hernández-Cristóbal, O., Zárraga-Núñez, R. A., Obregón-Herrera, A., López-Romero, E., & Pedraza-Reyes, M. (2019). Synthesis of bimetallic nanoparticles of Cd4HgS5 by *Candida* species. *Crystals*, *9*(2), 61.

Rónavári, A., Igaz, N., Adamecz, D. I., Szerencsés, B., Molnar, C., Kónya, Z., Pfeiffer, I., & Kiricsi, M. (2021). Green silver and gold nanoparticles: Biological synthesis approaches and potentials for biomedical applications. *Molecules (Basel, Switzerland)*, *26*(4), 844.

Roy, K., Sarkar, C. K., & Ghosh, C. K. (2015). Photocatalytic activity of biogenic silver nanoparticles synthesized using yeast (*Saccharomyces cerevisiae*) extract. *Applied Nanoscience*, *5*(8), 953−959. Available from https://doi.org/10.1007/s13204-014-0392-4.

Rassas, I., Braiek, M., Bonhomme, A., Bessueille, F., Raffin, G., Majdoub, H., & Jaffrezic-Renault, N. (2019). Highly sensitive voltammetric glucose biosensor based on glucose oxidase encapsulated in a chitosan/kappa-carrageenan/gold nanoparticle bionanocomposite. *Sensors*, *19*(1), 154.

Russell, S. L., McCartney, E., & Cavanaugh, C. M. (2018). Transmission strategies in a chemosynthetic symbiosis: Detection and quantification of symbionts in host tissues and their environment. *Proceedings of the Royal Society B: Biological Sciences*, *285*(1890), 20182157. Available from https://doi.org/10.1098/rspb.2018.2157.

Saha, D., Thorpe, R., Van Bramer, S. E., Alexander, N., Hensley, D. K., Orkoulas, G., & Chen, J. (2018). Synthesis of nitrogen and sulfur codoped nanoporous carbons from algae: Role in CO_2 separation. *ACS Omega*, *3*(12), 18592−18602. Available from https://doi.org/10.1021/acsomega.8b02892.

Sahoo, D., Seckbach, J., Sahoo, D., & Baweja, P. (2015). General characteristics of algae. *The Algae World* (pp. 3−29). Springer Netherlands. Available from https://doi.org/10.1007/978-94-017-7321-8_1.

Saif Hasan, S., Singh, S., Parikh, R. Y., Dharne, M. S., Patole, M. S., Prasad, B. L. V., & Shouche, Y. S. (2008). Bacterial synthesis of copper/copper oxide nanoparticles. *Journal of Nanoscience and Nanotechnology*, *8*(6), 3191−3196. Available from https://doi.org/10.1166/jnn.2008.095.

Saim, A. K., Kumah, F. N., & Oppong, M. N. (2021). Extracellular and intracellular synthesis of gold and silver nanoparticles by living plants: a review. *Nanotechnology for Environmental Engineering*, *6*(1), 1−11.

Salunke, B. K., Sawant, S. S., Lee, S.-I., & Kim, B. S. (2015). Comparative study of MnO_2 nanoparticle synthesis by marine bacterium *Saccharophagus degradans* and yeast *Saccharomyces cerevisiae*. *Applied Microbiology and Biotechnology*, *99*(13), 5419−5427. Available from https://doi.org/10.1007/s00253-015-6559-4.

Salvadori, M. R., Ando, R. A., Oller Nascimento, C. A., & Corrêa, B. (2015). Extra and intracellular synthesis of nickel oxide nanoparticles mediated by dead fungal biomass. *PLoS One*, *10*(6), e0129799. Available from https://doi.org/10.1371/journal.pone.0129799.

Sanghi, R., & Verma, P. (2009). Biomimetic synthesis and characterisation of protein capped silver nanoparticles. *Bioresource Technology*, *100*(1), 501−504. Available from https://doi.org/10.1016/j.biortech.2008.05.048.

Saravanakumar, K., Chelliah, R., MubarakAli, D., Jeevithan, E., Oh, D.-H., Kathiresan, K., & Wang, M.-H. (2018). Fungal enzyme-mediated synthesis of chitosan nanoparticles and its biocompatibility, antioxidant and bactericidal properties. *International Journal of Biological Macromolecules*, *118*, 1542−1549. Available from https://doi.org/10.1016/j.ijbiomac.2018.06.198.

Sastry, M., Ahmad, A., Khan, M. I., & Kumar, R. (2003). Biosynthesis of metal nanoparticles using fungi and actinomycete. *Current Science*, *85*(2), 162−170.

Scognamiglio, V. (2013). Nanotechnology in glucose monitoring: Advances and challenges in the last 10 years. *Biosensors and Bioelectronics*, *47*, 12−25. Available from https://doi.org/10.1016/j.bios.2013.02.043.

Scognamiglio, V., Aurilia, V., Cennamo, N., Ringhieri, P., Iozzino, L., Tartaglia, M., Staiano, M., Ruggiero, G., Orlando, P., & Labella, T. (2007). D-Galactose/D-glucose-binding protein from *Escherichia coli* as probe for a non-consuming glucose implantable fluorescence biosensor. *Sensors*, *7*(10), 2484−2491.

Semyonov, D., Ramon, O., Shoham, Y., & Shimoni, E. (2014). Enzymatically synthesized dextran nanoparticles and their use as carriers for nutraceuticals. *Food & Function*, *5*(10), 2463−2474.

Senapati, S., Syed, A., Moeez, S., Kumar, A., & Ahmad, A. (2012). Intracellular synthesis of gold nanoparticles using alga *Tetraselmis kochinensis*. *Materials Letters*, *79*, 116−118. Available from https://doi.org/10.1016/j.matlet.2012.04.009.

Shah, M., Fawcett, D., Sharma, S., Tripathy, S., & Poinern, G. (2015). Green synthesis of metallic nanoparticles via biological entities. *Materials, 8*(11), 7278−7308.

Shah, M., Nawaz, S., Jan, H., Uddin, N., Ali, A., Anjum, S., Giglioli-Guivarc'h, N., Hano, C., & Abbasi, B. H. (2020). Synthesis of bio-mediated silver nanoparticles from Silybum marianum and their biological and clinical activities. *Materials Science and Engineering: C, 112,* 110889.

Shankar, P. D., Shobana, S., Karuppusamy, I., Pugazhendhi, A., Ramkumar, V. S., Arvindnarayan, S., & Kumar, G. (2016). A review on the biosynthesis of metallic nanoparticles (gold and silver) using biocomponents of microalgae: Formation mechanism and applications. *Enzyme and Microbial Technology, 95,* 28−44. Available from https://doi.org/10.1016/j.enzmictec.2016.10.015.

Sharma, M., Behl, K., Nigam, S., & Joshi, M. (2018). TiO$_2$-GO nanocomposite for photocatalysis and environmental applications: A green synthesis approach. *Vacuum, 156,* 434−439. Available from https://doi.org/10.1016/j.vacuum.2018.08.009.

Shivaji, S., Madhu, S., & Singh, S. (2011). Extracellular synthesis of antibacterial silver nanoparticles using psychrophilic bacteria. *Process Biochemistry, 46*(9), 1800−1807. Available from https://doi.org/10.1016/j.procbio.2011.06.008.

Shwetha, U. R., Latha, M. S., Rajith Kumar, C. R., Kiran, M. S., & Betageri, V. S. (2020). Facile synthesis of zinc oxide nanoparticles using novel *Areca catechu* leaves extract and their in vitro antidiabetic and anticancer studies. *Journal of Inorganic and Organometallic Polymers and Materials, 30,* 4876−4883.

Siddiqi, K. S., Husen, A., & Rao, R. A. K. (2018). A review on biosynthesis of silver nanoparticles and their biocidal properties. *Journal of Nanobiotechnology, 16*(1), 14. Available from https://doi.org/10.1186/s12951-018-0334-5.

Singaravelu, G., Arockiamary, J. S., Kumar, V. G., & Govindaraju, K. (2007). A novel extracellular synthesis of monodisperse gold nanoparticles using marine alga, *Sargassum wightii* Greville. *Colloids and Surfaces B: Biointerfaces, 57*(1), 97−101. Available from https://doi.org/10.1016/j.colsurfb.2007.01.010.

Singh, A. V., Patil, R., Anand, A., Milani, P., & Gade, W. N. (2010). Biological synthesis of copper oxide nano particles using *Escherichia coli*. *Current Nanoscience, 6*(1), 365−369. Available from https://doi.org/10.2174/157341310791659062.

Singh, M., Kalaivani, R., Manikandan, S., Sangeetha, N., & Kumaraguru, A. K. (2013). Facile green synthesis of variable metallic gold nanoparticle using *Padina gymnospora*, a brown marine macroalga. *Applied Nanoscience, 3*(2), 145−151. Available from https://doi.org/10.1007/s13204-012-0115-7.

Singh, P., Kim, Y. J., Wang, C., Mathiyalagan, R., & Yang, D. C. (2016). Microbial synthesis of flower-shaped gold nanoparticles. *Artificial Cells, Nanomedicine, and Biotechnology, 44*(6), 1469−1474.

Singh, P., Kim, Y.-J., Zhang, D., & Yang, D.-C. (2016). Biological synthesis of nanoparticles from plants and microorganisms. *Trends in Biotechnology, 34*(7), 588−599.

Singh, R., Shedbalkar, U. U., Wadhwani, S. A., & Chopade, B. A. (2015). Bacteriagenic silver nanoparticles: Synthesis, mechanism, and applications. *Applied Microbiology and Biotechnology, 99*(11), 4579−4593.

Solairaj, D., Rameshthangam, P., Muthukumaran, P., & Wilson, J. (2017). Studies on electrochemical glucose sensing, antimicrobial activity and cytotoxicity of fabricated copper nanoparticle immobilized chitin nanostructure. *International Journal of Biological Macromolecules, 101,* 668−679. Available from https://doi.org/10.1016/j.ijbiomac.2017.03.147.

Solgi, M., & Taghizadeh, M. (2020). Biogenic synthesis of metal nanoparticles by plants. *Biogenic nanoparticles and their use in agro-ecosystems* (pp. 593−606). Springer.

Song, S., Cong, W., Zhou, S., Shi, Y., Dai, W., Zhang, H., Wang, X., He, B., & Zhang, Q. (2019). Small GTPases: Structure, biological function and its interaction with nanoparticles. *Asian Journal of Pharmaceutical Sciences, 14*(1), 30−39.

Srivastava, M., Singh, J., Yashpal, M., Gupta, D. K., Mishra, R. K., Tripathi, S., & Ojha, A. K. (2012). Synthesis of superparamagnetic bare Fe$_3$O$_4$ nanostructures and core/shell (Fe$_3$O$_4$/alginate) nanocomposites. *Carbohydrate Polymers, 89*(3), 821−829. Available from https://doi.org/10.1016/j.carbpol.2012.04.016.

Srivastava, P., Bragança, J., Ramanan, S. R., & Kowshik, M. (2013). Synthesis of silver nanoparticles using haloarchaeal isolate *Halococcus salifodinae* BK3. *Extremophiles: Life Under Extreme Conditions, 17*(5), 821−831. Available from https://doi.org/10.1007/s00792-013-0563-3.

Suh, S., Jo, A., Traore, M. A., Zhan, Y., Coutermarsh-Ott, S. L., Ringel-Scaia, V. M., Allen, I. C., Davis, R. M., & Behkam, B. (2019). Nanoscale bacteria-enabled autonomous drug delivery system (NanoBEADS) enhances intratumoral transport of nanomedicine. *Advanced Science, 6*(3), 1801309. Available from https://doi.org/10.1002/advs.201801309.

Syed, A., & Ahmad, A. (2012). Extracellular biosynthesis of platinum nanoparticles using the fungus *Fusarium oxysporum. Colloids and Surfaces B: Biointerfaces, 97,* 27−31. Available from https://doi.org/10.1016/j.colsurfb.2012.03.026.

Tang, Y. S., Wang, D., Zhou, C., Ma, W., Zhang, Y. Q., Liu, B., & Zhang, S. (2012). Bacterial magnetic particles as a novel and efficient gene vaccine delivery system. *Gene Therapy, 19*(12), 1187.

Taran, M., Monazah, A., & Alavi, M. (2017). Using petrochemical wastewater for synthesis of cruxrhodopsin as an energy capturing nanoparticle by *Haloarcula* sp. IRU1. *Progress in Biological Sciences, 6*(2), 151−157. Available from https://doi.org/10.22059/PBS.2016.590017.

Taylor, N. S., Merrifield, R., Williams, T. D., Chipman, J. K., Lead, J. R., & Viant, M. R. (2016). Molecular toxicity of cerium oxide nanoparticles to the freshwater alga *Chlamydomonas reinhardtii* is associated with supra-environmental exposure concentrations. *Nanotoxicology, 10*(1), 32−41. Available from https://doi.org/10.3109/17435390.2014.1002868.

Timko, M., Dzarova, A., Zavisova, V., Koneracka, M., Sprincova, A., Kopcansky, P., Kovac, J., Vavra, I., & Szlaferek, A. (2008). Magnetic properties of bacterial magnetosomes and chemosynthesized magnetite nanoparticles. *Magnetohydrodynamics, 44,* 113−120.

Tsai, Y.-H., Yang, Y.-N., Ho, Y.-C., Tsai, M.-L., & Mi, F.-L. (2018). Drug release and antioxidant/antibacterial activities of silymarin-zein nanoparticle/bacterial cellulose nanofiber composite films. *Carbohydrate Polymers, 180,* 286−296. Available from https://doi.org/10.1016/j.carbpol.2017.09.100.

Tsukamoto, R., Muraoka, M., Seki, M., Tabata, H., & Yamashita, I. (2007). Synthesis of CoPt and FePt3 nanowires using the central channel of tobacco mosaic virus as a biotemplate. *Chemistry of Materials, 19*(10), 2389−2391. Available from https://doi.org/10.1021/cm062187k.

Tugarova, A. V., & Kamnev, A. A. (2017). Proteins in microbial synthesis of selenium nanoparticles. *Talanta, 174,* 539−547. Available from https://doi.org/10.1016/j.talanta.2017.06.013.

Tuo, Y., Liu, G., Dong, B., Yu, H., Zhou, J., Wang, J., & Jin, R. (2017). Microbial synthesis of bimetallic PdPt nanoparticles for catalytic reduction of 4-nitrophenol. *Environmental Science and Pollution Research, 24*(6), 5249−5258. Available from https://doi.org/10.1007/s11356-016-8276-7.

Ullah, H., Wahid, F., Santos, H. A., & Khan, T. (2016). Advances in biomedical and pharmaceutical applications of functional bacterial cellulose-based nanocomposites. *Carbohydrate Polymers, 150,* 330−352. Available from https://doi.org/10.1016/j.carbpol.2016.05.029.

Unden, G., Strecker, A., Kleefeld, A., & Kim, O. B. (2016). C4-Dicarboxylate utilization in aerobic and anaerobic growth. *EcoSal Plus, 7*(1). Available from https://doi.org/10.1128/ecosalplus.esp-0021-2015.

Vaidyanathan, S., Cherng, J.-Y., Sun, A.-C., & Chen, C.-Y. (2016). Bacteria-templated NiO nanoparticles/microstructure for an enzymeless glucose sensor. *International Journal of Molecular Sciences, 17*(7), 1104.

Varghese, R. J., Zikalala, N., & Oluwafemi, O. S. (2020). Green synthesis protocol on metal oxide nanoparticles using plant extracts. *Colloidal metal oxide nanoparticles* (pp. 67−82). Elsevier.

Velusamy, P., Kumar, G. V., Jeyanthi, V., Das, J., & Pachaiappan, R. (2016). Bio-inspired green nanoparticles: Synthesis, mechanism, and antibacterial application. *Toxicological Research, 32*(2), 95−102. Available from https://doi.org/10.5487/TR.2016.32.2.095.

Vigneshwaran, N., Ashtaputre, N. M., Varadarajan, P. V., Nachane, R. P., Paralikar, K. M., & Balasubramanya, R. H. (2007). Biological synthesis of silver nanoparticles using the fungus *Aspergillus flavus. Materials Letters, 61*(6), 1413−1418. Available from https://doi.org/10.1016/j.matlet.2006.07.042.

Vinotha, V., Iswarya, A., Thaya, R., Govindarajan, M., Alharbi, N. S., Kadaikunnan, S., Khaled, J. M., Al-Anbr, M. N., & Vaseeharan, B. (2019). Synthesis of ZnO nanoparticles using insulin-rich leaf extract: Anti-diabetic, antibiofilm and anti-oxidant properties. *Journal of Photochemistry and Photobiology B: Biology, 197,* 111541. Available from https://doi.org/10.1016/j.jphotobiol.2019.111541.

Vitta, S., & Thiruvengadam, V. (2012). Multifunctional bacterial cellulose and nanoparticle-embedded composites. *Current Science, 102*(10), 1398−1405. Available from http://www.jstor.org/stable/24107797.

Wahab, A. W., Karim, A., La Nafie, N., Satrimafitrah, P., Triana., & Sutapa, I. W. (2019). Production of the nanoparticles using leaf of *Muntingia calabura* L. as bioreductor and potential as a blood sugar nanosensor. *Journal of Physics: Conference Series, 1242*, 012004. Available from https://doi.org/10.1088/1742-6596/1242/1/012004.

Wang, L., Zhang, C., Gao, F., Mailhot, G., & Pan, G. (2017). Algae decorated TiO_2/Ag hybrid nanofiber membrane with enhanced photocatalytic activity for Cr(VI) removal under visible light. *Chemical Engineering Journal, 314*, 622−630. Available from https://doi.org/10.1016/j.cej.2016.12.020.

Wang, W., Li, H.-Y., Zhang, D.-W., Jiang, J., Cui, Y.-R., Qiu, S., Zhou, Y.-L., & Zhang, X.-X. (2010). Fabrication of bienzymatic glucose biosensor based on novel gold nanoparticles-bacteria cellulose nanofibers nanocomposite. *Electroanalysis, 22*(21), 2543−2550. Available from https://doi.org/10.1002/elan.201000235.

Wang, Y., Wei, W., Zeng, J., Liu, X., & Zeng, X. (2008). Fabrication of a copper nanoparticle/chitosan/carbon nanotube-modified glassy carbon electrode for electrochemical sensing of hydrogen peroxide and glucose. *Microchimica Acta, 160*(1), 253−260. Available from https://doi.org/10.1007/s00604-007-0844-6.

Ward, A., Ball, A., & Lewis, D. (2015). Halophytic microalgae as a feedstock for anaerobic digestion. *Algal Research, 7*, 16−23. Available from https://doi.org/10.1016/j.algal.2014.11.008.

Webster, J., & Weber, R. (2007). *Introduction to fungi.* Cambridge University Press.

Wege, C., Lomonossoff, G. P., Vieweger, S. E., Tsvetkova, I. B., & Dragnea, B. G. (2018). In vitro assembly of virus-derived designer shells around inorganic nanoparticles. *Virus-derived nanoparticles for advanced technologies: methods and protocols* (pp. 279−294). Springer New York. Available from https://doi.org/10.1007/978-1-49397808-3_19.

Wu, J., Zheng, Y., Liu, M., Shan, W., Zhang, Z., & Huang, Y. (2018). Biomimetic viruslike and charge reversible nanoparticles to sequentially overcome mucus and epithelial barriers for oral insulin delivery. *ACS Applied Materials & Interfaces, 10*(12), 9916−9928. Available from https://doi.org/10.1021/acsami.7b16524.

Wu, J., Zheng, Y., Song, W., Luan, J., Wen, X., Wu, Z., Chen, X., Wang, Q., & Guo, S. (2014). In situ synthesis of silver-nanoparticles/bacterial cellulose composites for slow-released antimicrobial wound dressing. *Carbohydrate Polymers, 102*, 762−771. Available from https://doi.org/10.1016/j.carbpol.2013.10.093.

Wu, W., He, Q., & Jiang, C. (2008). Magnetic iron oxide nanoparticles: synthesis and surface functionalization strategies. *Nanoscale Research Letters, 3*(11), 397. Available from https://doi.org/10.1007/s11671-008-9174-9.

Yahyaei, B., Manafi, S., Fahimi, B., Arabzadeh, S., & Pourali, P. (2018). Production of electrospun polyvinyl alcohol/microbial synthesized silver nanoparticles scaffold for the treatment of fungating wounds. *Applied Nanoscience, 8*(3), 417−426. Available from https://doi.org/10.1007/s13204-018-0711-2.

Yamamoto, E., & Kuroda, K. (2016). Colloidal mesoporous silica nanoparticles. *Bulletin of the Chemical Society of Japan, 89*(5), 501−539. Available from https://doi.org/10.1246/bcsj.20150420.

Yao, J., Ji, P., Wang, B., Wang, H., & Chen, S. (2018). Color-tunable luminescent macrofibers based on CdTe QDs-loaded bacterial cellulose nanofibers for pH and glucose sensing. *Sensors and Actuators B: Chemical, 254*, 110−119. Available from https://doi.org/10.1016/j.snb.2017.07.071.

Yoon, H., Ahn, J.-H., Barone, P. W., Yum, K., Sharma, R., Boghossian, A. A., Han, J.-H., & Strano, M. S. (2011). Periplasmic binding proteins as optical modulators of single-walled carbon nanotube fluorescence: Amplifying a nanoscale actuator. *Angewandte Chemie, 123*(8), 1868−1871. Available from https://doi.org/10.1002/ange.201006167.

Yulizar, Y., Sudirman., Apriandanu, D. O. B., & Al Jabbar, J. L. (2021). Facile one-pot preparation of V_2O_5-Fe_2O_3 nanocomposites using *Foeniculum vulgare* extracts and their catalytic property. *Inorganic Chemistry Communications, 123*, 108320. Available from https://doi.org/10.1016/j.inoche.2020.108320.

Zaib, M., Akhtar, A., Maqsood, F., & Shahzadi, T. (2021). Green synthesis of carbon dots and their application as photocatalyst in dye degradation studies. *Arabian Journal for Science and Engineering, 46*(1), 437−446.

Zare, B., Babaie, S., Setayesh, N., & Shahverdi, A. R. (2013). Isolation and characterization of a fungus for extracellular synthesis of small selenium nanoparticles. *Nanomedicine Journal*, *1*(1), 13−19. Available from https://doi.org/10.7508/nmj.2013.01.002.

Zhang, C., Sun, R., & Xia, T. (2020). Adaption/resistance to antimicrobial nanoparticles: Will it be a problem? *Nano Today*, *34*, 100909.

Zhang, H., Zhou, H., Bai, J., Li, Y., Yang, J., Ma, Q., & Yuanyuan, Q. (2019). Biosynthesis of selenium nanoparticles mediated by fungus *Mariannaea* sp. HJ and their characterization. *Colloids and Surfaces A: Physicochemical and Engineering Aspects*. Available from https://doi.org/10.1016/j.colsurfa.2019.02.070.

Zhang, N., Ping, Q., Huang, G., Xu, W., Cheng, Y., & Han, X. (2006). Lectin-modified solid lipid nanoparticles as carriers for oral administration of insulin. *International Journal of Pharmaceutics*, *327*(1), 153−159. Available from https://doi.org/10.1016/j.ijpharm.2006.07.026.

Zhao, B., Shao, Q., Hao, L., Zhang, L., Liu, Z., Zhang, B., Ge, S., & Guo, Z. (2018). Yeast-template synthesized Fe-doped cerium oxide hollow microspheres for visible photodegradation of acid orange 7. *Journal of Colloid and Interface Science*, *511*, 39−47. Available from https://doi.org/10.1016/j.jcis.2017.09.077.

Zhao, X., Rajashekar, C. B., Carey, E. E., & Wang, W. (2006). Does organic production enhance phytochemical content of fruit and vegetables? Current knowledge and prospects for research. *Horttechnology*, *16*(3), 449−456.

Zonaro, E., Lampis, S., Turner, R. J., Qazi, S. J. S., & Vallini, G. (2015). Biogenic selenium and tellurium nanoparticles synthesized by environmental microbial isolates efficaciously inhibit bacterial planktonic cultures and biofilms. *Frontiers in Microbiology*, *6*(584). Available from https://doi.org/10.3389/fmicb.2015.00584.

CHAPTER 7

Cytotoxicity of nanoparticles toward diabetic cell models

Introduction

Cytotoxicity analysis is used to evaluate the toxicity levels of drug compounds toward cultured cells to viability (Casey et al., 2007). Cytotoxicity analysis is broadly classified into in vitro and in vivo methods, depending on the origin of cells that are used for the analysis (Nguyen et al., 2001). Generally, in vivo analysis relies on cells, that are obtained from live animals, and are analyzed for viability after the uptake of the test material (drugs or biomolecules) (Shoemaker et al., 2006). The *in vivo* analysis involves obtaining relevant cells from animal tissues and subjecting them to the test. MTT (3-(4, 5-dimethylthiazol-2-yl)-2, 5-diphenyl tetrazolium bromide) assay is one of the first cytotoxicity analysis methods for testing a large number of samples (Mosmann, 1983). In this method, viable cells reduce the MTT dye into formazan precipitate with intense color and later solubilized into a stable and uniform colored solution analyzed spectrophotometrically at 570 nm optical absorbance. The color intensity is directly proportional to the cultured cell viability (Riss et al., 2011). 3-(4,5-dimethylthiazol-2-yl)-5-(3-carboxymethoxyphenyl)-2-(4-sulfophenyl)-2H-tetrazolium (MTS tetrazolium), 2,3-bis-(2-methoxy-4-nitro-5-sulfophenyl)-2H-tetrazolium-5-carboxanilide (XTT), and water-soluble tetrazolium salts are some other dyes that are used to evaluate cell viability (Arab-Bafrani et al., 2016; Atmaca et al., 2016; Peskin & Winterbourn, 2017). These dyes have the advantage of being directly converted into aqueous soluble formazan by cells to eliminate the solubilization step in the MTT assay (Riss et al., 2016). A redox-sensitive dye called resazurin is also used in cytotoxicity analysis. Viable cells convert the dye into a fluorescent compound known as resorufin for analysis. The test exhibits a greater sensitivity for high-throughput screening (Elshikh et al., 2016).

Nanoparticles are extensively used pharmaceutical and biomedical applications (Ojha et al., 2021). Nanoparticles can cause toxic effects on cells due to their smaller size and ease of cellular internalization (Lewinski et al., 2008; Sengul & Asmatulu, 2020). Nanoparticles display distinct mechanisms of exhibiting toxic reactions to living organisms, and cytotoxicity analysis can be used to probe the effects of these mechanisms on cells directly exposed to nanoparticles (Bayal et al., 2019; Favi et al., 2015). Cytotoxicity analysis of nanoparticles can also be used to recommend acceptable drug dosages (LD50 and LC50) for delivery (Azevedo et al., 2016). In previous chapters, various nanoparticles are mentioned as useful for the treatment of diabetes. However, in addition to toxic effects on diabetic cells, some of these nanoparticles may lead to significant toxic effects on normal cells. Thus it is highly necessary to perform a comprehensive evaluation of the cytotoxicity of nanoparticles toward diabetic cells and normal cells. The present chapter discusses the cytotoxicity of various nanoparticles, nanomedicines, and nanocomposites against diabetic cells.

Diabetic cell models

The selection of specific diabetic cell models is important in evaluating the cytotoxicity of nanomedicines intended for diabetes treatment. Radioimmunoassay techniques were used to study insulin secretion by incubating pieces of rabbit pancreas (Feldman & Chapman, 1975; Zhong et al., 2019). This method was further refined by Hellerstrom using micro-dissected islets from the pancreas (Hellerström & Swenne, 1991; Jeevanandam et al., 2018). Later, Hellman demonstrated in his seminal studies that microdissection of large beta cell–rich islets isolated from ob/ob mice can yield up to 40 islets in an hour, and this provided a good opportunity to evaluate the mechanism of insulin secretion (Hellman, 1959a, 1959b; Irles et al., 2015). Islet research received rapid developments after the introduction of the collagenase digestion method for the isolation of islets from the pancreas, a method that enables the generation of many functional mouse or rat islets in few hours (Farney et al., 2016). However, this method requires exceptional skills and techniques to isolate islets from the pancreas. Moreover, it presents difficulties in obtaining the required islet numbers for cytotoxicity analysis (Carter et al., 2009). Isolated islets may also be denervated without vascular supply, hypoxic and lose its function, and exhibit reduction in glucose-stimulated secretion of insulin (Baeyens et al., 2018). Immortal pancreatic beta cells have also been extensively used for in vitro diabetic studies due to their enhanced ability to grow rapidly in tissue culture. However, they are not considered as a complete replica of healthy beta cells present within the pancreas with cellular, neural, and vascular regulators (Green et al., 2018). Insulin-secreting cells derived from rodents were also introduced as in vitro diabetic cell models due to their advantages in evaluating the biology and physiology of beta cells, especially when antidiabetic drugs are administered (Jansson et al., 2016). However, differences in the arrangement of cells in rodent and human islets exist as a hurdle to their use as in vitro diabetic cell models (Carter et al., 2009). Rodent beta cell lines used as diabetic cell models include rodent insulinoma cell lines (He & Mei, 2018), and simian vacuolating virus 40 (SV40) expressing rodent cell lines, such as HIT-T15 (Kaneto, 2015), beta-TC (Lawlor et al., 2017), beta-HC (Hohmeier et al., 2000), NIT-1 (Wang et al., 2018), and MIN6 mouse beta cell lines (Kobayashi et al., 2016; Oshima et al., 2020). In addition, RIN (Huang et al., 2018), RINr (Skelin et al., 2010), RINm5f CRI-G1 (Tadayyon et al., 2000), INS-1 (Ho et al., 2019), RINm (Bugliani et al., 2018), and INS-2 (Besseiche et al., 2018) have also been used as diabetic cell models in various studies. Further, special cell lines, such as In-111, where insulinomas are induced via poly-oncogenic BK viral infection in hamsters

(Eter et al., 2017), super clone INS-1 832/13 (Lawlor et al., 2017), and BRIN–BD11 are used as diabetic models to evaluate specific drugs (Chen et al., 2016). The variation in rodent and human cells is a drawback to performing accurate assessments of antidiabetic drugs. To this end, a variety of cell lines obtained from humans have been cultured as diabetic cell models for the evaluation of diabetic drugs. Spontaneous human insulinoma CM (Abdelmessih et al., 2018), HIN-D8 cell line (McCluskey et al., 2011), NES2Y from the islets of persistent hyperinsulinemic hypoglycemic patients (Šrámek et al., 2016), and beta-lox-5 cell lines are the examples of widely used human diabetic cell models (Mullapudi et al., 2010). Also, NAKT-15 cell lines (Pellegrini et al., 2016), electrofusion of isolated cadaveric human beta cells with immortal PANC-1 epithelial cells to form 1.1B4, 1.4E7, and 1.1E7 insulin–releasing cell lines (Collier & Burke, 2018), EndoC-betaH1, EndoC-betaH2, and EndoC-betaH3 are some human diabetic cell models used for the evaluation of drugs (Benazra et al., 2015; Ravassard et al., 2011; Scharfmann et al., 2014).

3T3-L1 adipose cell lines are widely used to evaluate diabetes-related obesity complications and drug efficacy for their treatment (Armani et al., 2010; Yi et al., 2020). Martineau et al. (2006) evaluated the antidiabetic properties of *Vaccinium angustifolium* (Canadian low-bush blueberry) by assessing the uptake of deoxyglucose in differentiated 3T3-L1 adipocytes. The result showed that the stem and root extracts of the plant enhanced deoxyglucose uptake after 1 h of exposure toward 3T3-L1 cells and the uptake increased to 75% in the absence of insulin (Martineau et al., 2006). 3T3-L1 has also been used as an in vitro diabetic cell model in determining the LD50 of a polyherbal extract formulation (Diabecine) with antidiabetic activity (Wong et al., 2017). Circulation endothelial progenitor cells (Eleftheriadou et al., 2020; Ingram et al., 2008), isolated human skin fibroblast cells obtained from the site of diabetic wounds (Ayuk et al., 2016), BACE2 suppressed beta cells (Alcarraz-Vizán et al., 2017), and D3T as a prooxidant in diabetes-induced peripheral neuropathy cell culture models are additional cell lines used as in vitro diabetes models (Stochelski et al., 2019). Stem cells, such as mouse embryonic stem cells with knocked down *pdx-1* and *Irs-1* genes (Saito et al., 2013), mouse bone marrow–derived stem cells (Mohammadi et al., 2020; Tang et al., 2004), stem cell–derived beta cells (Millman et al., 2016), stem cell–derived extracellular vesicles as diabetic nephropathy model (Grange et al., 2019), and induced pluripotent stem cells (Stepniewski et al., 2015) are some newer diabetic cell models. It is important that the cytotoxicity of nanoparticles and nanomedicines with antidiabetic activity is accurately assessed using correct diabetic cell models for further preclinical and clinical assessments.

Cytotoxicity of nanoparticles toward diabetic cell models

The cytotoxicity of nanoparticles that are intended for use as medicines, antidiabetic agents, or drug carriers has to be evaluated. Variations in the properties of nanoparticles make it challenging to evaluate and designate a specific margin of safety and therapeutic value for all nanoparticles as each nanoparticle is distinct. The size, morphology, source of precursors, synthesis methods, and surface charge of the nanoparticles are some parameters that affect the degree of cytotoxicity (Andra et al., 2019). Here, we discuss the cytotoxicity of different types of nanoparticles such as metal, metal oxide, carbon-based, and polymer nanoparticles as shown in Table 7.1.

Metal nanoparticles

Gold and silver nanoparticles are common metal nanoparticles that are widely researched for biomedical and pharmaceutical applications. For example, Kalpana et al. (2013) synthesized gold nanoparticles from *Cinnamomum japonicum*, *Torreya nucifera*, and *Nerium indicum* as reducing agents and evaluated their in vitro cytotoxicity toward

Table 7.1 Summary of nanoparticle cytotoxicity toward in vitro diabetic cell models.

Nanoparticles	Cell lines	Cytotoxic effect	References
Metal nanoparticles			
Gold nanoparticles from leaf extract of *Cinnamomum japonicum, Torreya nucifera*, and *Nerium indicum*	3T3-L1 cell lines	Low cytotoxicity at high 10 µg/mL concentration	Kalpana et al. (2013)
Gold nanoparticles from freshwater fern powder	3T3-L1 cell lines	100 µM of concentration exhibited 71.23% ± 1.56% cell viability	Chowdhury et al. (2017)
Gold nanoparticles from gymnemic acid of *Gymnema sylvestre*	3T3-L1 cells	56.67% cell viability at 1000 µM concentration	Rajarajeshwari et al. (2014)
Panax ginseng synthesized gold nanoparticles	3T3-L1 cells	Less toxic by downregulating PPARγ/CEBPα signals	Singh et al. (2017)
α-Helix peptide conjugated gold nanoparticles	3T3-L1 cells	Less toxic by downregulating PPARγ/CEBPα signals	Park et al. (2013) and Simu et al. (2019)
Gold nanoparticles	MIN6 mouse beta cells	Low-toxic contrasting agent	Saudek et al. (2008)
Gold nanoparticles	MIN6 mouse beta cells	Cell death via cold atmospheric pressure plasma	Iuchi et al. (2018)

(Continued)

Table 7.1 (Continued)

Nanoparticles	Cell lines	Cytotoxic effect	References
DNA-functionalized gold nanoparticles	Insulin-secreting pancreatic islet cell line	Nontoxic and increase insulin secretion up to 1.69-fold	Chan et al. (2019)
Gold complexes	INS-1 insulinoma cells	Toxic via binding with human islet amyloid polypeptide	He et al. (2015)
Gold nanoclusters	Endothelial progenitor cells	Cell viability not affected below 500 nmol/L	Wang et al. (2011)
Gold nanoparticles	Human embryonic stem cells (hESCs)-derived neural progenitor cells	Size-dependent cytotoxicity	Senut et al. (2016)
FM19G11-loaded gold nanoparticles	Ependymal stem progenitor cells	Proliferation and regeneration of cells	Marcuzzo et al. (2019)
Silver nanoparticles	Human skin keratinocytes	Low cytotoxicity after long time exposure	Habas and Shang, (2019)
Silver nanoparticles	Cochlear cell lines	Cytotoxicity and genotoxicity	Perde-Schrepler et al. (2019)
Galactomannan extracted *from Punica granatum* fruit rind synthesized silver nanoparticles	Red blood cells, human and murine cancer cells	Biocompatible with red blood cells, cytotoxic to human and murine cancer cells	Padinjarathil et al. (2018)
Aloe vera gel extracted silver nanoparticles	Normal healthy cells	Dose-dependent antioxidant activity lead to reduced cytotoxicity	Sohal et al. (2019)
Silver nanoparticles	Human adipose-derived stem cells	Minimal toxicity at higher concentration	Samberg et al. (2012)
Silver nanoparticles	Beige adipocytes	Cytotoxicity by reducing differentiation ability, mitochondrial activity and thermogenic response	Yue et al. (2019)
Silicon nanoparticles	3T3-L1 cells	Less toxic at high concentration	Wang et al. (2012)

(Continued)

Table 7.1 (Continued)

Nanoparticles	Cell lines	Cytotoxic effect	References
Metal oxide nanoparticles			
Zinc oxide nanoparticles	3T3–L1 adipocytes	Less toxic toward adipocytes	Chandrasekaran and Pandurangan, (2016)
Lemon juice-synthesized zinc oxide nanoparticles	Normal mammalian L929 and 3T3-L1 cell lines	Nontoxic with selectivity index of >10	Gopala Krishna et al. (2017)
ZnO nanoparticles	MIN6 and THP-1 cell lines	Nanoparticle supports insulin fibrillation, insulin amyloid growth and reduce toxicity	Asthana et al. (2019)
ZnO nanoparticles	In vitro diabetic cell lines	Endocrine disruption and pathogenesis	Priyam et al. (2018)
ZnO nanoparticles	Human adipose–derived stromal cells	Nontoxic up to concentration of 0.2 µg/mL	Radeloff et al. (2019)
ZnO nanoparticles	BEAS-2B bronchial epithelial cells, A549 alveolar adenocarcinoma cells, and RKO colon carcinoma cells	Toxic	Vandebriel and De Jong, (2012)
ZnO nanoparticles	Human hepatocyte (L02) and human embryonic kidney cell lines (HEK293)	Cytotoxic effect via lipid peroxidation and oxidative stress	Guan et al. (2012)
Silicon dioxide nanoparticles	3T3-L1 murine cell lines	Size-dependent toxicity	Stępnik et al. (2012)
Gadolinium-labeled mesoporous silicon dioxide nanoparticles	Mouse Insulinoma 6 beta cells	Less toxic to islets	Sarkis et al. (2017)
Peptide conjugate-functionalized maghemite (γ-Fe$_2$O$_3$)	Diabetic cells	Nontoxic	Munder et al. (2017)
Magnesium oxide nanoparticles	3T3-L1 cell lines	Nontoxic and reverses insulin resistance	Jeevanandam (2017)
Titanium oxide	Diabetic cell lines	Less toxicity	Hu et al. (2016)
Cerium oxide	Diabetic cell lines	Less toxicity	Lopez-Pascual et al. (2019)

(Continued)

Table 7.1 (Continued)

Nanoparticles	Cell lines	Cytotoxic effect	References
Carbon-based nanoparticles			
Graphite nanoparticles	Glioblastoma (brain cancer) and hepatoma (liver cancer) cells	Highly toxic to glioblastoma	Zakrzewska et al. (2015)
Graphite nanoparticles	HepG2 and HepaRG liver cells	Toxic by inhibiting cytochrome P450 isoforms	Strojny et al. (2018)
Graphene oxide nanoparticles	Insulin-secreting NIT-1 pancreatic beta cells	Less toxic to beta cells	Nedumpully-Govindan et al. (2016)
Graphene oxide nanoparticles	Leydig (TM3) and Sertoli (TM4) cells	Size-dependent toxicity toward germ cells	Gurunathan et al. (2019)
Graphene nanoparticles	Cancer and certain healthy cells	Toxic	Zhang et al. (2014)
F127-coated multiwalled carbon nanotubes	NIH-3T3-L1 preadipocytes	Less toxic	Bardi et al. (2009)
Multiwalled CNTs	Normal human dermal fibroblast cells	Induce cytotoxicity, genotoxicity, and apoptosis	Patlolla et al. (2010)
Single-walled CNTs	Islets and beta cells	Oxidative stress-mediated toxicity	Ahangarpour et al. (2018)
Water-soluble fullerene	Pancreatic cancer cells	Toxic	Serda, Ware, Newton et al. (2018) and Serda, Ware, Newton, Sachdeva et al. (2018)
Colloidal fullerene	Pancreatic cell	Morphology alteration	Dziubenko et al. (2019)
Graphene quantum dots	Pancreatic cancer cells	Toxic	Joshi et al. (2017) and Nigam et al. (2014)
Graphene quantum dots	Adipocytes	Less toxic and regulates insulin receptor traffic	Zheng et al. (2013)
Diamond nanoparticles	Healthy human cells	Genotoxic	Dworak et al. (2014) and Schrand et al. (2007)

(Continued)

Table 7.1 (Continued)

Nanoparticles	Cell lines	Cytotoxic effect	References
Polymeric nanoparticles			
Alginate-coated chitosan core-shell nanoparticles	Diabetic cells	Less toxicity and reduces oxidative stress	Maity et al. (2017)
OH-terminated poly (amidoamine) dendrimer	Pancreatic MIN6 and NIT-1 cells	Aggregation of hIAPP and reduce toxicity	Gurzov et al. (2016)
Polymeric nanoparticles	Diabetic in vitro cell models	Less toxic	Fonte et al. (2015)
Nanocomposites			
Neuroligin-2-derived poly amidoamine-based dendrimer polymer nanocomposites	Pancreatic beta cells	Increase proliferation	Munder et al. (2019)
PEG nanocomposites	GLP-1R MIN6 cells and CHL-GLP-1R positive cells	Less toxic to beta cells	Babič et al. (2018)
Taurocholic acid microencapsulation	Beta cells	Reduced toxicity	Mooranian et al. (2016)
Cerium oxide nanocomposites	Beta cells	Cytoprotection of beta cells from oxidative stress	Abuid et al. (2019)
Eysenhardtia polystachya coated silver nanoparticles	INS-1 pancreatic beta cell lines	Nontoxic	Garcia Campoy et al. (2018)
Nanoformulations			
Dextran nanoformulation	Caco-2 cell monolayer and Caco-2/HT29 coculture	Less toxic with greater insulin permeation	Woitiski et al. (2011)
Red blood cell membrane-coated nanoparticle formulation	Blood cells	Nontoxic and reduce blood glucose level	Wang et al. (2017)
Zwitterionic Janus dendrimer	Blood cells	Less toxic and control blood glucose level	Wang et al. (2019)

3T3-L1 cell lines. The result showed that the nanoparticles exhibited a low-level toxicity toward the cells even at a concentration of 10 µg/mL (Kalpana et al., 2013). Chowdhury et al. (2017) also reported the synthesis of gold nanoparticles from *Marsilea quadrifolia* and performed a cytotoxicity analysis using 3T3-L1 cell lines, and the cell viability was determined via MTT assay. The result showed that the phyto-chemical synthesized gold nanoparticles at 100 µM concentration exhibited 71.23% ± 1.56% of 3T3-L1 adipose cell viability (Chowdhury et al., 2017). Similarly, the cyto-toxicity of gold nanoparticles synthesized by a secondary metabolite of *Gymnema sylvestre* called gymnemic acid was evaluated using 3T3-L1 cells. The cell viability was about 56.67% even after exposure to 1000 µM gold nanoparticles (Rajarajeshwari et al., 2014). Gold nanoparticles synthesized using *Panax ginseng* (Singh et al., 2017), biogenic gold nanoparticles synthesized via *M. quadrifolia* (Chowdhury et al., 2017), and α-helix peptide conjugated gold nanoparticles (Park et al., 2013) are other reported nanosized gold particles that are less toxic to 3T3-L1 cells and support adipo-genesis by downregulating the signaling of PPARγ/CEBPα (Simu et al., 2019).

Gold nanoparticles have also been demonstrated to be less toxic toward MIN6 mouse beta cells and are used as a contrasting agent for the imaging of beta cells (Mohan et al., 2020; Saudek et al., 2008). However, some reports have indicated that gold nanoparticles can lead to MIN6 cell death (Iuchi et al., 2018). Gold nanoparticles functionalized with DNA was exposed to insulin-secreting pancreatic islet cell line (RIN-5F) and incubated for 24 h to evaluate toxicity. The result showed that the gold nanoparticles were nontoxic and enhanced the secretion of functional insulin up to 1.69-fold with the capability of binding to insulin receptors (Chan et al., 2019). Wang et al. (2011) investigated the cytotoxicity of fluorescent gold nanoclusters toward endothelial progenitor cells used as a diabetic cell model. The result via MTT assay showed that cell viability was not affected below 500 nmol/L and the nanocluster did not result in impaired angiogenesis, indicating that the nanoclusters were biocompatible and useful for imaging cardiac diabetic cells (Wang et al., 2011). Silver nanoparticles are widely researched for pharmaceutical especially as antimicrobial agents due to their toxicity to microbes (Kailasa et al., 2019). However, there are some concerns in the utilization of silver nanoparticles as nanomedicines or drug carriers due to limited capability to preferentially tune their antimicrobial activity to target cells (Chaloupka et al., 2010). The introduction of biogenic synthesis methods has promoted the application of silver nanoparticles for pharmaceutical applications (Sumitha et al., 2019). Habas and Shang (2019) showed that silver nanoparticles can lead to increased oxidative stress levels toward human skin keratinocytes (Habas & Shang, 2019). Also, Perde-Schrepler et al. (2019) reported that silver nanoparticles possess size-dependent cytotoxicity and genotoxicity toward cochlear cell lines through an in vitro study (Perde-Schrepler et al., 2019). These studies indicate that silver nanoparticles can be toxic to normal cells. However, silver nanoparticles synthesized by Padinjarathil et al. (2018) via Galactomannan extracted from *Punica granatum* fruit rind were found to be biocompatible with red blood cells and cytotoxic toward human and murine cancer cells (Padinjarathil et al., 2018). Sohal et al. (2019) also reported that silver nanoparticles synthesized via an aqueous extract of *Aloe vera* gel possessed dose-dependent antioxidant activity and this can be used to control cytotoxicity toward normal cells (Sohal et al., 2019). Samberg et al. (2012) evaluated the toxicity of 10- to 20-nm-sized silver nanoparticles toward human adipose-derived stem cells. The result showed that the silver nanoparticles did not influence the differentiation of stem cells. However, they caused minimal toxicity at higher concentrations (Samberg et al., 2012). More research effort is needed to fully uncover the specific mechanisms of metal nanoparticles responsible for their toxicity to diabetic and normal cells.

Metal oxide nanoparticles

Metal oxide nanoparticles are recommended for biomedical applications due to their high stability in body fluids compared to nanosized metal particles (Moore et al., 2015; Sengul & Asmatulu, 2020). Several metal oxide nanoparticles have been subjected to cytotoxicity analysis using normal and diabetic cells. Zinc oxide (ZnO) nanoparticles are widely assessed for cytotoxicity toward diabetic cells due to their antidiabetic activity (Bala et al., 2015). The cytotoxicity of zinc oxide nanoparticles in the size range of 80—100 nm was evaluated using 3T3-L1 adipocytes and cocultured C2C12 myoblastoma cancer cells. The results demonstrated that the ZnO nanoparticles were less toxic toward the adipocytes with increased caspase-3 enzyme activity and reactive oxygen species production (Chandrasekaran & Pandurangan, 2016). The cytotoxicity of solution combustion synthesized ZnO nanoparticles with lemon juice as fuel was evaluated using normal mammalian cell lines such as L929 and 3T3-L1. The results showed that the ZnO nanoparticles were nontoxic toward the diabetic adipose cells with a selectivity index of >10 (Gopala Krishna et al., 2017). Asthana et al. (2019) revealed that the interaction of ZnO nanoparticles and insulin improves the fibrillation of proteins at optimum pH. Even though amyloid-like fibrils are toxic to MIN6 and THP-1 cell lines, the fresh mix of nanoparticles and insulin played a significant role in cell proliferation. Thus the nanoparticles act as a nucleation template to support the fibrillation of insulin and insulin amyloid growth (Asthana et al., 2019). The cytotoxicity of ZnO nanoparticles was also examined using human adipose-derived stromal cells (Radeloff et al., 2019). Subtoxic concentration of ZnO nanoparticles at 0.2 μg/mL was exposed to the cell lines for 24 h. The results showed that cell functionality was not affected by nanoparticles exposure. The findings of the study indicated that while the initial concentration of ZnO nanoparticles is safe and nontoxic, long-term disposition of the nanoparticles may lead to toxic effects (Radeloff et al., 2019). ZnO nanoparticles have been reported to show toxic effects toward BEAS-2B bronchial epithelial cells, A549 alveolar adenocarcinoma cells, and RKO colon carcinoma cells (Vandebriel & De Jong, 2012), as well as human hepatocyte (L02) and human embryonic kidney cell lines (HEK293) via lipid peroxidation and oxidative stress (Guan et al., 2012).

Other than ZnO nanoparticles, the cytotoxicity of commercial silicon dioxide nanoparticles, Ludox CL and CL-X with sizes of 21 and 30 nm, respectively, was evaluated using 3T3-L1 murine cell lines. The results showed that the 21-nm-sized nanoparticles were not toxic to the cells, whereas the 30-nm-sized nanoparticles induced toxic effects. The findings of the study demonstrated that the size of the nanoparticles can play a significant role in cytotoxicity (Stępnik et al., 2012). In addition, gadolinium-labeled mesoporous silicon dioxide nanoparticles in alginate beads (with less toxicity) were used to encapsulate mouse Insulinoma 6 beta cells. These less toxic encapsulations were reported to be highly beneficial in MRI-mediated tracking of transplanted pancreatic islets (Sarkis et al., 2017). Also, a nontoxic neuroligin (NL-2)-derived peptide conjugate-functionalized maghemite (γ-Fe$_2$O$_3$) has been reported as a novel antidiabetic therapeutic agent (Munder et al., 2017). The authors of this book have previously reported that magnesium oxide (MgO) nanoparticles are highly useful in reversing insulin resistance in 3T3-L1 cell lines. They synthesized distinct morphology and sizes of MgO nanoparticles via chemical-based sol—gel method and plant leaf extract-mediated green synthesis approach. The results revealed that the hexagonal-shaped MgO nanoparticles synthesized using *Amaranthus tricolor* leaf extract were less toxic to diabetic, insulin-resistant 3T3-L1, and normal VERO cell lines. In addition, biochemical analysis demonstrated that the MgO nanoparticles possessed the ability to reverse insulin resistance and can promote type 2 diabetes treatment (Jeevanandam, 2017; Jeevanandam, Chan et al., 2020; Jeevanandam, San Chan et al., 2020). Titanium oxide (Hu et al., 2016) and cerium oxide (Lopez-Pascual et al., 2019) nanoparticles have been reported to be less toxic toward diabetic cell lines.

Carbon-based nanoparticles

Graphite, graphene, carbon nanotube (CNT), carbon dots, and diamond nanoparticles are carbon-based nanoparticles that have been researched extensively for biomedical applications (Grodzik et al., 2011). Various studies have reported that carbon-based nanoparticles are toxic to normal cells (Jia et al., 2005; Madannejad et al., 2019; Magrez et al., 2006). However, it has been shown that the toxicity of carbon-based nanoparticles is dependent on their size, surface charge, and stability (Magrez et al., 2006). The cytotoxicity of graphite nanoparticles has been evaluated using glioblastoma (brain cancer) and hepatoma (liver cancer) cells and showed high toxicity to glioblastoma and low toxicity to hepatoma cells (Afshar et al., 2020; Zakrzewska et al., 2015). Graphene is a thin two-dimensional layer of graphite that is stable in the form of graphene oxide and has been explored in various pharma and biomedical applications (Zhu et al., 2010). Nedumpully-Govindan et al. (2016) studied the cytotoxicity of graphene oxide nanoparticles in the form of two-dimensional sheets toward insulin-secreting NIT-1 pancreatic beta cells. The results showed that the nanosheets were less toxic to beta cells, inhibiting the aggregation of human islet amyloid polypeptide (amylin/hIAPP) (Nedumpully-Govindan et al., 2016). However, it has also been reported that graphene oxide nanoparticles in the form of sheets of 20 and 100 nm in size were toxic to Leydig (TM3) and Sertoli (TM4) cells. The nanosheets with 20 nm size were highly toxic toward the cells compared to the 100 nm nanoparticles, probably due to the regulation of EGFR/AKT pathway (Gurunathan et al., 2019). However, the toxicity of graphene oxide nanoparticles toward diabetic cells have not yet been reported.

CNTs are one-dimensional carbon allotropes with enormous applications in the biomedical and pharmaceutical fields (Alshehri et al., 2016). There are two types of CNTs: single and multiwalled based on their structures (Mehra et al., 2015). Both single- and multiwalled CNTs have been demonstrated to have varying degrees of toxicity to normal cells (Cammisuli et al., 2018; Rubio et al., 2016). Bardi et al. (2009) showed that pluronic F127-coated multiwalled CNTs were less toxic to NIH-3T3-L1 preadipocytes which turn into white adipose tissues with endocrine roles, including maturation into diabetic cells (Bardi et al., 2009). However, it has been reported by Patlolla et al. (2010) that multiwalled CNTs are toxic to normal human dermal fibroblast cells in terms of cytotoxicity, genotoxicity, and apoptosis (Patlolla et al., 2010). Ahangarpour et al. (2018) studied the toxicity of single-walled CNTs toward islets and beta cells. The results indicated that CNTs induced oxidative stress to the pancreatic islets and elevated diabetic complications (Ahangarpour et al., 2018). The mechanism of the toxicity of CNTs is still under investigation. However, the most accepted mechanism is their DNA methylation ability upon cellular internalization (Mohanta et al., 2019).

In addition to CNTs, there are zero-dimensional carbon nanomaterials such as fullerenes and graphene quantum dots. Fullerenes have been reported to be toxic to healthy cells (Kolosnjaj et al., 2007). Serda, Ware, Newton et al. (2018) and Serda, Ware, Newton, Sachdeva et al. (2018) reported that water-soluble fullerene shows toxicity toward pancreatic cancer cells (Serda, Ware, Newton et al., 2018; Serda, Ware, Newton, Sachdeva et al., 2018). However, the toxicity of fullerenes toward diabetic cells has not yet been reported. Several studies have demonstrated that graphene quantum dots possess the ability to inhibit pancreatic cancer cells and act as anticancer agents (Joshi et al., 2017; Nigam et al., 2014). Graphene quantum dots act as universal fluorophores and are highly beneficial in regulating insulin receptor trafficking in adipocytes (Zheng et al., 2013). Diamond nanoparticles are reported to be genotoxic by exhibiting mutagenic activity to healthy human cells (Dworak et al., 2014; Schrand et al., 2007). In brief, carbon-based nanomaterials are toxic to healthy human cells. Research works on the cytotoxicity of carbon-based nanomaterials on diabetic cells are yet to be investigated widely to support emerging applications as drug delivery agents for diabetes treatments.

Polymer nanoparticles

Polymer nanoparticles are extensively used as diabetic drug delivery agents, especially to deliver insulin in diabetic patients (Lim et al., 2018). Polymer nanoparticles can be synthesized using natural and synthetic polymers (Abbasi et al., 2001). Polymeric nanoparticles have been explored for targeted and controlled insulin delivery (Jamwal et al., 2019; Li et al., 2018; Sgorla et al., 2018). Natural polymeric nanoparticles such as chitin, chitosan, and dextran are widely used in insulin delivery applications as they are less toxic toward healthy cells and are bioavailable and biocompatible (Mukhopadhyay & Prajapati, 2015). Maity et al. (2017) fabricated novel mucoadhesive alginate-coated chitosan core-shell nanoparticles for the oral delivery of naringenin, a diabetes drug in the form of flavonoid. The nanoparticles showed 90% drug entrapment with controlled drug release. The nanoparticles showed promise for the treatment of type 1 diabetes-related complications such as oxidative stress-mediated by iron in hemoglobin, dyslipidemia, and hyperglycemia via oral delivery (Maity et al., 2017). Polymeric nanoparticles such as hyaluronic acid, dextran, acrylic polymers, alginate, polycaprolactone (PCL), polyallylamine, poly (lactic acid), poly (γ-glutamic acid), and poly (lactide-*co*-glycolic acid) have been reported to be less toxic toward diabetic in vitro cell models with enhanced drug delivery efficacy (Fonte et al., 2015). While there have been significant research applications of polymeric nanoparticles for insulin delivery, progress toward clinical approval has not been encouraging due to reports on toxic reactions toward to normal cells (Fonte et al., 2015; Masood, 2016). Thus more research effort is needed to address the cytotoxicity issues to enable clinical applications as a potential diabetes drug or insulin delivery agent.

Other novel nanoparticles and nanocomposites

There are numerous nonconventional and novel nanoparticles as well as nanocomposites that are reported to possess less or no toxicity. A novel derivative of exendin-4 was synthesized using poly (ethylene glycol) and a complex nanocomposite was formed by adding amphiphilic squalene and a Cy5 fluorescence probe. The nanocomposite was added to GLP-1R MIN6 cells and CHL-GLP-1R positive cells transfected via recombination to evaluate their cytotoxicity and bioimaging ability. The results revealed that the nanocomposites were less toxic to beta cells with enhanced biodistribution in the pancreas and liver and a prolonged half-life of 3.8 h in the blood (Babič et al., 2018). Abuid et al. (2019) demonstrated that cerium oxide nanoparticles arranged in layers over alginate microbeads possessed the ability to act as an antioxidant agent and protect beta cells via encapsulation. The results showed that 12 layers of coating were required for cytoprotection of beta cells from oxidative stress. The nanocomposites were proposed to be useful in beta cell transplantation for the treatment of diabetes (Abuid et al., 2019). Garcia Campoy et al. (2018) synthesized silver nanoparticles protected with phytochemical extracts of *Eysenhardtia polystachya*. The cytotoxicity analysis of the nanoparticles' biogenic nanocomposites revealed that they are nontoxic to INS-1 pancreatic beta cell lines with protection against oxidative injury induced by hydrogen peroxide (Garcia Campoy et al., 2018). These studies indicate the potential of nanocomposites to exhibit less cytotoxicity.

Cytotoxicity of nanoformulations toward diabetic cell models

Nanoformulations are widely used in controlled delivery of insulin to diabetic cells. For example, Woitiski et al. (2011) synthesized a nanoformulation by encapsulating insulin within the core of alginate and dextran sulfate with calcium-bounded poloxamer which was chitosan-stabilized and coated with albumin. The study reported that the nanoformulation facilitated 2.1-fold permeation of insulin into Caco-2 cell monolayer without cytotoxicity, compared to nonformulated insulin (Woitiski et al., 2011). Wang et al. (2017) fabricated novel red blood cell membrane-coated nanoparticles loaded with glucose derivative-modified insulin as a biomimetic nanoformulation for smart insulin delivery. The nanoformulation assisted in smart delivery of insulin with no notifiable toxic effects (Wang et al., 2017). Modified chitosan nanoparticle formulations of insulin such as TMC, native chitosan, PEG-graft-TMC copolymer, TMC-γ-PGA (poly glutamic acid), triethyl chitosan, insulin-dimethyl ethyl chitosan (DMEC), diethyl methyl chitosan (DEMC), TMC/cysteine conjugate, and anionic poly-γ-glutamic acid chitosan are under extensive research to evaluate their cytotoxicity and insulin delivery efficacy in diabetic cells (Mukhopadhyay et al., 2012).

Nanocarriers represent another form of nanoformulations that are useful as insulin delivery systems. Solid—lipid insulin, dextran-insulin, chitosan-insulin, poly-lactic-*co*-glycolic acid (PLGA)-insulin, and polyalkyl cyanoacrylated-insulin are novel nanocarriers with less toxicity that are used for controlled delivery of insulin (Khursheed et al., 2021; Sharma et al., 2015). Even though several nanoformulations are reported to be less or nontoxic with enhanced efficacy for insulin delivery, most of these nanoformulations are still in the lab and preclinical stages.

Cytotoxicity mechanism of nanoparticles

Generally, the cytotoxicity of nanoparticles is dependent on synergistic effects of their size, morphology, surface charge, stability, synthesis approaches, and functional groups. However, each class of nanoparticles possesses a specific cytotoxicity mechanism (Andra et al., 2019) as shown in Fig. 7.1. Diabetic cells are usually weaker due to their metabolic malfunctions, causing irregular insulin levels and ATP production. Thus diabetic cells are more prone to toxic reactions compared to normal cells (Nakayama et al., 2015). Metal nanoparticles mostly bind onto the diabetic cell membrane to

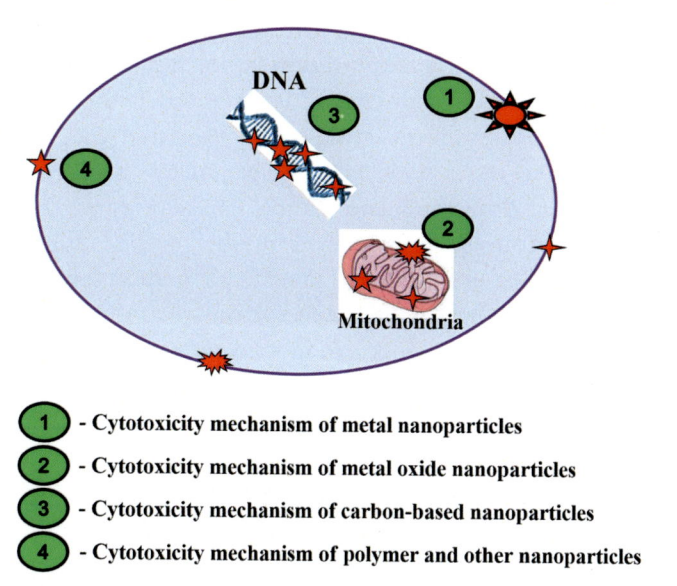

- **1** - Cytotoxicity mechanism of metal nanoparticles
- **2** - Cytotoxicity mechanism of metal oxide nanoparticles
- **3** - Cytotoxicity mechanism of carbon-based nanoparticles
- **4** - Cytotoxicity mechanism of polymer and other nanoparticles

Figure 7.1 Cytotoxicity mechanisms of different nanoparticles.

cause damage and cytotoxic effects (Choudhury et al., 2020; Fard et al., 2015). Smaller sized metal nanoparticles may also enter the cells, disintegrate into ions, and increase ion potential in cells to cause cytotoxicity in diabetic cells (Pan et al., 2007). Metal oxide nanoparticles bind to the cell membrane or internalize into diabetic cells, disintegrate, release reactive oxygen species to cause oxidative stress, and inhibit cell growth (Hu et al., 2009). Particularly, they attack the mitochondria, which leads to immediate cell death (Gajewicz et al., 2015). It is noteworthy that carbon-based nanoparticles can exhibit a cytotoxic mechanism, similar to metal and metal oxide nanoparticles (Yang et al., 2017). In addition, they can enter the cell, bind with cellular DNA, and cause DNA methylation to shut down cell function (Sierra et al., 2017). The cytotoxic mechanism of polymers, nanocomposites, and nanoformulations may be due to a combination of the mechanisms discussed early (Kim et al., 2015). Understanding the cytotoxicity mechanism of nanoparticles will help to develop techniques or formulations for reducing cytotoxicity.

Future perspective

The cytotoxicity evaluation of nanoparticles is essential to assess their efficacy in the treatment of diabetes (Vauthier, 2019). However, it is difficult to correlate the cytotoxicity of nanoparticles to different cell lines in complex systems. There is a reduced enthusiasm in the use of cell line-based cytotoxicity analysis in recent times for diabetes studies as researchers prefer in vivo analysis in animals as a direct method. In silico methods are also available to evaluate the cytotoxicity of nanoparticles (Cho et al., 2013). In silico methods are widely employed in biomedical and pharmaceutical applications to evaluate the toxicity of drugs and compounds without using live cells or animal models. It helps to circumvent ethical issues associated with live animal models and offers opportunities to test a wide range of parameters to have a comprehensive cytotoxicity evaluation of drug candidates (Abdel-Moneim et al., 2020; Lavecchia & Cerchia, 2016). While cytotoxicity results from in silico analyses still require validation with animal models (Hodos et al., 2016), it helps to reduce the number of experiments involving animal models (Raies & Bajic, 2016).

Zebrafish is another innovative inclusion in in vitro cytotoxicity analysis of nanoparticles (Asharani et al., 2008). Zebrafish is classified as an in vitro model instead of in vivo due to their size and ability to multiply in a shorter period (Bai & Tang, 2020; Sukardi et al., 2011). Zebrafish possesses organs that are similar to humans, which play significant roles in the cytotoxicity analysis of nanoparticles (Jeevanandam et al., 2019; Spitsbergen & Kent, 2003). Further, zebrafish does not require stringent animal ethics approval, compared to other live animal models, which is an added advantage to their use (Dametto et al., 2018). However, the use of zebrafish as a regular in vitro cytotoxicity analysis model still requires research validation. Development of a toxicological repository for various nanoparticles against different cell lines will help determine the cytotoxicity profiles of common nanomaterials to advance their applications in drug formulation for diabetes treatment.

Conclusion

This chapter presents an overview of the cytotoxicity properties of various nanoparticles toward different cell lines. This is important to support research and development of nanomedicines for diabetes treatment. A large number of cell line-based in vitro studies for a wide range of nanoparticles, such as metal, metal oxide, carbon, polymer, and composites, have been performed to determine their cytotoxicity. However, there is currently a limited understanding of the specific cytotoxicity mechanisms of nanoparticles, particularly nanocomposites and polymeric nanoparticles. Such knowledge is important to support the development of efficacious nanoformulations with reduced cytotoxicity effects. With further research, in silico and zebrafish-based cytotoxicity analyses have the potential to play a significant role in the development of nanoparticles for diabetes treatment.

References

Abbasi, F., Mirzadeh, H., & Katbab, A. (2001). Modification of polysiloxane polymers for biomedical applications: A review. *Polymer International, 50*(12), 1279–1287.

Abdelmessih, S., Giesecke, Y., Pascher, A., Wiedenmann, B., Mergler, S., Strowski, M., & Grötzinger, C. (2018). TRP channel expression in neuroendocrine tumors and cell lines-characterization of TRPV4 function in neuroendocrine CM cells. *Frontiers in Endocrinology, 9*, 385.

Abdel-Moneim, A., El-Shahawy, A., Yousef, A. I., Abd El-Twab, S. M., Elden, Z. E., & Taha, M. (2020). Novel polydatin-loaded chitosan nanoparticles for safe and efficient type 2 diabetes therapy: In silico, in vitro and in vivo approaches. *International Journal of Biological Macromolecules, 154,* 1496−1504.

Abuid, N. J., Gattás-Asfura, K. M., Schofield, E. A., & Stabler, C. L. (2019). Layer-by-layer cerium oxide nanoparticle coating for antioxidant protection of encapsulated beta cells. *Advanced Healthcare Materials, 0*(0), 1801493. Available from https://doi.org/10.1002/adhm.201801493.

Afshar, E. G., Zarrabi, A., Dehshahri, A., Ashrafizadeh, M., Dehghannoudeh, G., Behnam, B., Mandegary, A., Pardakhty, A., Mohammadinejad, R., & Tavakol, S. (2020). Graphene as a promising multifunctional nanoplatform for glioblastoma theranostic applications. *FlatChem, 22,* 100173.

Ahangarpour, A., Alboghobeish, S., Oroojan, A. A., & Dehghani, M. A. (2018). Mice pancreatic islets protection from oxidative stress induced by single-walled carbon nanotubes through naringin. *Human & Experimental Toxicology, 37*(12), 1268−1281. Available from https://doi.org/10.1177/0960327118769704.

Alcarraz-Vizán, G., Castaño, C., Visa, M., Montane, J., Servitja, J.-M., & Novials, A. (2017). BACE2 suppression promotes β-cell survival and function in a model of type 2 diabetes induced by human islet amyloid polypeptide overexpression. *Cellular and Molecular Life Sciences, 74*(15), 2827−2838.

Alshehri, R., Ilyas, A. M., Hasan, A., Arnaout, A., Ahmed, F., & Memic, A. (2016). Carbon nanotubes in biomedical applications: Factors, mechanisms, and remedies of toxicity: Miniperspective. *Journal of Medicinal Chemistry, 59*(18), 8149−8167.

Andra, S., Balu, S. K., Jeevanandham, J., Muthalagu, M., Vidyavathy, M., San Chan, Y., & Danquah, M. K. (2019). Phytosynthesized metal oxide nanoparticles for pharmaceutical applications. *Naunyn-Schmiedeberg's Archives of Pharmacology,* 1−17.

Arab-Bafrani, Z., Shahbazi-Gahrouei, D., Abbasian, M., & Fesharaki, M. (2016). Multiple MTS assay as the alternative method to determine survival fraction of the irradiated HT-29 colon cancer cells. *Journal of Medical Signals and Sensors, 6*(2), 112.

Armani, A., Mammi, C., Marzolla, V., Calanchini, M., Antelmi, A., Rosano, G. M. C., Fabbri, A., & Caprio, M. (2010). Cellular models for understanding adipogenesis, adipose dysfunction, and obesity. *Journal of Cellular Biochemistry, 110*(3), 564−572.

Asharani, P. V., Wu, Y. L., Gong, Z., & Valiyaveettil, S. (2008). Toxicity of silver nanoparticles in zebrafish models. *Nanotechnology, 19*(25), 255102.

Asthana, S., Hazarika, Z., Nayak, P. S., Roy, J., Jha, A. N., Mallick, B., & Jha, S. (2019). Insulin adsorption onto zinc oxide nanoparticle mediates conformational rearrangement into amyloid-prone structure with enhanced cytotoxic propensity. *Biochimica et Biophysica Acta (BBA) - General Subjects, 1863* (1), 153−166. Available from https://doi.org/10.1016/j.bbagen.2018.10.004.

Atmaca, H., Bozkurt, E., Kısım, A., & Uslu, R. (2016). Comparative analysis of XTT assay and xCELLigence system by measuring cytotoxicity of resveratrol in human cancer cell lines. *Turkish Journal of Biochemistry, 41*(6), 413−421.

Ayuk, S. M., Abrahamse, H., & Houreld, N. N. (2016). The role of photobiomodulation on gene expression of cell adhesion molecules in diabetic wounded fibroblasts in vitro. *Journal of Photochemistry and Photobiology B: Biology, 161,* 368−374. Available from https://doi.org/10.1016/j.jphotobiol.2016.05.027.

Azevedo, S. L., Ribeiro, F., Jurkschat, K., Soares, A. M. V. M., & Loureiro, S. (2016). Co-exposure of ZnO nanoparticles and UV radiation to *Daphnia magna* and *Danio rerio*: Combined effects rather than protection. *Environmental Toxicology and Chemistry, 35*(2), 458−467.

Babič, A., Vinet, L., Chellakudam, V., Janikowska, K., Allémann, E., & Lange, N. (2018). Squalene-PEG-exendin as high-affinity constructs for pancreatic beta-cells. *Bioconjugate Chemistry, 29*(8), 2531−2540. Available from https://doi.org/10.1021/acs.bioconjchem.8b00186.

Baeyens, L., Lemper, M., Staels, W., De Groef, S., De Leu, N., Heremans, Y., German, M. S., & Heimberg, H. (2018). (Re) generating human beta cells: Status, pitfalls, and perspectives. *Physiological Reviews, 98*(3), 1143−1167.

Bai, C., & Tang, M. (2020). Toxicological study of metal and metal oxide nanoparticles in zebrafish. *Journal of Applied Toxicology, 40*(1), 37−63.

Bala, N., Saha, S., Chakraborty, M., Maiti, M., Das, S., Basu, R., & Nandy, P. (2015). Green synthesis of zinc oxide nanoparticles using *Hibiscus subdariffa* leaf extract: effect of temperature on synthesis, antibacterial activity and anti-diabetic activity. *RSC Advances, 5*(7), 4993–5003.

Bardi, G., Vittorio, O., Maffei, M., Pizzorusso, T., & Costa, M. (2009). Adipocytes differentiation in the presence of Pluronic F127–coated carbon nanotubes. *Nanomedicine: Nanotechnology, Biology and Medicine, 5*(4), 378–381. Available from https://doi.org/10.1016/j.nano.2009.01.016.

Bayal, M., Janardhanan, P., Tom, E., Chandran, N., Devadathan, S., Ranjeet, D., Unnikrishnan, U., Rajendra, P., & Nair, S. S. (2019). Cytotoxicity of nanoparticles - Are the size and shape only matters? or the media parameters too?: A study on band engineered ZnS nanoparticles and calculations based on equivolume stress model. *Nanotoxicology*, 1–16.

Benazra, M., Lecomte, M.-J., Colace, C., Müller, A., Machado, C., Pechberty, S., Bricout-Neveu, E., Grenier-Godard, M., Solimena, M., & Scharfmann, R. (2015). A human beta cell line with drug inducible excision of immortalizing transgenes. *Molecular Metabolism, 4*(12), 916–925.

Besseiche, A., Riveline, J. P., Delavallée, L., Foufelle, F., Gautier, J. F., & Blondeau, B. (2018). Oxidative and energetic stresses mediate beta-cell dysfunction induced by PGC-1α. *Diabetes & Metabolism, 44*(1), 45–54.

Bugliani, M., Syed, F., Paula, F. M. M., Omar, B. A., Suleiman, M., Mossuto, S., Grano, F., Cardarelli, F., Boggi, U., & Vistoli, F. (2018). DPP-4 is expressed in human pancreatic beta cells and its direct inhibition improves beta cell function and survival in type 2 diabetes. *Molecular and Cellular Endocrinology, 473*, 186–193.

Cammisuli, F., Giordani, S., Gianoncelli, A., Rizzardi, C., Radillo, L., Zweyer, M., Da Ros, T., Salomé, M., Melato, M., & Pascolo, L. (2018). Iron-related toxicity of single-walled carbon nanotubes and crocidolite fibres in human mesothelial cells investigated by Synchrotron XRF microscopy. *Scientific Reports, 8*(1), 706.

Carter, J. D., Dula, S. B., Corbin, K. L., Wu, R., & Nunemaker, C. S. (2009). A practical guide to rodent islet isolation and assessment. *Biological Procedures Online, 11*(1), 3.

Casey, A., Herzog, E., Davoren, M., Lyng, F. M., Byrne, H. J., & Chambers, G. (2007). Spectroscopic analysis confirms the interactions between single walled carbon nanotubes and various dyes commonly used to assess cytotoxicity. *Carbon, 45*(7), 1425–1432.

Chaloupka, K., Malam, Y., & Seifalian, A. M. (2010). Nanosilver as a new generation of nanoproduct in biomedical applications. *Trends in Biotechnology, 28*(11), 580–588.

Chan, K. P., Chao, S.-H., & Kah, J. C. Y. (2019). Enhanced secretion of functional insulin with DNA-functionalized gold nanoparticles in cells. *ACS Biomaterials Science & Engineering, 5*(3), 1602–1610. Available from https://doi.org/10.1021/acsbiomaterials.9b00032.

Chandrasekaran, M., & Pandurangan, M. (2016). In vitro selective anti-proliferative effect of zinc oxide nanoparticles against co-cultured C2C12 myoblastoma cancer and 3T3-L1 normal cells. *Biological Trace Element Research, 172*(1), 148–154. Available from https://doi.org/10.1007/s12011-015-0562-6.

Chen, Y., Carlessi, R., Walz, N., Cruzat, V. F., Keane, K., John, A. N., Jiang, F.-X., Carnagarin, R., Dass, C. R., & Newsholme, P. (2016). Pigment epithelium-derived factor (PEDF) regulates metabolism and insulin secretion from a clonal rat pancreatic beta cell line BRIN-BD11 and mouse islets. *Molecular and Cellular Endocrinology, 426*, 50–60.

Cho, C. W., Ranke, J., Arning, J., Thöming, J., Preiss, U., Jungnickel, C., Diedenhofen, M., Krossing, I., & Stolte, S. (2013). In silico modelling for predicting the cationic hydrophobicity and cytotoxicity of ionic liquids towards the Leukemia rat cell line, *Vibrio fischeri* and *Scenedesmus vacuolatus* based on molecular interaction potentials of ions. *SAR and QSAR in Environmental Research, 24*(10), 863–882.

Choudhury, H., Pandey, M., Lim, Y. Q., Low, C. Y., Lee, C. T., Marilyn, T. C. L., Loh, H. S., Lim, Y. P., Lee, C. F., & Bhattamisra, S. K. (2020). Silver nanoparticles: advanced and promising technology in diabetic wound therapy. *Materials Science and Engineering: C, 112*, 110925.

Chowdhury, A., Kunjiappan, S., Bhattacharjee, C., Somasundaram, B., & Panneerselvam, T. (2017). Biogenic synthesis of *Marsilea quadrifolia* gold nanoparticles: A study of improved glucose utilization efficiency on 3T3-L1 adipocytes. *In Vitro Cellular & Developmental Biology - Animal, 53*(6), 483–493. Available from https://doi.org/10.1007/s11626-017-0136-3.

Collier, J. J., & Burke, S. J. (2018). Pancreatic islet beta-cell replacement strategies. *Cell Engineering and Regeneration*, 1−23.

Dametto, F. S., Fior, D., Idalencio, R., Rosa, J. G. S., Fagundes, M., Marqueze, A., Barreto, R. E., Piato, A., & Barcellos, L. J. G. (2018). Feeding regimen modulates zebrafish behavior. *PeerJ*, 6, e5343.

Dworak, N., Wnuk, M., Zebrowski, J., Bartosz, G., & Lewinska, A. (2014). Genotoxic and mutagenic activity of diamond nanoparticles in human peripheral lymphocytes in vitro. *Carbon*, 68, 763−776.

Dziubenko, N. V., Kuznietsova, H. M., Lynchak, O. V., & Rybalchenko, V. K. (2019). Influence of c 60-fullerene aqueous colloid solution on liver and pancreas morphological state and blood amino-transferases of rats with experienced acute cholangitis. *Biotechnologia Acta*, 12(1), 66−74.

Eleftheriadou, I., Dimitrakopoulou, N., Kafasi, N., Tentolouris, A., Dimitrakopoulou, A., Anastasiou, I. A., Mourouzis, I., Jude, E., & Tentolouris, N. (2020). Endothelial progenitor cells and peripheral neuropathy in subjects with type 2 diabetes mellitus. *Journal of Diabetes and Its Complications*, 34(4), 107517.

Elshikh, M., Ahmed, S., Funston, S., Dunlop, P., McGaw, M., Marchant, R., & Banat, I. M. (2016). Resazurin-based 96-well plate microdilution method for the determination of minimum inhibitory concentration of biosurfactants. *Biotechnology Letters*, 38(6), 1015−1019.

Eter, W. A., Van der Kroon, I., Andralojc, K., Buitinga, M., Willekens, S. M. A., Frielink, C., Bos, D., Joosten, L., Boerman, O. C., & Brom, M. (2017). Non-invasive in vivo determination of viable islet graft volume by 111 In-exendin-3. *Scientific Reports*, 7(1), 7232.

Fard, J. K., Jafari, S., & Eghbal, M. A. (2015). A review of molecular mechanisms involved in toxicity of nanoparticles. *Advanced Pharmaceutical Bulletin*, 5(4), 447.

Farney, A. C., Sutherland, D. E. R., & Opara, E. C. (2016). Evolution of islet transplantation for the last 30 years. *Pancreas*, 45(1), 8−20.

Favi, P. M., Gao, M., Johana Sepúlveda Arango, L., Ospina, S. P., Morales, M., Pavon, J. J., & Webster, T. J. (2015). Shape and surface effects on the cytotoxicity of nanoparticles: Gold nanospheres vs gold nanostars. *Journal of Biomedical Materials Research Part A*, 103(11), 3449−3462.

Feldman, J. M., & Chapman, B. (1975). Monoamine oxidase inhibitors: nature of their interaction with rabbit pancreatic islets to alter insulin secretion. *Diabetologia*, 11(6), 487−494.

Fonte, P., Araújo, F., Silva, C., Pereira, C., Reis, S., Santos, H. A., & Sarmento, B. (2015). Polymer-based nanoparticles for oral insulin delivery: Revisited approaches. *Biotechnology Advances,* 33(6, Part, 3), 1342−1354. Available from https://doi.org/10.1016/j.biotechadv.2015.02.010.

Gajewicz, A., Schaeublin, N., Rasulev, B., Hussain, S., Leszczynska, D., Puzyn, T., & Leszczynski, J. (2015). Towards understanding mechanisms governing cytotoxicity of metal oxides nanoparticles: Hints from nano-QSAR studies. *Nanotoxicology*, 9(3), 313−325.

Garcia Campoy, A. H., Perez Gutierrez, R. M., Manriquez-Alvirde, G., & Muñiz Ramirez, A. (2018). Protection of silver nanoparticles using *Eysenhardtia polystachya* in peroxide-induced pancreatic β-Cell damage and their antidiabetic properties in zebrafish. *International Journal of Nanomedicine*, 13, 2601−2612. Available from https://doi.org/10.2147/IJN.S163714.

Gopala Krishna, P., Paduvarahalli Ananthaswamy, P., Trivedi, P., Chaturvedi, V., Bhangi Mutta, N., Sannaiah, A., Erra, A., & Yadavalli, T. (2017). Antitubercular activity of ZnO nanoparticles prepared by solution combustion synthesis using lemon juice as bio-fuel. *Materials Science and Engineering: C*, 75, 1026−1033. Available from https://doi.org/10.1016/j.msec.2017.02.093.

Grange, C., Tritta, S., Tapparo, M., Cedrino, M., Tetta, C., Camussi, G., & Brizzi, M. F. (2019). Stem cell-derived extracellular vesicles inhibit and revert fibrosis progression in a mouse model of diabetic nephropathy. *Scientific Reports*, 9(1), 4468.

Green, A. D., Vasu, S., & Flatt, P. R. (2018). Cellular models for beta-cell function and diabetes gene therapy. *Acta Physiologica*, 222(3), e13012.

Grodzik, M., Sawosz, E., Wierzbicki, M., Orlowski, P., Hotowy, A., Niemiec, T., Szmidt, M., Mitura, K., & Chwalibog, A. (2011). Nanoparticles of carbon allotropes inhibit glioblastoma multiforme angiogenesis in ovo. *International Journal of Nanomedicine*, 6, 3041.

Guan, R., Kang, T., Lu, F., Zhang, Z., Shen, H., & Liu, M. (2012). Cytotoxicity, oxidative stress, and genotoxicity in human hepatocyte and embryonic kidney cells exposed to ZnO nanoparticles. *Nanoscale Research Letters*, 7(1), 602. Available from https://doi.org/10.1186/1556-276X-7-602.

Gurunathan, S., Kang, M.-H., Jeyaraj, M., & Kim, J.-H. (2019). Differential cytotoxicity of different sizes of graphene oxide nanoparticles in Leydig (TM3) and Sertoli (TM4) cells. *Nanomaterials, 9*(2). Available from https://doi.org/10.3390/nano9020139.

Gurzov, E. N., Wang, B., Pilkington, E. H., et al. (2016). Inhibition of hIAPP amyloid aggregation and pancreatic β-cell toxicity by OH-terminated PAMAM dendrimer. *Small (Weinheim an der Bergstrasse, Germany), 12*(12), 1615–1626.

Habas, K., & Shang, L. (2019). Silver nanoparticle-mediated cellular responses in human keratinocyte cell line HaCaT in vitro. *Nanoscale Reports, 2*(2), 1–9.

He, L., Zhu, D., Zhao, C., et al. (2015). Effects of gold complexes on the assembly behavior of human islet amyloid polypeptide. *Journal of Inorganic Biochemistry, 152*, 114–122.

He, Q., & Mei, X. (2018). *Tools for diabetes research and their use in drug and gene therapy.*

Hellerström, C., & Swenne, I. (1991). Functional maturation and proliferation of fetal pancreatic β-cells. *Diabetes, 40*(Supplement 2), 89–93.

Hellman, B. (1959a). The relation between age and the B/A cell ratio in the islet tissue of the rat. *European Journal of Endocrinology, 31*(I), 80–90.

Hellman, B. (1959b). The volumetric distribution of the pancreatic islet tissue in young and old rats. *European Journal of Endocrinology, 31*(I), 91–106.

Ho, H.-J., Shirakawa, H., Hirahara, K., Sone, H., Kamiyama, S., & Komai, M. (2019). Menaquinone-4 amplified glucose-stimulated insulin secretion in isolated mouse pancreatic islets and INS-1 rat insulinoma cells. *International Journal of Molecular Sciences, 20*(8), 1995.

Hodos, R. A., Kidd, B. A., Shameer, K., Readhead, B. P., & Dudley, J. T. (2016). In silico methods for drug repurposing and pharmacology. *Wiley Interdisciplinary Reviews: Systems Biology and Medicine, 8*(3), 186–210.

Hohmeier, H. E., Mulder, H., Chen, G., Henkel-Rieger, R., Prentki, M., & Newgard, C. B. (2000). Isolation of INS-1-derived cell lines with robust ATP-sensitive K + channel-dependent and-independent glucose-stimulated insulin secretion. *Diabetes, 49*(3), 424–430.

Hu, H., Li, L., Guo, Q., Jin, S., Zhou, Y., Oh, Y., Feng, Y., Wu, Q., & Gu, N. (2016). A mechanistic study to increase understanding of titanium dioxide nanoparticles-increased plasma glucose in mice. *Food and Chemical Toxicology, 95*, 175–187. Available from https://doi.org/10.1016/j.fct.2016.07.010.

Hu, X., Cook, S., Wang, P., & Hwang, H. (2009). In vitro evaluation of cytotoxicity of engineered metal oxide nanoparticles. *Science of the Total Environment, 407*(8), 3070–3072.

Huang, C.-F., Yang, C.-Y., Tsai, J.-R., Wu, C.-T., Liu, S.-H., & Lan, K.-C. (2018). Low-dose tributyltin exposure induces an oxidative stress-triggered JNK-related pancreatic β-cell apoptosis and a reversible hypoinsulinemic hyperglycemia in mice. *Scientific Reports, 8*(1), 5734.

Ingram, D. A., Lien, I. Z., Mead, L. E., Estes, M., Prater, D. N., Derr-Yellin, E., DiMeglio, L. A., & Haneline, L. S. (2008). In vitro hyperglycemia or a diabetic intrauterine environment reduces neonatal endothelial colony-forming cell numbers and function. *Diabetes, 57*(3), 724. Available from https://doi.org/10.2337/db07-1507.

Irles, E., Ñeco, P., Lluesma, M., Villar-Pazos, S., Santos-Silva, J. C., Vettorazzi, J. F., Alonso-Magdalena, P., Carneiro, E. M., Boschero, A. C., & Nadal, Á. (2015). Enhanced glucose-induced intracellular signaling promotes insulin hypersecretion: Pancreatic beta-cell functional adaptations in a model of genetic obesity and prediabetes. *Molecular and Cellular Endocrinology, 404*, 46–55.

Iuchi, K., Morisada, Y., Yoshino, Y., Himuro, T., Saito, Y., Murakami, T., & Hisatomi, H. (2018). Cold atmospheric-pressure nitrogen plasma induces the production of reactive nitrogen species and cell death by increasing intracellular calcium in HEK293T cells. *Archives of Biochemistry and Biophysics, 654*, 136–145. Available from https://doi.org/10.1016/j.abb.2018.07.015.

Jamwal, S., Ram, B., Ranote, S., Dharela, R., & Chauhan, G. S. (2019). New glucose oxidase-immobilized stimuli-responsive dextran nanoparticles for insulin delivery. *International Journal of Biological Macromolecules, 123*, 968–978.

Jansson, E. T., Comi, T. J., Rubakhin, S. S., & Sweedler, J. V. (2016). Single cell peptide heterogeneity of rat islets of Langerhans. *ACS Chemical Biology, 11*(9), 2588–2595.

Jeevanandam, J. (2017). *Enhanced synthesis and delivery of magnesium oxide nanoparticles for reverse insulin resistance in type 2 diabetes mellitus.*

Jeevanandam, J., Barhoum, A., Chan, Y. S., Dufresne, A., & Danquah, M. K. (2018). Review on nanoparticles and nanostructured materials: history, sources, toxicity and regulations. *Beilstein Journal of Nanotechnology, 9*(1), 1050−1074.

Jeevanandam, J., Chan, Y. S., & Danquah, M. K. (2020). Cytotoxicity and insulin resistance reversal ability of biofunctional phytosynthesized MgO nanoparticles. *3 Biotech, 10*, 489.

Jeevanandam, J., San Chan, Y., & Danquah, M. K. (2019). Zebrafish as a model organism to study nanomaterial toxicity. *Emerging Science Journal, 3*(3), 195−208.

Jeevanandam, J., San Chan, Y., Danquah, M. K., & Law, M. C. (2020). Cytotoxicity analysis of morphologically different sol-gel-synthesized MgO nanoparticles and their in vitro insulin resistance reversal ability in adipose cells. *Applied Biochemistry and Biotechnology, 190*(4), 1385−1410.

Jia, G., Wang, H., Yan, L., Wang, X., Pei, R., Yan, T., Zhao, Y., & Guo, X. (2005). Cytotoxicity of carbon nanomaterials: single-wall nanotube, multi-wall nanotube, and fullerene. *Environmental Science & Technology, 39*(5), 1378−1383.

Joshi, P. N., Agawane, S., Athalye, M. C., Jadhav, V., Sarkar, D., & Prakash, R. (2017). Multifunctional inulin tethered silver-graphene quantum dots nanotheranostic module for pancreatic cancer therapy. *Materials Science and Engineering: C, 78*, 1203−1211.

Kailasa, S. K., Park, T.-J., Rohit, J. V., & Koduru, J. R. (2019). *Antimicrobial activity of silver nanoparticles. Nanoparticles in Pharmacotherapy* (pp. 461−484). Elsevier.

Kalpana, D., Pichiah, P. B. T., Sankarganesh, A., Park, W. S., Lee, S. M., Wahab, R., Cha, Y. S., & Lee, Y. S. (2013). Biogenesis of gold nanoparticles using plant powders and assessment of in vitro cytotoxicity in 3T3-L1 cell line. *Journal of Pharmaceutical Innovation, 8*(4), 265−275. Available from https://doi.org/10.1007/s12247-013-9166-x.

Kaneto, H. (2015). Pancreatic β-cell glucose toxicity in type 2 diabetes mellitus. *Current Diabetes Reviews, 11*(1), 2−6.

Khursheed, R., Singh, S. K., Wadhwa, S., Gulati, M., Kapoor, B., Awasthi, A., Kr, A., Kumar, R., Pottoo, F. H., & Kumar, V. (2021). Opening eyes to therapeutic perspectives of bioactive polyphenols and their nanoformulations against diabetic neuropathy and related complications. *Expert Opinion on Drug Delivery, 18*(4), 427−448.

Kim, C. S., Mout, R., Zhao, Y., Yeh, Y.-C., Tang, R., Jeong, Y., Duncan, B., Hardy, J. A., & Rotello, V. M. (2015). Co-delivery of protein and small molecule therapeutics using nanoparticle-stabilized nanocapsules. *Bioconjugate Chemistry, 26*(5), 950−954.

Kobayashi, M., Yamato, E., Tanabe, K., Tashiro, F., Miyazaki, S., & Miyazaki, J. (2016). Functional analysis of novel candidate regulators of insulin secretion in the min6 mouse pancreatic β cell line. *PloS One, 11*(3), e0151927.

Kolosnjaj, J., Szwarc, H., & Moussa, F. (2007). *Toxicity studies of fullerenes and derivatives. Bio-Applications of Nanoparticles* (pp. 168−180). Springer.

Lavecchia, A., & Cerchia, C. (2016). In silico methods to address polypharmacology: Current status, applications and future perspectives. *Drug Discovery Today, 21*(2), 288−298.

Lawlor, N., Youn, A., Kursawe, R., Ucar, D., & Stitzel, M. L. (2017). Alpha TC1 and Beta-TC-6 genomic profiling uncovers both shared and distinct transcriptional regulatory features with their primary islet counterparts. *Scientific Reports, 7*(1), 11959.

Lewinski, N., Colvin, V., & Drezek, R. (2008). Cytotoxicity of nanoparticles. *Small, 4*(1), 26−49. Available from https://doi.org/10.1002/smll.200700595.

Li, H., Zhang, Z., Bao, X., Xu, G., & Yao, P. (2018). Fatty acid and quaternary ammonium modified chitosan nanoparticles for insulin delivery. *Colloids and Surfaces B: Biointerfaces, 170*, 136−143.

Lim, E.-K., Chung, B. H., & Chung, S. J. (2018). Recent advances in pH-sensitive polymeric nanoparticles for smart drug delivery in cancer therapy. *Current Drug Targets, 19*(4), 300−317.

Lopez-Pascual, A., Urrutia-Sarratea, A., Lorente-Cebrián, S., Martinez, J. A., & González-Muniesa, P. (2019). Cerium oxide nanoparticles regulate insulin sensitivity and oxidative markers in 3T3-L1 adipocytes and C2C12 myotubes. *Oxidative Medicine and Cellular Longevity*, 2019.

Madannejad, R., Shoaie, N., Jahanpeyma, F., Darvishi, M. H., Azimzadeh, M., & Javadi, H. (2019). Toxicity of carbon-based nanomaterials: Reviewing recent reports in medical and biological systems. *Chemico-Biological Interactions, 307*, 206−222.

Magrez, A., Kasas, S., Salicio, V., Pasquier, N., Seo, J. W., Celio, M., Catsicas, S., Schwaller, B., & Forró, L. (2006). Cellular toxicity of carbon-based nanomaterials. *Nano Letters, 6*(6), 1121–1125.

Maity, S., Mukhopadhyay, P., Kundu, P. P., & Chakraborti, A. S. (2017). Alginate coated chitosan core-shell nanoparticles for efficient oral delivery of naringenin in diabetic animals—An in vitro and in vivo approach. *Carbohydrate Polymers, 170*, 124–132. Available from https://doi.org/10.1016/j.carbpol.2017.04.066.

Marcuzzo, S., Isaia, D., Bonanno, S., et al. (2019). FM19G11-loaded gold nanoparticles enhance the proliferation and self-renewal of ependymal stem progenitor cells derived from ALS mice. *Cells, 8*(3), 279.

Martineau, L. C., Couture, A., Spoor, D., Benhaddou-Andaloussi, A., Harris, C., Meddah, B., Leduc, C., Burt, A., Vuong, T., Mai Le, P., Prentki, M., Bennett, S. A., Arnason, J. T., & Haddad, P. S. (2006). Anti-diabetic properties of the Canadian lowbush blueberry *Vaccinium angustifolium* Ait. *Phytomedicine, 13*(9), 612–623. Available from https://doi.org/10.1016/j.phymed.2006.08.005.

Masood, F. (2016). Polymeric nanoparticles for targeted drug delivery system for cancer therapy. *Materials Science and Engineering: C, 60*, 569–578.

McCluskey, J. T., Hamid, M., Guo-Parke, H., McClenaghan, N. H., Gomis, R., & Flatt, P. R. (2011). Development and functional characterization of insulin-releasing human pancreatic beta cell lines produced by electrofusion. *Journal of Biological Chemistry, 286*(25), 21982–21992.

Mehra, N. K., Jain, K., & Jain, N. K. (2015). Pharmaceutical and biomedical applications of surface engineered carbon nanotubes. *Drug Discovery Today, 20*(6), 750–759.

Millman, J. R., Xie, C., Van Dervort, A., Gürtler, M., Pagliuca, F. W., & Melton, D. A. (2016). Generation of stem cell-derived β-cells from patients with type 1 diabetes. *Nature Communications, 7*, 11463.

Mohammadi, N., Mardomi, A., Hassannia, H., Enderami, S. E., Ranjbaran, H., Rafiei, A., & Abediankenari, S. (2020). Mouse bone marrow-derived mesenchymal stem cells acquire immunogenicity concurrent with differentiation to insulin-producing cells. *Immunobiology, 225*(5), 151994.

Mohan, P. K., Moola, A. K., Kumar, T. S., & Kumari, B. D. (2020). A comprehensive review of the phytochemical and pharmacological properties of *Desmodium gangeticum* (L.) DC. *Journal of Advanced Scientific Research, 11*.

Mohanta, D., Patnaik, S., Sood, S., & Das, N. (2019). Carbon nanotubes: Evaluation of toxicity at biointerfaces. *Journal of Pharmaceutical Analysis*.

Mooranian, A., Negrulj, R., & Al-Salami, H. (2016). Viability and topographical analysis of microencapsulated β-cells exposed to a biotransformed tertiary bile acid: an ex vivo study. *International Journal of Nano and Biomaterials, 6*(2), 74–82.

Moore, T. L., Rodriguez-Lorenzo, L., Hirsch, V., Balog, S., Urban, D., Jud, C., Rothen-Rutishauser, B., Lattuada, M., & Petri-Fink, A. (2015). Nanoparticle colloidal stability in cell culture media and impact on cellular interactions. *Chemical Society Reviews, 44*(17), 6287–6305.

Mosmann, T. (1983). Rapid colorimetric assay for cellular growth and survival: Application to proliferation and cytotoxicity assays. *Journal of Immunological Methods, 65*(1–2), 55–63.

Mukhopadhyay, P., Mishra, R., Rana, D., & Kundu, P. P. (2012). Strategies for effective oral insulin delivery with modified chitosan nanoparticles: A review. *Progress in Polymer Science, 37*(11), 1457–1475. Available from https://doi.org/10.1016/j.progpolymsci.2012.04.004.

Mukhopadhyay, P., & Prajapati, A. K. (2015). Quercetin in anti-diabetic research and strategies for improved quercetin bioavailability using polymer-based carriers—A review. *RSC Advances, 5*(118), 97547–97562.

Mullapudi, B., Ding, Y., Ding, X., & Grippo, P. (2010). *Drug evaluations in pancreatic cancer culture systems. Drug discovery in pancreatic cancer* (pp. 1–27). Springer.

Munder, A., Israel, L. L., Kahremany, S., Ben-Shabat-Binyamini, R., Zhang, C., Kolitz-Domb, M., Viskind, O., Levine, A., Senderowitz, H., Chessler, S., Lellouche, J.-P., & Gruzman, A. (2017). Mimicking Neuroligin-2 functions in β-cells by functionalized nanoparticles as a novel approach for antidiabetic therapy. *ACS Applied Materials & Interfaces, 9*(2), 1189–1206. Available from https://doi.org/10.1021/acsami.6b10568.

Munder, A., Moskovitz, Y., Meir, A., et al. (2019). Neuroligin-2-derived peptide-covered polyamidoamine-based (PAMAM) dendrimers enhance pancreatic β-cells' proliferation and functions. *MedChemComm, 10*(2), 280–293.

Nakayama, M., McDaniel, K., Fitzgerald-Miller, L., Kiekhaefer, C., Snell-Bergeon, J. K., Davidson, H. W., Rewers, M., Yu, L., Gottlieb, P., & Kappler, J. W. (2015). Regulatory vs. inflammatory cytokine T-cell responses to mutated insulin peptides in healthy and type 1 diabetic subjects. *Proceedings of the National Academy of Sciences, 112*(14), 4429–4434.

Nedumpully-Govindan, P., Gurzov, E. N., Chen, P., Pilkington, E. H., Stanley, W. J., Litwak, S. A., Davis, T. P., Ke, P. C., & Ding, F. (2016). Graphene oxide inhibits hIAPP amyloid fibrillation and toxicity in insulin-producing NIT-1 cells. *Physical Chemistry Chemical Physics, 18*(1), 94–100. Available from https://doi.org/10.1039/C5CP05924K.

Nguyen, D. M., Lorang, D., Chen, G. A., Stewart Iv, J. H., Tabibi, E., & Schrump, D. S. (2001). Enhancement of paclitaxel-mediated cytotoxicity in lung cancer cells by 17-allylamino geldanamycin: in vitro and in vivo analysis. *The Annals of Thoracic Surgery, 72*(2), 371–379.

Nigam, P., Waghmode, S., Louis, M., Wangnoo, S., Chavan, P., & Sarkar, D. (2014). Graphene quantum dots conjugated albumin nanoparticles for targeted drug delivery and imaging of pancreatic cancer. *Journal of Materials Chemistry B, 2*(21), 3190–3195.

Ojha, B., Jain, V. K., Mehra, N. K., & Jain, K. (2021). *Nanotechnology: Introduction and basic concepts. Dendrimers in nanomedicine* (pp. 1–17). CRC Press.

Oshima, M., Pechberty, S., Bellini, L., Göpel, S. O., Campana, M., Rouch, C., Dairou, J., Cosentino, C., Fantuzzi, F., & Toivonen, S. (2020). Stearoyl CoA desaturase is a gatekeeper that protects human beta cells against lipotoxicity and maintains their identity. *Diabetologia, 63*(2), 395–409.

Padinjarathil, H., Joseph, M. M., Unnikrishnan, B. S., Preethi, G. U., Shiji, R., Archana, M. G., Maya, S., Syama, H. P., & Sreelekha, T. T. (2018). Galactomannan endowed biogenic silver nanoparticles exposed enhanced cancer cytotoxicity with excellent biocompatibility. *International Journal of Biological Macromolecules, 118*, 1174–1182. Available from https://doi.org/10.1016/j.ijbiomac.2018.06.194.

Pan, Y., Neuss, S., Leifert, A., Fischler, M., Wen, F., Simon, U., Schmid, G., Brandau, W., & Jahnen-Dechent, W. (2007). Size-dependent cytotoxity of gold nanoparticles. *Small, 3*(11), 1941–1949.

Park, H., Tsutsumi, H., & Mihara, H. (2013). Cell penetration and cell-selective drug delivery using α helix peptides conjugated with gold nanoparticles. *Biomaterials, 34*(20), 4872–4879.

Patlolla, A., Knighten, B., & Tchounwou, P. (2010). Multi-walled carbon nanotubes induce cytotoxicity, genotoxicity and apoptosis in normal human dermal fibroblast cells. *Ethnicity & Disease, 20*(1 Suppl 1), S1–S72. Available from https://www.ncbi.nlm.nih.gov/pubmed/20521388.

Pellegrini, S., Cantarelli, E., Sordi, V., Nano, R., & Piemonti, L. (2016). The state of the art of islet transplantation and cell therapy in type 1 diabetes. *Acta Diabetologica, 53*(5), 683–691.

Perde-Schrepler, M., Florea, A., Brie, I., Virag, P., Fischer-Fodor, E., Vâlcan, A., Gurzău, E., Lisencu, C., & Maniu, A. (2019). Size-dependent cytotoxicity and genotoxicity of silver nanoparticles in cochlear cells in vitro. *Journal of Nanomaterials, 2019*.

Peskin, A. V., & Winterbourn, C. C. (2017). Assay of superoxide dismutase activity in a plate assay using WST-1. *Free Radical Biology and Medicine, 103*, 188–191.

Priyam, A., Singh, P. P., & Gehlout, S. (2018). Role of endocrine-disrupting engineered nanomaterials in the pathogenesis of type 2 diabetes mellitus. *Frontiers in Endocrinology, 9*, 704, 704.

Radeloff, K., Radeloff, A., Tirado, M. R., Scherzad, A., Hagen, R., Kleinsasser, N. H., & Hackenberg, S. (2019). Long-term impact of zinc oxide nanoparticles on differentiation and cytokine secretion of human adipose-derived stromal cells. *Materials, 12*(11), 1823. Available from https://doi.org/10.3390/ma12111823.

Raies, A. B., & Bajic, V. B. (2016). In silico toxicology: Computational methods for the prediction of chemical toxicity. *Wiley Interdisciplinary Reviews: Computational Molecular Science, 6*(2), 147–172.

Rajarajeshwari, T., Shivashri, C., & Rajasekar, P. (2014). Synthesis and characterization of biocompatible gymnemic acid–gold nanoparticles: A study on glucose uptake stimulatory effect in 3T3-L1 adipocytes. *RSC Advances, 4*(108), 63285–63295. Available from https://doi.org/10.1039/C4RA07087A.

Ravassard, P., Hazhouz, Y., Pechberty, S., Bricout-Neveu, E., Armanet, M., Czernichow, P., & Scharfmann, R. (2011). A genetically engineered human pancreatic β cell line exhibiting glucose-inducible insulin secretion. *The Journal of Clinical Investigation, 121*(9).

Riss, T. L., Moravec, R. A., & Niles, A. L. (2011). Cytotoxicity testing: Measuring viable cells, dead cells, and detecting mechanism of cell death. *Methods in Molecular Biology, 740*, 103–114. Available from https://doi.org/10.1007/978-1-61779-108-6_12.

Riss, Terry L., Moravec, R.A., Niles, A.L., Duellman, S., Benink, H.A., Worzella, T.J., & Minor, L. (2016). Cell viability assays.

Rubio, L., El Yamani, N., Kazimirova, A., Dusinska, M., & Marcos, R. (2016). Multi-walled carbon nanotubes (NM401) induce ROS-mediated HPRT mutations in Chinese hamster lung fibroblasts. *Environmental Research, 146,* 185−190.

Saito, M., Hayakawa, A., Inagaki, N., & Matsuoka, H. (2013). Development of novel cell lines of diabetic dysfunction model fit for cell-based screening tests of medicinal materials. *Cytotechnology, 65*(1), 105−118. Available from https://doi.org/10.1007/s10616-012-9466-x.

Samberg, M. E., Loboa, E. G., Oldenburg, S. J., & Monteiro-Riviere, N. A. (2012). Silver nanoparticles do not influence stem cell differentiation but cause minimal toxicity. *Nanomedicine, 7*(8), 1197−1209. Available from https://doi.org/10.2217/nnm.12.18.

Sarkis, S., Silencieux, F., Markwick, K. E., Fortin, M.-A., & Hoesli, C. A. (2017). Magnetic resonance imaging of alginate beads containing pancreatic beta cells and paramagnetic nanoparticles. *ACS Biomaterials Science & Engineering, 3*(12), 3576−3587. Available from https://doi.org/10.1021/acsbiomaterials.7b00404.

Saudek, F., Brogren, C.-H., & Manohar, S. (2008). Imaging the beta-cell mass: Why and how. *The Review of Diabetic Studies: RDS, 5*(1), 6.

Scharfmann, R., Pechberty, S., Hazhouz, Y., Von Bülow, M., Bricout-Neveu, E., Grenier-Godard, M., Guez, F., Rachdi, L., Lohmann, M., & Czernichow, P. (2014). Development of a conditionally immortalized human pancreatic β cell line. *The Journal of Clinical Investigation, 124*(5), 2087−2098.

Schrand, A. M., Huang, H., Carlson, C., Schlager, J. J., Ōsawa, E., Hussain, S. M., & Dai, L. (2007). Are diamond nanoparticles cytotoxic? *The Journal of Physical Chemistry B, 111*(1), 2−7.

Sengul, A. B., & Asmatulu, E. (2020). Toxicity of metal and metal oxide nanoparticles: A review. *Environmental Chemistry Letters, 18*(5), 1659−1683.

Senut, M.-C., Zhang, Y., Liu, F., et al. (2016). Size-dependent toxicity of gold nanoparticles on human embryonic stem cells and their neural derivatives. *Small (Weinheim an der Bergstrasse, Germany), 12*(5), 631−646.

Serda, M., Ware, M., Newton, J., et al. (2018). PO-515 Novel water-solube [60] fullerene nanotherapeutic agent for pancreatic cancer treatment. *BMJ Publishing Group Limited.*

Serda, M., Ware, M. J., Newton, J. M., Sachdeva, S., Krzykawska-Serda, M., Nguyen, L., Law, J., Anderson, A. O., Curley, S. A., & Wilson, L. J. (2018). Development of photoactive Sweet-C60 for pancreatic cancer stellate cell therapy. *Nanomedicine, 13*(23), 2981−2993.

Sgorla, D., Lechanteur, A., Almeida, A., Sousa, F., Melo, E., Bunhak, É., Mainardes, R., Khalil, N., Cavalcanti, O., & Sarmento, B. (2018). Development and characterization of lipid-polymeric nanoparticles for oral insulin delivery. *Expert Opinion on Drug Delivery, 15*(3), 213−222.

Sharma, G., Sharma, A. R., Nam, J.-S., Doss, G. P. C., Lee, S.-S., & Chakraborty, C. (2015). Nanoparticle based insulin delivery system: The next generation efficient therapy for Type 1 diabetes. *Journal of Nanobiotechnology, 13*(1), 74. Available from https://doi.org/10.1186/s12951-015-0136-y.

Shoemaker, A. R., Oleksijew, A., Bauch, J., Belli, B. A., Borre, T., Bruncko, M., Deckwirth, T., Frost, D. J., Jarvis, K., & Joseph, M. K. (2006). A small-molecule inhibitor of Bcl-XL potentiates the activity of cytotoxic drugs in vitro and in vivo. *Cancer Research, 66*(17), 8731−8739.

Sierra, M. I., Rubio, L., Bayón, G. F., Cobo, I., Menendez, P., Morales, P., Mangas, C., Urdinguio, R. G., Lopez, V., & Valdes, A. (2017). DNA methylation changes in human lung epithelia cells exposed to multi-walled carbon nanotubes. *Nanotoxicology, 11*(7), 857−870.

Simu, S. Y., Ahn, S., Castro-Aceituno, V., Singh, P., Mathiyalagan, R., Jiménez-Pérez, Z. E., Hurh, J., Oi, L. Z., Hun, N. J., & Kim, Y.-J. (2019). Gold nanoparticles synthesized with fresh *Panax ginseng* leaf extract suppress Adipogenesis by downregulating PPARγ/CEBPα signaling in 3T3-L1 mature adipocytes. *Journal of Nanoscience and Nanotechnology, 19*(2), 701−708.

Singh, P., Singh, H., Ahn, S., Castro-Aceituno, V., Jiménez, Z., Simu, S. Y., Kim, Y. J., & Yang, D. C. (2017). Pharmacological importance, characterization and applications of gold and silver nanoparticles synthesized by Panax ginseng fresh leaves. *Artificial Cells, Nanomedicine, and Biotechnology, 45*(7), 1415−1424. Available from https://doi.org/10.1080/21691401.2016.1243547.

Skelin, M., Rupnik, M., & Cencič, A. (2010). Pancreatic beta cell lines and their applications in diabetes mellitus research. *ALTEX-Alternatives to Animal Experimentation, 27*(2), 105−113.

Sohal, J. K., Saraf, A., Shukla, K., & Shrivastava, M. (2019). Determination of antioxidant potential of biochemically synthesized silver nanoparticles using aloe vera gel extract. *Plant Science Today, 2*, 208−217%V 6. Available from https://doi.org/10.14719/pst.2019.6.2.532.

Spitsbergen, J. M., & Kent, M. L. (2003). The state of the art of the zebrafish model for toxicology and toxicologic pathology research—Advantages and current limitations. *Toxicologic Pathology, 31(1_suppl)*, 62−87.

Šrámek, J., Němcová-Fürstová, V., & Kovář, J. (2016). Kinase signaling in apoptosis induced by saturated fatty acids in pancreatic β-cells. *International Journal of Molecular Sciences, 17*(9), 1400.

Stepniewski, J., Kachamakova-Trojanowska, N., Ogrocki, D., Szopa, M., Matlok, M., Beilharz, M., Dyduch, G., Malecki, M. T., Józkowicz, A., & Dulak, J. (2015). Induced pluripotent stem cells as a model for diabetes investigation. *Scientific Reports, 5*, 8597.

Stępnik, M., Arkusz, J., Smok-Pieniążek, A., Bratek-Skicki, A., Salvati, A., Lynch, I., Dawson, K. A., Gromadzińska, J., De Jong, W. H., & Rydzyński, K. (2012). Cytotoxic effects in 3T3-L1 mouse and WI-38 human fibroblasts following 72 hour and 7 day exposures to commercial silica nanoparticles. *Toxicology and Applied Pharmacology, 263*(1), 89−101. Available from https://doi.org/10.1016/j.taap.2012.06.002.

Stochelski, M. A., Wilmanski, T., Walters, M., & Burgess, J. R. (2019). D3T acts as a pro-oxidant in a cell culture model of diabetes-induced peripheral neuropathy. *Redox Biology, 21*, 101078.

Strojny, B., Sawosz, E., Grodzik, M., et al. (2018). Nanostructures of diamond, graphene oxide and graphite inhibit CYP1A2, CYP2D6 and CYP3A4 enzymes and downregulate their genes in liver cells. *International Journal of Nanomedicine, 13*, 8561−8575.

Sukardi, H., Chng, H. T., Chan, E. C. Y., Gong, Z., & Lam, S. H. (2011). Zebrafish for drug toxicity screening: bridging the in vitro cell-based models and in vivo mammalian models. *Expert Opinion on Drug Metabolism & Toxicology, 7*(5), 579−589.

Sumitha, S., Vasanthi, S., Shalini, S., Chinni, S. V., Gopinath, S. C. B., Kathiresan, S., Anbu, P., & Ravichandran, V. (2019). *Durio zibethinus* rind extract mediated green synthesis of silver nanoparticles: Characterization and biomedical applications. *Pharmacognosy Magazine, 15*(60), 52.

Tadayyon, M., Welters, H. J., Haynes, A. C., Cluderay, J. E., & Hervieu, G. (2000). Expression of melanin-concentrating hormone receptors in insulin-producing cells: MCH stimulates insulin release in RINm5F and CRI-G1 cell-lines. *Biochemical and Biophysical Research Communications, 275*(2), 709−712.

Tang, D.-Q., Cao, L.-Z., Burkhardt, B. R., Xia, C.-Q., Litherland, S. A., Atkinson, M. A., & Yang, L.-J. (2004). In vivo and in vitro characterization of insulin-producing cells obtained from murine bone marrow. *Diabetes, 53*(7), 1721. Available from https://doi.org/10.2337/diabetes.53.7.1721.

Vandebriel, R. J., & De Jong, W. H. (2012). A review of mammalian toxicity of ZnO nanoparticles. *Nanotechnology, Science and Applications, 5*, 61−71. Available from https://doi.org/10.2147/NSA. S23932.

Vauthier, C. (2019). A journey through the emergence of nanomedicines with poly (alkylcyanoacrylate) based nanoparticles. *Journal of Drug Targeting*, 1−23.

Wang, C., Ye, Y., Sun, W., Yu, J., Wang, J., Lawrence, D. S., Buse, J. B., & Gu, Z. (2017). Red blood cells for glucose-responsive insulin delivery. *Advanced Materials, 29*(18), 1606617.

Wang, H.-H., Lin, C.-A. J., Lee, C.-H., Lin, Y.-C., Tseng, Y.-M., Hsieh, C.-L., Chen, C.-H., Tsai, C.-H., Hsieh, C.-T., Shen, J.-L., Chan, W.-H., Chang, W. H., & Yeh, H.-I. (2011). Fluorescent gold nanoclusters as a biocompatible marker for in vitro and in vivo tracking of endothelial cells. *ACS Nano, 5*(6), 4337−4344. Available from https://doi.org/10.1021/nn102752a.

Wang, L., Shi, C., Wang, X., et al. (2019). Zwitterionic Janus dendrimer with distinct functional disparity for enhanced protein delivery. *Biomaterials, 215*, 119233.

Wang, Q., Bao, Y., Zhang, X., et al. (2012). Uptake and toxicity studies of poly-acrylic acid functionalized silicon nanoparticles in cultured mammalian cells. *Advanced Healthcare Materials, 1*(2), 189−198.

Wang, Y., Liu, J., Liu, Z., Chen, J., Hu, X., Hu, Y., Yuan, Y., Wu, G., Dai, Z., & Xu, Y. (2018). Sall2 knockdown exacerbates palmitic acid induced dysfunction and apoptosis of pancreatic NIT-1 beta cells. *Biomedicine & Pharmacotherapy, 104*, 375−382.

Woitiski, C. B., Sarmento, B., Carvalho, R. A., Neufeld, R. J., & Veiga, F. (2011). Facilitated nanoscale delivery of insulin across intestinal membrane models. *International Journal of Pharmaceutics, 412*(1), 123−131. Available from https://doi.org/10.1016/j.ijpharm.2011.04.003.

Wong, T. S., Hashim, Z., Zulkifli, R. M., Ismail, H. F., Zainol, S. N., Rajib, N. S. M., Teh, L. C., & Majid, F. A. A. (2017). LD 50 estimations for diabecine tm polyherbal extracts based on in vitro diabetic models of 3T3-L1, WRL-68 and 1.1 B4 cell lines. *Chemical Engineering Transactions*, *56*, 1567–1572.

Yang, F., Jiang, Q., Xie, W., & Zhang, Y. (2017). Effects of multi-walled carbon nanotubes with various diameters on bacterial cellular membranes: cytotoxicity and adaptive mechanisms. *Chemosphere*, *185*, 162–170.

Yi, M. H., Simu, S. Y., Ahn, S., Aceituno, V. C., Wang, C., Mathiyalagan, R., Hurh, J., Batjikh, I., Ali, H., & Kim, Y.-J. (2020). Anti-obesity effect of gold nanoparticles from *Dendropanax morbifera* Léveille by suppression of triglyceride synthesis and downregulation of PPARγ and CEBPα signaling pathways in 3T3-L1 mature adipocytes and HepG2 cells. *Current Nanoscience*, *16*(2), 196–203.

Yue, L., Zhao, W., Wang, D., et al. (2019). Silver nanoparticles inhibit beige fat function and promote adiposity. *Molecular Metabolism*, *22*, 1–11.

Zakrzewska, K. E., Samluk, A., Wierzbicki, M., Jaworski, S., Kutwin, M., Sawosz, E., Chwalibog, A., Pijanowska, D. G., & Pluta, K. D. (2015). Analysis of the cytotoxicity of carbon-based nanoparticles, diamond and graphite, in human glioblastoma and hepatoma cell lines. *PLOS ONE*, *10*(3), e0122579. Available from https://doi.org/10.1371/journal.pone.0122579.

Zhang, Y., Petibone, D., Xu, Y., et al. (2014). Toxicity and efficacy of carbon nanotubes and graphene: The utility of carbon-based nanoparticles in nanomedicine. *Drug Metabolism Reviews*, *46*(2), 232–246.

Zheng, X. T., Than, A., Ananthanaraya, A., Kim, D.-H., & Chen, P. (2013). Graphene quantum dots as universal fluorophores and their use in revealing regulated trafficking of insulin receptors in adipocytes. *ACS Nano*, *7*(7), 6278–6286. Available from https://doi.org/10.1021/nn4023137.

Zhong, B., Ma, S., & Wang, D. H. (2019). TRPV1 mediates glucose-induced insulin secretion through releasing neuropeptides. *In Vivo*, *33*(5), 1431–1437.

Zhu, Y., Murali, S., Cai, W., Li, X., Suk, J. W., Potts, J. R., & Ruoff, R. S. (2010). Graphene and graphene oxide: Synthesis, properties, and applications. *Advanced Materials*, *22*(35), 3906–3924.

CHAPTER 8

In vivo studies of nanoparticles in diabetic models

Introduction

Toxicological analyses are essential to evaluate the side effects of drugs in patients at the in cellular, molecular, organ, and/or psychological levels (Flanagan et al., 2020; Stephens et al., 2016). In vitro and in vivo toxicity analyses are the two major approaches that are used to evaluate the toxic behavior of drugs (Suter et al., 2004). In vitro cyto-models used for toxicity analysis are beneficial in evaluating toxic effects corresponding to specific cells (Drasler et al., 2017; Tice et al., 2000). However, drugs that are developed for disease treatment must be evaluated using organisms to obtain comprehensive toxicity profiles rather than specific toxic profiles for cells (Khlebtsov & Dykman, 2011; Patnaik et al., 2021). Thus in vivo toxicity analyses are crucial in demonstrating the efficacy of drugs without toxic or side effects (Winner et al., 2011). It is noteworthy that successful in vivo evaluations using animal models is a requirement for clinical trials in drug development (Mak et al., 2014; Sukhanova et al., 2018). Thus animal models are important in in vivo toxicity and efficacy analysis of drugs.

Nanomedicines, particularly those synthesized via chemical approaches, are likely to exhibit toxic effects on target or nontarget cells (Fard et al., 2015). Also, some bio-synthesized nanomedicines can generate toxic reactions and side effects (Jahangirian et al., 2017; Krishnaraj et al., 2014). Hence, it is necessary to properly assess the in vivo toxicological profiles of nanodrugs and nanomedicines for efficient treatment of diseases (Monteiro-Riviere et al., 2009; Sayes et al., 2007). The possibility to manipulate healthy in vivo models into disease-affected models helps in determining the efficacy of nanomedicines for the treatment of specific diseases (Lieschke & Currie, 2007). This chapter provides an overview of various diabetic in vivo animal models used to assess the toxicity and efficacy of nanomedicines in diabetes treatment.

Diabetic animal models

The first diabetic animal model that was used for diabetes research was pancreatomized dog in 1880s to examine their intestinal function (Rees & Alcolado, 2005). Drug-induced animal models were introduced later to replace conventional pancreatomized models. In 1838 a pyrimidine derivative, called alloxan, as a diabetes drug was

Emerging Nanomedicines for Diabetes Mellitus Theranostics
DOI: https://doi.org/10.1016/B978-0-323-85396-5.00006-3

synthesized (Wöhler & Liebig, 1838) and used to induce necrosis in endocrine beta cells (Dunn et al., 1943; Dunn & McLetchie, 1943). In 1963 STZ, a N–nitroso derivative of glucosamine, also called streptozotocin, was reported to act as a potential drug to induce diabetes in animal models (Rakeiten, 1963). Several spontaneous diabetic animal models were later introduced to evaluate both type 1 and 2 diabetes conditions, which led to the emergence of diabetic mouse models. Mice models are extensively used in diabetes studies to evaluate pharmacokinetic conditions. There are several classes of mice used in diabetes research, depending on the type of diabetes complication and the drug to be evaluated. Nonobese diabetic mouse from JcI-ICR strain of female mice (Makino et al., 1980), biobreeding rat from outbred Wistar rats (Nakhooda et al., 1977), LEW.1AR1/-iddm from congenic Lewis rats with MHC haplotype (Lenzen et al., 2001), and Akita mouse with *C57BL/6NSlc* (Mathews et al., 2002) are some of the mouse models used for in vivo type 1 diabetes studies. Also, mouse models such as ob/ob, db/db, Zuckar fa/fa rats, Goto-kakizaki rats, and Otsuka Long–Evans Tokushima fatty rats have been used as in vivo type 2 diabetic models (Sakata et al., 2012). In addition, hyperleptinemic strains of wild-derived ddY mice (KK and KK-A^Y), New Zealand obese (NZO), TallyHo/Jng, NoncNZO 10/LtJ, and hlAPP are other mice models used for type 2 diabetes studies (King, 2012). Transgenic mice models were introduced in 1990s with insulin, IRS1, and IRS2 gene knockouts for in vivo analysis of both types 1 and 2 diabetes complications (Joshi et al., 1996; Kubota et al., 2000; Tamemoto et al., 1994). Kimba and Akimba mouse models have also been used to study diabetic retinopathy conditions (Okamoto et al., 1997; Tee et al., 2008). Fig. 8.1 shows the historic timeline of various animal and mouse models used for in vivo diabetes studies.

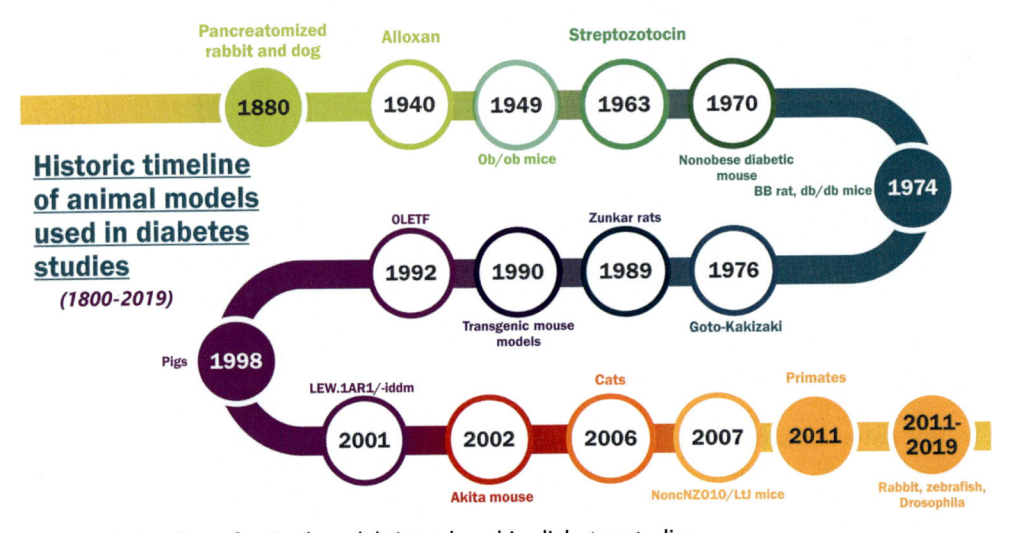

Figure 8.1 Timeline of animal models introduced in diabetes studies.

Rats are more commonly used as pancreas injury models compared to mice (Leahy et al., 1988). Neonatal rats are used as type 2 diabetes models via streptozotocin (STZ) administration (Portha et al., 1974; Tourrel et al., 2001). Desert Gerbil (*Psammomys obesus*) (Bödvarsdóttir et al., 2010) and Nile grass rat (*Arvicanthis niloticus*) (Noda et al., 2010) have been employed as diet-induced rodent models for type 2 diabetes research. Apart from mice and rats, animal models such as pigs (Mellert et al., 1998), dogs (Fisher et al., 2001), and primates (He et al., 2011) have been used to evaluate the complications of pancreatectomy in type 1 diabetes. Cynomolgus monkey has also been used to evaluate type 1 diabetic condition via chemical ablation of beta cells (Dufrane et al., 2006). Animals such as cats (Feline) (Henson & O'Brien, 2006), dog (canine) (Ionut et al., 2009), and several pigs (swine) (Bellinger et al., 2006) strains have been used as in vivo models to study type 2 diabetes conditions.

More recently, rabbits are gaining attention as in vivo animal models to study diabetes conditions. Diabetes complications such as nephropathy (Zhao et al., 2016), atherosclerosis (Matsuura et al., 2017), hind limb ischemia (Sligar et al., 2019), vertebral microvascular permeability, and fat fraction (Hu et al., 2017) can be studied using rabbits as in vivo animal models. Zebrafish and drosophila are newer models for diabetes studies. Zebrafish (*Dania rerio*) are a transparent fish species that are widely used to screen drugs based on their toxicity, especially genotoxicity. Further, they possess a rapid multiplication property along with 69% organ physiology, cell development, and genetic similarity with humans (Jeevanandam et al., 2019). Thus they have been employed in diabetes research to evaluate obesity and hyperglycemia (Morningstar et al., 2017). Zebrafish have been used to evaluate microbiome alteration (Okazaki et al., 2019), protein X amelioration (Zang et al., 2019), obesity-mediated vascular damage (Wiggenhauser & Kroll, 2019), and beta cell excitability in islets (Emfinger et al., 2019) due to diabetes complications. Similar to zebrafish, *Drosophila melanogaster* (fruit fly) has been used to study diabetes (Teleman et al., 2012). This model has been used to study glucose homeostasis (Haselton & Fridell, 2010), insulin resistance (Pick et al., 2017), and obesity (Morris et al., 2012). Animal models such as mouse, *Drosophila*, and zebrafish are major in vivo organisms used to evaluate the toxicity of nanomedicines. *Daphnia magna* crustacean, *Poecilia reticulata* fish, *Oreochromis mossambicus* fish, and mussels *Mytilus galloprovincialis* are some minor animal models used for in vivo toxicity evaluation of nanomedicines (Andra et al., 2019; Shakya et al., 2020).

Toxicity of nanoparticles in animal models
Mice and rats

Metal, metal oxides, carbon, polymer, and nanocomposites are some nanoparticle classes that have been reported to possess antidiabetic ability.

Metal nanoparticles

Various mouse models have been used to study the toxicity and in vivo mechanism of nanoparticles for diabetes treatment. Leu et al. (2012) evaluated the toxicity and diabetic wound healing efficacy of gold nanoparticles synthesized via molecular beam epitaxy method. The nanoparticles size was in the range of 1−30 nm. The nanoparticles were formulated using epigallocatechin gallate and α-lipoic acid with antioxidant property to increase wound healing capability in diabetic mice. In vitro analysis of the gold nanoparticles mixture in Hs68 (Human foreskin fibroblasts) and HaCaT (human keratinocyte cells) showed that the addition of nanoparticles helped to increase cell proliferation and demonstrated wound healing properties. Further, BALB/c mice were used in in vivo analysis, and they demonstrated that the gold nanoparticles possessed enhanced wound healing capability in the skin (Leu et al., 2012). Gold nanoparticles have also been used as a carrier and targeted delivery system for miR155 antagonist macrophage (Jia et al., 2017). The gold nanocarriers possessed efficient macrophage delivery ability at targeted site to restore cardiac function in ovariectomized C57BL/6 diabetic mouse model. Tian et al. (2007) demonstrated that silver nanoparticles possess swift wound healing property using genetically diabetic C57BLKs/J-m + /db, db/db male mice. The silver nanoparticles with 14 ± 9.8 nm diameter were coated in the wound dressing and improved the wound healing process (Tian et al., 2007). Silver nanoparticles have also been used to obtain in vivo biochemical information from female Sprague-Dawley rat tissues via a high spatial resolution−based near-infrared surface-enhanced Raman scattering approach. The study showed that silver nanoparticles can serve as a bioimaging agent to analyze in vivo diabetes complications in tissues (Huang et al., 2013). Copper metal organic framework nanoparticles fabricated and stabilized with folic acid have been demonstrated to improve wound healing ability in db/db diabetic mice (Xiao et al., 2018). Palladium and selenium nanoparticles are additional metal nanoparticles that have been shown to possess antidiabetic activity using mice models (El-Borady et al., 2020; Lushchak et al., 2018).

Metal oxide nanoparticles

Metal oxide nanoparticles such as zinc oxide, iron oxide, titanium dioxide, silicon and cerium oxides have been shown to possess antidiabetic activity via in vivo analysis using mice, rat, and rodent. Luyts et al. (2014) used Bmal1 gene knockout mice with altered procoagulant phenotype and circadian rhythm to evaluate the hemostatic effect of zinc oxide (ZnO) nanoparticles. The results showed that the ZnO nanoparticles improved procoagulant activity, inflammation, and oxidative stress in mice models, and are thus beneficial in enhancing hemostatic and pulmonary activity in diabetic conditions (Luyts et al., 2014). Also, ZnO nanoparticles synthesized using *Hibiscus subdariffa* leaf extract demonstrated to possess antidiabetic activity using streptozotocin-induced diabetic mice by expressing insulin receptors and other pancreatic genes

(Bala et al., 2015). Zgheib et al. (2019) conjugated cerium oxide nanoparticles with microRNA-146a to reduce reactive oxygen species (ROS) and improve antiinflammation of swift dorsal skin wound healing in diabetic db/db mice (Zgheib et al., 2019).

The antidiabetic activity of ZnO nanoparticles has also been evaluated using diabetic rats. Wahba et al. (2016) investigated alterations in the function and histopathology of streptozotocin-induced diabetic rat model and showed that ZnO nanoparticles possess the ability to reverse pancreatic injury by stabilizing serum insulin and blood glucose levels in diabetic rats (Wahba et al., 2016). The ameliorative effect of ZnO nanoparticles has been evaluated using male streptozotocin-induced diabetic albino rats. The results showed that ZnO nanoparticles have the potential to elevate sperm count and act as a shield to protect testicular tissues of diabetic rats from oxidative stress (Afifi et al., 2015). Ali et al. (2020) demonstrated the effects of superparamagnetic iron oxide nanoparticles on glucose homeostasis of type 2 diabetic experimental rat model. The study showed that the magnetic nanoparticles significantly elevated glucose sensing of insulin signaling pathways (Ali et al., 2020). Cerium oxide nanoparticles have also been investigated using streptozotocin-induced diabetic rats and demonstrated the ability to cure diabetes induced sperm and testicular damage via reduction in DNA fragmentation and increment in DNA integrity (Artimani et al., 2018). Contrarily, Mao et al. (2019) showed that titanium dioxide (TiO_2) nanoparticles possess the ability to alter the maternal gut microbiota of Sprague-Dawley female rats during pregnancy and demonstrated to increase blood glucose level with the potential to increase the risk of gestational diabetes (Mao et al., 2019). Silicon oxide nanoparticles have been used to deliver flightless I neutralizing antibody to improve wound healing in streptozotocin-induced diabetic rats for the treatment of diabetic ulcer (Turner et al., 2017).

Carbon-based nanoparticles

Graphene, carbon nanotubes (CNTs), and carbon dots are the main carbon-based nanoparticles that are evaluated using mice/rat models to show their antidiabetic properties. Fu et al. (2019) reported that acellular dermal composite scaffolds incorporated with reduced graphene oxide nanoparticles possess the ability to deliver mesenchymal stem cells to accelerate wound healing in diabetes. The results showed that the graphene oxide nanoparticles supported collagen deposition, robust vascularization, and swift re-epithelialization in healing diabetic wounds (Fu et al., 2019). Polydopamine-reduced graphene oxide nanoparticles incorporated chitosan (in silk fibroin), which possessed enhanced mechanical, antioxidative, and electroactive properties, was demonstrated to be an excellent wound dressing material using a Sprawgue-Dawley rat model (Tang et al., 2019). Graphene quantum dots have demonstrated burn wounds healing potential by increasing angiogenesis, collagen synthesis, and skin layer thickness in Wistar rat models (Haghshenas et al., 2019).

Single/multiwalled CNT embedded in chitosan has shown to possess the ability to improve wound healing in CC-72-line mice models (Kittana et al., 2018). Using a diabetic rat model, it was demonstrated that fullerene (C60) possesses the ability to reduce diet-induced obesity by regulating glucose level (Halenova et al., 2018) and inhibit reactive gliosis in retina (Nedzvetskii et al., 2016). Also, gadofullerene functionalized with water soluble amino acid was shown to improve glucose tolerance and insulin sensitivity in diabetic db/db mice models (Li et al., 2019).

Polymer nanoparticles

Different types of polymeric nanoparticles have shown antidiabetic properties in mice/ rat models. For examples, chitosan/poly(γ-glutamic acid) nanoparticle capsules with encapsulated insulin have been investigated as a controlled oral insulin delivery agent. The study, using streptozotocin-induced diabetic male Wistar rats, showed the potential of the nanoparticles for oral insulin delivery to provide enhanced hypoglycemic effects (Sonaje et al., 2010). Natural curcumin polymer nanoparticles showed the potential to delay cataract by blocking relevant biochemical pathways in streptozotocin-induced diabetic cataract rat models (Grama et al., 2013).

Amphiphilic polymer poly[oligo(ethylene glycol) methacrylate]-2-(dimethylamino) ethyl methacrylate-Methylaluminoxane (PEGMA-DMAEMA-MAO) was used to encapsulate curcumin to generate nanoparticles. The polymeric nanoparticles showed enhanced targeted delivery of curcumin in diabetic male Sprague-Dawley rats and showed the ability to inhibit diabetic neuropathic pain (Jia et al., 2018). Also, nano thermogels, prepared using polyethylene glycol/polyester copolymers to encapsulate a fatty acid—modified antidiabetic polypeptide (called Liraglutide), demonstrated the ability to deliver the polypeptide to the targeted site in a controlled manner (Chen, Hu, et al., 2016; Chen, Li, et al., 2016). A glucagon-like peptide-1 (GLP-1) called Exendin-4 was encapsulated using chondroitin sulfate-g-glycocholic acid as nano-liposomes. *In vivo* analysis using diabetic male Sprague-Dawley rats revealed that the liposomes improved body weight as well as blood lipid concentration (Suzuki et al., 2019). An immunosuppressant (FK506) was encapsulated with 4-dihydroxyphenethylamine (DOPA)-conjugated polymers [such as poly(lactide-co-glycolide)-poly(ethylene glycol)] to generate nanoparticles. The nanoparticles were reported to be tissue-adhesive and demonstrated to be beneficial in multilayered pancreatic islet nano-shielding to improve xenograft survival in a streptozotocin-induced type 1 diabetic C57BL/6 mice model (Pham et al., 2018). In addition, self-assembled chitosan nanoparticles (Mukhopadhyay et al., 2013), B12-coated dextran nanoparticles (Chalasani et al., 2007), polysaccharide nanoparticles (Sarmento et al., 2007), polyurethane-alginate nanoparticles (Bhattacharyya et al., 2016), and nanosized mucoadhesive polymers (Bouttefeux et al., 2016) are other polymeric nanoparticles evaluated using rat/mice models to deliver insulin and antidiabetic drugs.

Nanocomposites

Nanocomposites are gaining importance in diabetes research as antidiabetic agents for drug delivery applications. Shrestha et al. (2016) developed a nanocomposite system using porous silicon nanoparticles loaded with chitosan and enteric hydroxypropyl methylcellulose acetate succinate polymer to deliver GLP-1 incretin hormone and dipeptidylpeptidase-4 enzyme in diabetic rat models. The result showed that oral delivery of the hormone and enzyme via the nanocomposites resulted in 32% reduction in blood glucose level and approximately sixfold pancreatic insulin content increase (Shrestha et al., 2016). Curcumin-loaded poly(ε-caprolactone) nanofibers have shown antioxidative and antiinflammatory properties with the potential for wound healing using diabetic mice (Merrell et al., 2009). Recently, the antidiabetic effect of curcumin-zinc oxide nanocomposite was evaluated using diabetic rats induced by streptozotocin. In this study, adult Wistar albino rats were used for the analysis and showed a significant reduction in blood glucose levels and improved levels of insulin (Abd El-Aziz et al., 2021). Poly(sulfobetaine acrylamide)-silver nanoparticles formulated as zwitterionic nanocomposites have demonstrated to be a good wound dressing agent to cure infected chronic wounds in diabetic male Wistar rat models (Huang et al., 2017). Additionally, mice/rat in vivo models have been used to study doxycycline-amoxicillin/lactate dehydrogenase (El-Ela et al., 2019), chitosan-silver-sericin films (Shah et al., 2019), and copper-containing bioactive glass-egg shell membrane (Li et al., 2016) to evaluate wound healing in diabetic conditions.

Zebrafish

Zebrafish are extensively employed as a model for initial evaluation of the toxicity of nanoparticles as diabetes drug carriers or as antidiabetic agents. It is a transparent fish and particularly used to evaluate the genotoxicity of nanoparticles (Jeevanandam et al., 2019; Modarresi Chahardehi et al., 2020). The initial screening process using zebrafish helps to reduce the number of samples that are evaluated in preclinical studies, removing genotoxic nanoparticles as diabetic nanomedicine (Marins et al., 2019).

Metal and metal oxide nanoparticles

Gold and silver nanoparticles are widely evaluated for genotoxicity using zebrafish. The analysis has useful in characterizing specific morphologies of nanoparticles based on their genotoxicity properties. Zebrafish embryos have been used to evaluate the genotoxicity of gold nanoparticles with narrow core diameters of 0.8−5.8 nm and functionalized with positively charged N,N,N-trimethylammonium ethanethiol ligands. The developmental zebrafish assay showed that larger nanoparticles displayed more toxicity toward the embryos via a surface area−mediated size-dependent mechanism (Truong et al., 2019). Sivaji and Kannan (2019) prepared 70−90 nm monodispersed gold nanoparticles functionalized with polyethylene glycol and polysorbate as a therapeutic agent to cure neurodegenerative diseases. The nanoparticles were evaluated to bypass the blood—brain barrier, revealing that the gold nanoparticles led to focal cell aggregation in pia matter (Sivaji & Kannan, 2019). It can extrapolate from current literature show that in vivo toxicity evaluation using zebrafish will be useful in developing nontoxic gold nanoparticles as implantable glucose sensors in diabetic patients.

Silver nanoparticles have also been evaluated using zebrafish to screen toxic particles. Campoy et al. (2018) reported 10—12 nm spherical silver nanoparticles synthesized from the bark extract of *Eysenhardtia polystachya* using methanol and aqueous mixture as solvent. The nanoparticles promoted insulin secretion, the survival of pancreatic beta cells, hyperlipidemia, and enhanced hyperglycemia in a zebrafish diabetic model induced by glucose (Garcia Campoy et al., 2018). Gutierrez et al. (2018) demonstrated the synthesis of silver nanoparticles using *E. polystachya* methanolic bark extract and investigated their ability to inhibit the formation of advanced glycation end product (AGE) in diabetic models. Zebrafish was used as the in vivo model to test diabetic activity by inducing diabetes via high blood glucose concentration. The results revealed that the silver nanoparticles reduced AGE formation, amyloid cross-beta structure, amadorin or fructosamine product, protein carbonyl content, and N^{ε}-(carboxymethyl)-lysine, and elevated total thiol group in the hyperglycemic zebrafish (Gutierrez et al., 2018). In addition, nanoparticles of platinum (Labrador-Rached et al., 2018), selenium (Chandramohan et al., 2018), and copper (Yen et al., 2019) have also been evaluated using zebrafish to screen highly genotoxic nanoparticles.

The genotoxicity of metal oxide nanoparticles has also been evaluated extensively using zebrafish. Kteeba et al. (2018) reported that zinc oxide (ZnO) nanoparticles possess the ability to alter the development of neurons and the vascular system of zebrafish, potentially resulting in genotoxicity and facilitating the formation of secondary abnormal motoneurons phenotypes (Kteeba et al., 2018). Aksakal and Ciltas (2019) showed that copper oxide (CuO) nanoparticles possess the ability to cause severe developmental abnormalities, induce mortality, and affect oxidative stress as well as immune-related gene expression in zebrafish embryos (Aksakal & Ciltas, 2019). Iron oxide nanoparticles of size 100—250 nm were synthesized using spinach extract, and their toxicity was evaluated using zebrafish. The study showed that the iron oxide nanoparticles were toxic to zebrafish embryos, causing mortality and delay in hatching (Hafiz et al., 2018). The toxicity of titanium dioxide (TiO_2) nanoparticles synthesized using *Sesbania grandiflora* extract was evaluated using zebrafish embryos. The toxicity of nanosized oxides of cerium (Bhagat et al., 2018), zirconium (Samuel et al., 2019), and silicon (Vranic et al., 2019) particles have been evaluated using different stages of zebrafish to screen their toxicity.

Carbon and polymer nanoparticles

The toxicity and efficacy of various types of carbon-based nanoparticles have been evaluated using zebrafish. Li et al. (2015) reported that the exposure of carbon nanoparticles, including quantum dots and carbon tubes, can lead to brain and gonadal alterations in zebrafish. They analyzed Fourier-transform infrared spectra to evaluate toxic effects (Li et al., 2015). Similarly, Chen et al. (2015) demonstrated the in vivo toxicity of graphene oxide nanoparticles using embryos of zebrafish. The results from the study indicated that the toxicity of graphene oxide resulted from the envelopment of zebrafish embryo chorion via interactions of the hydroxyl group and pore canal blockage of chorionic membrane, leading to marked hypoxia and delay in hatching of eggs (Chen, Hu, et al., 2016; Chen, Li, et al., 2016). The toxicity analysis of both single- and multiwalled CNTs using zebrafish has shown that their toxic mechanism is based on surface charge (Gilbertson et al., 2016) and the medium in which they are suspended (Girardi et al., 2017). Carbon-based quantum dots have been used in fluorescence imaging (Kang et al., 2015) and as bio-probes (Zhang et al., 2016) in zebrafish. However, they have been reported to exhibit toxicity via selective antioxidant enzyme activity inhibition (Deng et al., 2019) and perturbation effects on hydrocarbon receptor pathway (Zhang et al., 2017). The toxicity of fullerenes has also been evaluated using zebrafish embryos. The results indicated that the presence of benzo(alpha) pyrene in fullerene triggers oxidative stress and influences their bioavailability as well as toxicity in zebrafish embryos (Della Torre et al., 2018).

Polymers are widely researched in the development of nanodrugs from diabetes treatment. The toxicity of various nano polymers has been assessed using zebrafish. Vong et al. (2016) reported the use of zebrafish embryos to evaluate the toxicity of ROS scavenging nanoparticles synthesized using methoxy-poly(ethylene glycol)-*b*-poly[4-(2,2,6,6-tetramethylpiperidine-1-pxyl)oxymethylstyrene] polymer. The result showed that the polymer possessed less toxicity and exhibited antioxidant property (Vong et al., 2016). Zebrafish has also been used to assess the toxicity of double-shell polymeric nanocapsules synthesized using hyaluronic acid and protamine. The result showed that the polymeric nanocapsules exhibited lower toxicity toward zebrafish and were able to cross the chorion and skin of zebrafish (Teijeiro-Valiño et al., 2017). These reported studies demonstrate the potential of zebrafish for in vivo toxicological evaluation of nanodrugs, facilitating their applications in evaluating diabetes nanomedicines.

Nanocomposites

In vivo toxicity of various nanocomposites has also been assessed using zebrafish. Prakash et al. (2019) investigated the toxicity of reduced graphene oxide-titanium dioxide nanocomposites using zebrafish embryo and larva. The results revealed that the nanocomposites exhibited concentration-dependent teratogenicity and cardiotoxicity toward zebrafish eggs and larva via generation of ROS (Prakash et al., 2019). Also, Khan et al. (2019) synthesized iron oxide–silicon dioxide nanocomposites that were coated with chitosan and gadolinium oxyfluoride doped with cerium and terbium ions. *In vivo* toxicity analysis in zebrafish showed that the nanocomposites were less toxic and uptake of the particles was mainly through oral exposure (Khan et al., 2019). Chitosan-zinc oxide (Younes et al., 2019), gold-iron oxide-MXene (Hussein et al., 2019), graphene oxide-titania-loaded nafion (Pecoraro et al., 2018), glycine-silica nanoparticles (Dumitrescu et al., 2017), and titanium dioxide-chondroitin-4-sulfate (Kandiah et al., 2015) are additional nanocomposites that have been evaluated using zebrafish.

Drosophila

D. melanogaster (fruit fly) belongs to the order Diptera and class Insecta, and has been used to evaluate the genotoxicity of nanodrug formulations. Alaraby et al. (2019) assessed the genotoxicity of silver nanoparticles using *Drosophila* and demonstrated that the nanoparticles can cross intestinal barriers depending on their size, affect hemocytes, release intracellular ROS, and lead to primary DNA damage (Alaraby et al., 2019). Carmona et al. (2015) evaluated the toxicity of copper oxide nanoparticles using somatic cells of *Drosophila* and revealed that the nanoparticles exhibited oxidative stress—mediated genotoxicity to cause DNA nicking, mutation, and recombination (Carmona et al., 2015). Anand et al. (2019) reported the toxicity of aluminum oxide nanoparticles investigated with *Drosophila*. The results showed that chronic exposure of the nanoparticles can lead to behavioral effects, phenotypic abnormalities in progeny flies, absence of appendages, expression of proteins related to alterations in the morphogenesis of the digestive tract, striated differentiation of muscle cells, chromatin organization regulation, phototransduction, and unwinding of DNA duplex (Anand et al., 2019). Dan et al. (2019) assessed the in vivo toxicity of hydroxyapatite nanoparticles using *Drosophila* and the result revealed that they are nontoxic even at high concentration and did not affect their climbing and crawling patterns in larvae and adult fruit flies (Dan et al., 2019). All the studies showed that *Drosophila* can be a potential in vivo model to evaluate the toxicity of nanomedicines for diabetes therapy.

Primates

Primates, including monkey, Chimpanzee, and others, can also be used to evaluate the toxicity of nanoparticles and their efficacy in diabetes treatment (Simon et al., 1993). Ye et al. (2012) reported the use of Rhesus macaques, a common nonhuman primate, as an animal model for in vivo toxicity evaluation of phospholipid micelle encapsulated cadmium selenide-cadmium sulfide-zinc sulfide quantum dots. The results showed that the quantum dots were less toxic without any abnormalities or side effects. However, the cadmium ion existed in the spleen, liver, and kidneys of the macaques for about 90 days, which could lead to some side effects (Ye et al., 2012). Kotb et al. (2016) showed that gadolinium nanoparticles are nontoxic to nonhuman primates and did not exhibit any dose, sex, or time-related toxicity effects (Kotb et al., 2016). Baboons have been used as a nonhuman primate model to evaluate the toxicity of DNA nanoparticles conjugated with polylysine and encapsulated with polyethylene glycol. The result showed that the nanoparticles were less toxic to the eyes of the primates and could be used as a potential drug for nonviral ocular therapy of retinal diseases (Kelley et al., 2018). *Cebus apella* has been used as a nonhuman primate model for the evaluation of organic effects associated with a lipid encapsulated paclitaxel nanoparticle system. The result showed that free paclitaxel was highly toxic and exhibited reactions such as skin flaking, diarrhea, vomiting, decreased physical activity, and 70% loss of body hair (Feio et al., 2017). These studies support the use of nonhuman primate models to evaluate the toxicity of nanoparticles. However, hurdles associated with obtaining the needed ethical approval for these higher animals, compared to mice, are a major limitation to their use in diabetes nanodrugs assessments. Even with these limitations, there are reports showing the use of nonhuman primate models for the evaluation of antidiabetic drugs with strict ethical procedures (Orlando et al., 2017; Song et al., 2016). Due to stringent ethical procedures, primate models are mostly used to evaluate islet graft transplantation and very rarely to evaluate the safety of antidiabetic drugs (Safley et al., 2018).

Other animal models

There are several other animal models that have been used to evaluate the toxicity of nanoparticles in vivo models. Mesak et al. (2018) used *Gallus gallus domesticus* (Phasianidae) chicks to predict the environmental toxicity of ZnO nanoparticles. The study investigated the behavioral alterations of chicks due to intraperitoneal administration of the nanoparticles. The results showed that low concentrations of nanoparticles possessed the ability to bypass the blood—brain barrier to affect the neuronal mechanism and structures, altering the defensive response (Mesak et al., 2018). Nazaktabar et al. (2017) used chicken embryos to assess the antiviral efficacy and toxicity of polyrhodanine nanoparticles against Newcastle disease. The results showed no potential toxic effects on the kidney, heart, liver, spleen, nerve tissues, and bursa of Fabricius in chicken embryo and exhibited elevated antiviral efficacy (Nazaktabar et al., 2017). *Columba livia* (Pigeon) has also been used as an animal model to evaluate the toxic effect of silver nanoparticles and the result showed dose-dependent toxicity, especially in the liver (Tashakori Miyanroudi & Arabi, 2016).

Rabbits are also a potential in vivo animal model. However, the requirement of stringent ethical clearance made rabbits less utilized for nanoparticle toxicity analysis (Rathore et al., 2020). Polyol-conjugated cobalt-zinc ferrite nanoparticles haves been determined to exhibit acute toxicity in the liver, kidneys, and lungs using New Zealand rabbits (Hanini et al., 2016). Also, bevacizumab drug, used for retinal and choroidal neovascularization treatment, was encapsulated with polymeric poly(lactic-co-glycolic acid) (PLGA) nanoparticles and their toxicity was assessed using rabbit. The results showed biocompatibility and less toxic effects (Varshochian et al., 2015). Moreover, silver nanoparticle incorporated in chitosan membranes (Shao et al., 2017), zinc oxide nanoparticles (Kim et al., 2016), chitosan-coated PLGA nanoparticles (Pandit et al., 2017), graphene oxide (Wu et al., 2016), and solid lipid nanoparticles (Leonardi et al., 2015) are other nanoparticles that have been evaluated using rabbits for their in vivo toxicity. Further, frog models such as bullfrog tadpoles (Oliveira et al., 2019), South African clawed frog (*Xenopus laevis*) (Coll et al., 2018), *Rhinella arenarum* (Lajmanovich et al., 2018), and marsh frog (*Pelophylax ridibundus*) (Falfushynska et al., 2016) have also been used as animal models for *in vivo* toxicity evaluation of nanoparticles. In addition, cockroaches (Khatami et al., 2019), land snails (Khatami et al., 2019), benthic aquatic snails (Oliveira-Filho et al., 2016), freshwater *Radix luteola* (Ali et al., 2016), freshwater Asian clam *Corbicula fluminea* (Cid et al., 2015), dogs (Ashwini & Mahalingam, 2020; Danmaigoro et al., 2018), cats (Abdoon et al., 2016), brine shrimp (Arumugam et al., 2019), and pigs (Lin et al., 2016; Nabofa et al., 2018) are other animals that have been used for the purpose of toxicity analysis of nanoparticles. Even though various animal models have been used to evaluate the toxicity of nanoparticles, only mice and zebrafish are used extensively to evaluate diabetic complications and antidiabetic activity of nanoparticles.

Toxicity of nanoformulations toward diabetic animal models

Nanoformulations can be developed from nanoparticles or nanosized drugs encapsulated in biodegradable nanomaterials for insulin delivery or for targeting diabetic cells. Similar to nanoparticles, the toxicity and antidiabetic efficacy of nanoformulations are evaluated using in vivo animal models. *Psoralea corylifolia* herbal seed extract formulated silver nanoparticles were determined to possess antidiabetic activity using female albino diabetic mice models. The results showed that the silver nanoformulation was safe and possessed antidiabetic properties (Shanker et al., 2017). Maity et al. (2017) demonstrated the antidiabetic efficacy of naringenin drug formulated with alginate-coated chitosan nanoparticles with a core—shell morphology using streptozotocin-induced diabetic rat models. They reported that the mucoadhesive nanoparticles helped in controlled oral delivery of the antidiabetic drug without any significant toxicity, demonstrating potential applications in dyslipidemia, hyperglycemia, and for iron in hemoglobin-mediated oxidative stress treatment in type 1 diabetes (Maity et al., 2017). The antidiabetic drug, methotrexate, formulated using lipid nanoparticles has been investigated to be beneficial for the treatment of diabetic cardiomyopathy using rat models. The study showed that the nanoformulated antidiabetic drug helped in reducing blood glucose level, increased insulin secretion, reduced cardiac hypertrophy, inflammation, and myocardial fibrosis without any adverse toxic effects (Cavalcante Maranhao et al., 2017). Zebrafish models have also been used to evaluate the toxicity of nanoformulations such as thioridazine-encapsulated PLGA nanoparticles to improve rifampicin therapy (Vibe et al., 2016), doxorubicin-loaded mixed micelles (Calienni et al., 2018), glycine-coated silica nanoparticles (Dumitrescu et al., 2017), and glyco-gold nanoparticles (Sangabathuni et al., 2017). It can be noted that mice/rats and zebrafish are commonly used in diabetes studies for toxicity and efficacy evaluations of nanoformulations. The requirements of animal ethical practices act as a hindrance other animal models. It is easier to induce diabetes in normal rats, compared to other animals, and this model has become a standard procedure for evaluating the efficacy of antidiabetic drugs. Rat models are used as a primary in vivo animal model to assess the antidiabetic efficacy of nanoparticles and zebrafish are used for initial screening of toxic nanoparticles.

Toxicity mechanism of nanoparticles toward diabetic animal models

Researchers are using animal models rather than cell lines to evaluate the biodistribution and toxicity of nanomedicines in various organs when administered via various routes (Johnson et al., 2001). The data from animal studies are authentic and reliable for extrapolation compared to cell cultures (Lanford & Bigger, 2002). Rat, rabbit, bird, primate, and frog have been used to assess the effects of nanoparticles on organs such as lung, trachea, spleen, liver, testis/ovary, brain, heart, and kidney (Groenink et al., 2015). Electron microscopes, Magnetic resonance imaging (MRI), hematoxylin and eosin staining methods, and immunohistochemistry are some of the approaches used to evaluate the in vivo effects of nanoparticles in animal models (Gambini et al., 2015). MRI can be used to evaluate the biocompatibility and bioavailability of nanoparticles in animal models (Haensel et al., 2015). The analysis can also provide data on the pharmacokinetics of nanomedicines in live animal models. Behavioral alterations and morphological changes in organs upon nanoparticles exposure can be determined using animal models (Mears et al., 2015). Fig. 8.2 shows a summary of key data that can be obtained from in vivo animal models to analyze toxicity and antidiabetic effects of nanoparticles.

A nanodrug or nanomedicine can be administered into an organism via oral, injection, intramuscular injection, topical medication, parenteral, inhalation, rectal, sublingual, nasal, buccal, systematic, intrathecal, transdermal, intravaginal, intraosseous infusion, epidural, insufflation, intravitreal and extra-amniotic approaches (Zunhammer et al., 2017). It is possible to evaluate the efficacy of nanodrugs via these administration routes using mice as animal models. Whilst there is no one specific mechanism that leads to the toxicity of all nanodrugs in animal models, the most common mechanism relates to the generation of ROS at the cellular level. This can

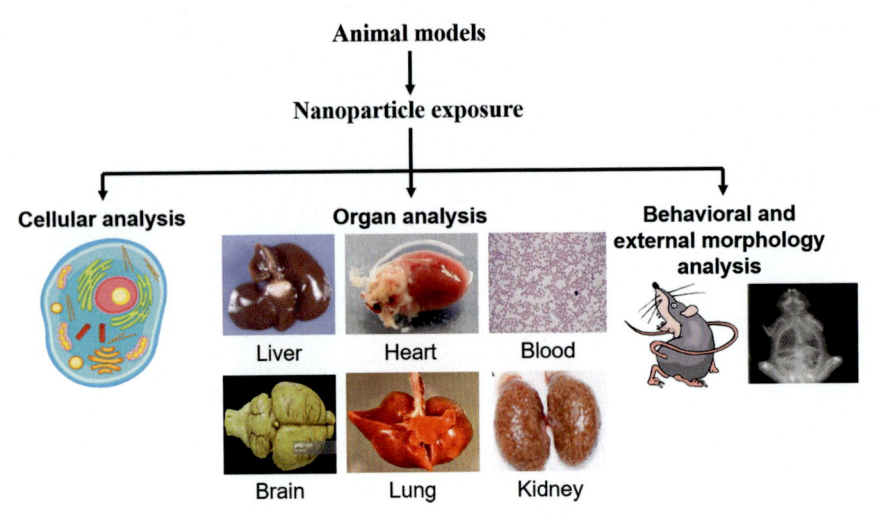

Figure 8.2 Summary of data that can be obtained from animal (mice) model for in vivo toxicity and antidiabetic evaluation of nanoparticles.

eventually lead to cell membrane disruption, cell organelle damage, nucleus degradation, and DNA breakage (Bahadar et al., 2016).

Future perspectives

In the perspective of pharmaceutical research, animal models are highly beneficial to evaluate the pharmacokinetics of nanomedicines and antidiabetic efficiency. However, ethical issues associated with the use of animals continue to be a major concern. An approach to circumvent this has been via in silico analysis. In silico animal models are being developed to evaluate the pharmacokinetics of newly developed drugs. Manshian et al. (2017) demonstrated that iron-doped zinc oxide nanoparticles can lead to cancer-specific toxicity in a preclinical rodent model using in silico optimal dissolution kinetics design (Manshian et al., 2017). These types of in silico animal models can be employed to predict the toxicity of new drug materials to screen drugs that are highly toxic (Raies & Bajic, 2016). In silico models can also be used to evaluate the treatment efficacy of diseases, especially against diabetes complications. Platania et al. (2019) have used an in silico animal model to evaluate the patterns of retinal and circulating miRNA expression in diabetic retinopathy (Platania et al., 2019). Also, Srinivasan et al. (2018) reported that quercetin extracted from *Phyllanthus emblica* fruit possess antidiabetic properties using in silico approaches (Srinivasan et al., 2018). However, in silico models and simulations are not yet recognized as a standard standalone approach for pharmaceutical evaluations. In silico approaches still require in vivo validation in animal models.

Conclusion

This chapter presents an overview of various in vivo animal models that are used for the evaluation of toxicity and antidiabetic potential of nanomedicines. Such in vivo studies are critical to nanodrugs discovery processes and will facilitate understanding of the toxicity and antidiabetic mechanisms of nanomedicines. With the current growth in nanomedicines, new strategies for in vivo evaluation of efficacy and toxicity, including in silico approaches, are being unearthed to address ethical concerns associated with the use of animals. In silico animal model−based computational simulation will play a significant role in evaluating the toxicity and antidiabetic properties of nanoparticles.

References

Abd El-Aziz, S. M., Raslan, M., Afify, M., Abdelmaksoud, M. D. E., & El-Nesr, K. A. (2021). Antidiabetic effects of curcumin/zinc oxide nanocomposite in streptozotocin-induced diabetic rats. *IOP Conference Series: Materials Science and Engineering, 1046*(1), 012023. Available from https://doi.org/10.1088/1757-899x/1046/1/012023.

Abdoon, A. S., Al-Ashkar, E. A., Kandil, O. M., Shaban, A. M., Khaled, H. M., El Sayed, M. A., El Shaer, M. M., Shaalan, A. H., Eisa, W. H., & Eldin, A. A. G. (2016). Efficacy and toxicity of plasmonic photothermal therapy (PPTT) using gold nanorods (GNRs) against mammary tumors in dogs and cats. *Nanomedicine: Nanotechnology, Biology and Medicine, 12*(8), 2291−2297.

Afifi, M., Almaghrabi, O. A., & Kadasa, N. M. (2015). Ameliorative effect of zinc oxide nanoparticles on antioxidants and sperm characteristics in streptozotocin-induced diabetic rat testes. *BioMed Research International, 2015*, 6. Available from https://doi.org/10.1155/2015/153573.

Aksakal, F. I., & Ciltas, A. (2019). Impact of copper oxide nanoparticles (CuO NPs) exposure on embryo development and expression of genes related to the innate immune system of zebrafish (*Danio rerio*). *Comparative Biochemistry and Physiology, Part C: Toxicology & Pharmacology, 223*, 78−87. Available from https://doi.org/10.1016/j.cbpc.2019.05.016.

Alaraby, M., Romero, S., Hernández, A., & Marcos, R. (2019). Toxic and genotoxic effects of silver nanoparticles in *Drosophila*. *Environmental and Molecular Mutagenesis, 60*(3), 277−285. Available from https://doi.org/10.1002/em.22262.

Ali, D., Ali, H., Alarifi, S., Masih, A. P., Manohardas, S., & Hussain, S. A. (2016). Eco-toxic efficacy of nano-sized magnesium oxide in freshwater snail *Radix leuteola* L. *Fresenius Environmental Bulletin, 25*(4), 1234−1242.

Ali, L. M. A., Shaker, S. A., Pinol, R., Millan, A., Hanafy, M. Y., Helmy, M. H., Kamel, M. A., & Mahmoud, S. A. (2020). Effect of superparamagnetic iron oxide nanoparticles on glucose homeostasis on type 2 diabetes experimental model. *Life Sciences, 245*, 117361. Available from https://doi.org/10.1016/j.lfs.2020.117361.

Anand, A. S., Gahlot, U., Prasad, D. N., Amitabh., & Kohli, E. (2019). Aluminum oxide nanoparticles mediated toxicity, loss of appendages in progeny of *Drosophila melanogaster* on chronic exposure. *Nanotoxicology, 13*(7), 977−989. Available from https://doi.org/10.1080/17435390.2019.1602680.

Andra, S., Balu, S. K., Jeevanandham, J., Muthalagu, M., Vidyavathy, M., San Chan, Y., & Danquah, M. K. (2019). Phytosynthesized metal oxide nanoparticles for pharmaceutical applications. *Naunyn-Schmiedeberg's Archives of Pharmacology, 392*, 755−771.

Artimani, T., Amiri, I., Soleimani Asl, S., Saidijam, M., Hasanvand, D., & Afshar, S. (2018). Amelioration of diabetes-induced testicular and sperm damage in rats by cerium oxide nanoparticle treatment. *Andrologia, 50*(9), e13089. Available from https://doi.org/10.1111/and.13089.

Arumugam, D. G., Sivaji, S., Dhandapani, K. V., Nookala, S., & Ranganathan, B. (2019). Panchagavya mediated copper nanoparticles synthesis, characterization and evaluating cytotoxicity in brine shrimp. *Biocatalysis and Agricultural Biotechnology (Reading, Mass.), 19*, 101132.

Ashwini, D., & Mahalingam, G. (2020). Green synthesized metal nanoparticles, characterization and its antidiabetic activities—A review. *Research Journal of Pharmacy and Technology, 13*(1), 468–474.

Bahadar, H., Maqbool, F., Niaz, K., & Abdollahi, M. (2016). Toxicity of nanoparticles and an overview of current experimental models. *Iranian Biomedical Journal, 20*(1), 1.

Bala, N., Saha, S., Chakraborty, M., Maiti, M., Das, S., Basu, R., & Nandy, P. (2015). Green synthesis of zinc oxide nanoparticles using *Hibiscus subdariffa* leaf extract: Effect of temperature on synthesis, antibacterial activity and anti-diabetic activity. *RSC Advances, 5*(7), 4993–5003. Available from https://doi.org/10.1039/C4RA12784F.

Bellinger, D. A., Merricks, E. P., & Nichols, T. C. (2006). Swine models of type 2 diabetes mellitus: Insulin resistance, glucose tolerance, and cardiovascular complications. *ILAR Journal, 47*(3), 243–258.

Bhagat, J., Greeshma, S. S., & Shyama, S. K. (2018). Genotoxicity of cerium oxide nanoparticle in zebrafish and green Mussel *Perna viridis* using alkaline comet assay. *LIFE: International Journal of Health and Life-Sciences, 4*(3).

Bhattacharyya, A., Mukherjee, D., Mishra, R., & Kundu, P. P. (2016). Development of pH sensitive polyurethane—alginate nanoparticles for safe and efficient oral insulin delivery in animal models. *RSC Advances, 6*(48), 41835–41846. Available from https://doi.org/10.1039/C6RA06749B.

Bödvarsdóttir, T. B., Hove, K. D., Gotfredsen, C. F., Pridal, L., Vaag, A., Karlsen, A. E., & Petersen, J. S. (2010). Treatment with a proton pump inhibitor improves glycaemic control in *Psammomys obesus*, a model of type 2 diabetes. *Diabetologia, 53*(10), 2220–2223.

Bouttefeux, O., Beloqui, A., & Preat, V. (2016). Delivery of peptides via the oral route: Diabetes treatment by peptide-loaded nanoparticles. *Current Pharmaceutical Design, 22*(9), 1161–1176.

Calienni, M. N., Cagel, M., Montanari, J., Moretton, M. A., Prieto, M. J., Chiappetta, D. A., & del Valle Alonso, S. (2018). Zebrafish (*Danio rerio*) model as an early stage screening tool to study the biodistribution and toxicity profile of doxorubicin-loaded mixed micelles. *Toxicology and Applied Pharmacology, 357*, 106–114.

Campoy, A. H. G., Gutierrez, R. M. P., Manriquez-Alvirde, G., & Ramirez, A. M. (2018). Protection of silver nanoparticles using Eysenhardtia polystachya in peroxide-induced pancreatic β-cell damage and their antidiabetic properties in zebrafish. *International journal of nanomedicine, 13*, 2601.

Carmona, E. R., Inostroza-Blancheteau, C., Obando, V., Rubio, L., & Marcos, R. (2015). Genotoxicity of copper oxide nanoparticles in *Drosophila melanogaster*. *Mutation Research/Genetic Toxicology and Environmental Mutagenesis, 791*, 1–11. Available from https://doi.org/10.1016/j.mrgentox.2015.07.006.

Cavalcante Maranhao, R., França Marques, A., Guido, M. C., Rufo Tavares, E., Lima Bispo, D., Dantas Tavares de Melo, M., & Maria Salemi, V. (2017). Effects of treatment with methotrexate associated to lipid nanoparticles on diabetic cardiomyopathy in rats. *Atherosclerosis, 263*, e48. Available from https://doi.org/10.1016/j.atherosclerosis.2017.06.165.

Chalasani, K. B., Russell-Jones, G. J., Jain, A. K., Diwan, P. V., & Jain, S. K. (2007). Effective oral delivery of insulin in animal models using vitamin B12-coated dextran nanoparticles. *Journal of Controlled Release, 122*(2), 141–150.

Chandramohan, S., Sundar, K., & Muthukumaran, A. (2018). Hollow selenium nanoparticles from potato extract and investigation of its biological properties and developmental toxicity in zebrafish embryos. *IET Nanobiotechnology, 13*(3), 275–281.

Chen, Y., Hu, X., Sun, J., & Zhou, Q. (2016b). Specific nanotoxicity of graphene oxide during zebrafish embryogenesis. *Nanotoxicology, 10*(1), 42–52. Available from https://doi.org/10.3109/17435390.2015.1005032.

Chen, Y., Li, Y., Shen, W., Li, K., Yu, L., Chen, Q., & Ding, J. (2016a). Controlled release of liraglutide using thermogelling polymers in treatment of diabetes. *Scientific Reports, 6*, 31593. Available from https://doi.org/10.1038/srep31593.

Chen, Y., Ren, C., Ouyang, S., Hu, X., & Zhou, Q. (2015). Mitigation in multiple effects of graphene oxide toxicity in zebrafish embryogenesis driven by humic acid. *Environmental Science & Technology, 49*(16), 10147–10154.

Cid, A., Picado, A., Correia, J. B., Chaves, R., Silva, H., Caldeira, J., de Matos, A. P. A., & Diniz, M. S. (2015). Oxidative stress and histological changes following exposure to diamond nanoparticles in the freshwater Asian clam *Corbicula fluminea* (Müller, 1774). *Journal of Hazardous Materials*, *284*, 27−34. Available from https://doi.org/10.1016/j.jhazmat.2014.10.055.

Coll, C. S. P., Pabón-Reyes, C., Meichtry, J. M., & Litter, M. I. (2018). Monitoring of toxicity of As (V) solutions by AMPHITOX test without and with treatment with zerovalent iron nanoparticles. *Environmental Toxicology and Pharmacology*, *60*, 138−145.

Dan, P., Sundararajan, V., Ganeshkumar, H., Gnanabarathi, B., Subramanian, A. K., Venkatasubu, G. D., Ichihara, S., Ichihara, G., & Sheik Mohideen, S. (2019). Evaluation of hydroxyapatite nanoparticles - induced in vivo toxicity in *Drosophila melanogaster*. *Applied Surface Science*, *484*, 568−577. Available from https://doi.org/10.1016/j.apsusc.2019.04.120.

Danmaigoro, A., Selvarajah, G. T., Noor, M., Hezmee, M., Mahmud, R., Bakar, A., & Zuki, M. (2018). Toxicity and safety evaluation of doxorubicin-loaded cockleshell-derived calcium carbonate nanoparticle in dogs. *Advances in Pharmacological and Pharmaceutical Sciences*, *2018*.

Della Torre, C., Maggioni, D., Ghilardi, A., Parolini, M., Santo, N., Landi, C., Madaschi, L., Magni, S., Tasselli, S., Ascagni, M., Bini, L., La Porta, C., Del Giacco, L., & Binelli, A. (2018). The interactions of fullerene C60 and benzo(α)pyrene influence their bioavailability and toxicity to zebrafish embryos. *Environmental Pollution*, *241*, 999−1008. Available from https://doi.org/10.1016/j.envpol.2018.06.042.

Deng, S., Fu, A., Junaid, M., Wang, Y., Yin, Q., Fu, C., Liu, L., Su, D.-S., Bian, W.-P., & Pei, D.-S. (2019). Nitrogen-doped graphene quantum dots (N-GQDs) perturb redox-sensitive system via the selective inhibition of antioxidant enzyme activities in zebrafish. *Biomaterials*, *206*, 61−72.

Drasler, B., Sayre, P., Steinhäuser, K. G., Petri-Fink, A., & Rothen-Rutishauser, B. (2017). In vitro approaches to assess the hazard of nanomaterials. *NanoImpact*, *8*, 99−116.

Dufrane, D., van Steenberghe, M., Guiot, Y., Goebbels, R.-M., Saliez, A., & Gianello, P. (2006). Streptozotocin-induced diabetes in large animals (pigs/primates): Role of GLUT2 transporter and β-cell plasticity. *Transplantation*, *81*(1), 36−45.

Dumitrescu, E., Karunaratne, D. P., Prochaska, M. K., Liu, X., Wallace, K. N., & Andreescu, S. (2017). Developmental toxicity of glycine-coated silica nanoparticles in embryonic zebrafish. *Environmental Pollution*, *229*, 439−447.

Dunn, J. S., Kirkpatrick, J., McLetchie, N. G. B., & Telfer, S. V. (1943). Necrosis of the islets of langerhans produced experimentally. *Journal of Pathology and Bacteriology*, *55*, 245−257.

Dunn, J. S., & McLetchie, N. G. B. (1943). Experimental alloxan diabetes in the rat. *The Lancet*, *242*(6265), 384−387.

El-Borady, O. M., Othman, M. S., Atallah, H. H., & Moneim, A. E. A. (2020). Hypoglycemic potential of selenium nanoparticles capped with polyvinyl-pyrrolidone in streptozotocin-induced experimental diabetes in rats. *Heliyon*, *6*(5), e04045.

El-Ela, F. I. A., Farghali, A. A., Mahmoud, R. K., Mohamed, N. A., & Moaty, S. A. A. (2019). New approach in ulcer prevention and wound healing treatment using doxycycline and amoxicillin/LDH nanocomposites. *Scientific Reports*, *9*(1), 6418.

Emfinger, C. H., Lőrincz, R., Wang, Y., York, N. W., Singareddy, S. S., Ikle, J. M., Tryon, R. C., McClenaghan, C., Shyr, Z. A., Huang, Y., Reissaus, C. A., Meyer, D., Piston, D. W., Hyrc, K., Remedi, M. S., & Nichols, C. G. (2019). Beta-cell excitability and excitability-driven diabetes in adult Zebrafish islets. *Physiological Reports*, *7*(11), e14101. Available from https://doi.org/10.14814/phy2.14101.

Falfushynska, H., Gnatyshyna, L., Fedoruk, O., Sokolova, I. M., & Stoliar, O. (2016). Endocrine activities and cellular stress responses in the marsh frog *Pelophylax ridibundus* exposed to cobalt, zinc and their organic nanocomplexes. *Aquatic Toxicology*, *170*, 62−71.

Fard, J. K., Jafari, S., & Eghbal, M. A. (2015). A review of molecular mechanisms involved in toxicity of nanoparticles. *Advanced Pharmaceutical Bulletin*, *5*(4), 447.

Feio, D. C. A., de Oliveira, N. C. L., Pereira, E. L. R., Morikawa, A. T., Muniz, J. A. P. C., Montenegro, R. C., Alves, A. P. N. N., de Lima, P. D. L., Maranhão, R. C., & Burbano, R. R. (2017). Organic effects of associating paclitaxel with a lipid-based nanoparticle system on a nonhuman primate, *Cebus apella*. *International Journal of Nanomedicine*, *12*, 3827−3837. Available from https://doi.org/10.2147/IJN.S129153.

Fisher, S. J., Shi, Z. Q., Lickley, H. L., Efendic, S., Vranic, M., & Giacca, A. (2001). Low-dose IGF-I has no selective advantage over insulin in regulating glucose metabolism in hyperglycemic depancreatized dogs. *Journal of Endocrinology, 168*(1), 49–58.

Flanagan, R. J., Cuypers, E., Maurer, H. H., & Whelpton, R. (2020). *Fundamentals of analytical toxicology: Clinical and forensic.* John Wiley & Sons.

Fu, J., Zhang, Y., Chu, J., Wang, X., Yan, W., Zhang, Q., & Liu, H. (2019). Reduced graphene oxide incorporated acellular dermal composite scaffold enables efficient local delivery of mesenchymal stem cells for accelerating diabetic wound healing. *ACS Biomaterials Science & Engineering.* Available from https://doi.org/10.1021/acsbiomaterials.9b00485.

Gambini, J., Inglés, M., Olaso, G., Lopez-Grueso, R., Bonet-Costa, V., Gimeno-Mallench, L., Mas-Bargues, C., Abdelaziz, K. M., Gomez-Cabrera, M. C., & Vina, J. (2015). Properties of resveratrol: In vitro and in vivo studies about metabolism, bioavailability, and biological effects in animal models and humans. *Oxidative Medicine and Cellular Longevity, 2015.*

Garcia Campoy, A. H., Perez Gutierrez, R. M., Manriquez-Alvirde, G., & Muñiz Ramirez, A. (2018). Protection of silver nanoparticles using *Eysenhardtia polystachya* in peroxide-induced pancreatic β-cell damage and their antidiabetic properties in zebrafish. *International Journal of Nanomedicine, 13*, 2601–2612. Available from https://doi.org/10.2147/IJN.S163714.

Gilbertson, L. M., Melnikov, F., Wehmas, L. C., Anastas, P. T., Tanguay, R. L., & Zimmerman, J. B. (2016). Toward safer multi-walled carbon nanotube design: Establishing a statistical model that relates surface charge and embryonic zebrafish mortality. *Nanotoxicology, 10*(1), 10–19.

Girardi, F. A., Bruch, G. E., Peixoto, C. S., Dal Bosco, L., Sahoo, S. K., Gonçalves, C. O. F., Santos, A. P., Furtado, C. A., Fantini, C., & Barros, D. M. (2017). Toxicity of single-wall carbon nanotubes functionalized with polyethylene glycol in zebrafish (*Danio rerio*) embryos. *Journal of Applied Toxicology, 37*(2), 214–221.

Grama, C. N., Suryanarayana, P., Patil, M. A., Raghu, G., Balakrishna, N., Kumar, M. N. V. R., & Reddy, G. B. (2013). Efficacy of biodegradable curcumin nanoparticles in delaying cataract in diabetic rat model. *PLoS One, 8*(10), e78217. Available from https://doi.org/10.1371/journal.pone.0078217.

Groenink, L., Folkerts, G., & Schuurman, H.-J. (2015). European journal of pharmacology, special issue on translational value of animal models: Introduction. *European Journal of Pharmacology, 759*, 1–2.

Gutierrez, R. M. P., Jeronimo, F. F. M., Campoy, A. H. G., & Vadillo, C. H. (2018). Silver nanoparticles synthesized using *Eysenhardtia polystachya* and assessment of the inhibition of glycation in multiple stages in vitro and in the zebrafish model. *Journal of Cluster Science, 29*(6), 1291–1303. Available from https://doi.org/10.1007/s10876-018-1448-5.

Haensel, J. X., Spain, A., & Martin, C. (2015). A systematic review of physiological methods in rodent pharmacological MRI studies. *Psychopharmacology, 232*(3), 489–499.

Hafiz, S. M., Kulkarni, S. S., & Thakur, M. K. (2018). In-vivo toxicity assessment of biologically synthesized iron oxide nanoparticles in zebrafish (*Danio rerio*). *Biosciences Biotechnology Research Asia, 15*(2), 419–425.

Haghshenas, M., Hoveizi, E., Mohammadi, T., & Kazemi Nezhad, S. R. (2019). Use of embryonic fibroblasts associated with graphene quantum dots for burn wound healing in Wistar rats. *In Vitro Cellular & Developmental Biology — Animal, 55*(4), 312–322. Available from https://doi.org/10.1007/s11626-019-00331-w.

Halenova, T., Raksha, N., Vovk, T., Savchuk, O., Ostapchenko, L., Prylutskyy, Y., Kyzyma, O., Ritter, U., & Scharff, P. (2018). Effect of C60 fullerene nanoparticles on the diet-induced obesity in rats. *International Journal of Obesity, 42*(12), 1987–1998. Available from https://doi.org/10.1038/s41366-018-0016-2.

Hanini, A., El Massoudi, M., Gavard, J., Kacem, K., Ammar, S., & Souilem, O. (2016). Nanotoxicological study of polyol-made cobalt-zinc ferrite nanoparticles in rabbit. *Environmental Toxicology and Pharmacology, 45*, 321–327.

Haselton, A. T., & Fridell, Y.-W. C. (2010). Adult *Drosophila melanogaster* as a model for the study of glucose homeostasis. *Aging, 2*(8), 523–526. Available from https://doi.org/10.18632/aging.100185.

He, S., Chen, Y., Wei, L., Jin, X., Zeng, L., Ren, Y., Zhang, J., Wang, L., Li, H., & Lu, Y. (2011). Treatment and risk factor analysis of hypoglycemia in diabetic rhesus monkeys. *Experimental Biology and Medicine, 236*(2), 212–218.

Henson, M. S., & O'Brien, T. D. (2006). Feline models of type 2 diabetes mellitus. *ILAR Journal, 47*(3), 234–242.

Hu, L., Zha, Y. F., Wang, L., Li, L., Xing, D., Gong, W., Wang, J., Lin, Y., Zeng, F. F., & Lu, X. S. (2017). Quantitative evaluation of vertebral microvascular permeability and fat fraction in alloxan-induced diabetic rabbits. *Radiology, 287*(1), 128–136. Available from https://doi.org/10.1148/radiol.2017170760.

Huang, H., Shi, H., Feng, S., Lin, J., Chen, W., Huang, Z., Li, Y., Yu, Y., Lin, D., Xu, Q., & Chen, R. (2013). Silver nanoparticle based surface enhanced Raman scattering spectroscopy of diabetic and normal rat pancreatic tissue under near-infrared laser excitation. *Laser Physics Letters, 10*(4), 045603. Available from https://doi.org/10.1088/1612-2011/10/4/045603.

Huang, K.-T., Fang, Y.-L., Hsieh, P.-S., Li, C.-C., Dai, N.-T., & Huang, C.-J. (2017). Non-sticky and antimicrobial zwitterionic nanocomposite dressings for infected chronic wounds. *Biomaterials Science, 5*(6), 1072–1081. Available from https://doi.org/10.1039/C7BM00039A.

Hussein, E. A., Zagho, M. M., Rizeq, B. R., Younes, N. N., Pintus, G., Mahmoud, K. A., Nasrallah, G. K., & Elzatahry, A. A. (2019). Plasmonic MXene-based nanocomposites exhibiting photothermal therapeutic effects with lower acute toxicity than pure MXene. *International Journal of Nanomedicine, 14*, 4529.

Ionut, V., Liu, H., Mooradian, V., Castro, A. V. B., Kabir, M., Stefanovski, D., Zheng, D., Kirkman, E. L., & Bergman, R. N. (2009). Novel canine models of obese prediabetes and mild type 2 diabetes. *American Journal of Physiology-Endocrinology and Metabolism, 298*(1), E38–E48.

Jahangirian, H., Lemraski, E. G., Webster, T. J., Rafiee-Moghaddam, R., & Abdollahi, Y. (2017). A review of drug delivery systems based on nanotechnology and green chemistry: Green nanomedicine. *International Journal of Nanomedicine, 12*, 2957.

Jeevanandam, J., Chan, Y. S., & Danquah, M. K. (2019). Zebrafish as a model organism to study nano-material toxicity. *Emerging Science Journal, 3*(3), 195–208.

Jia, C., Chen, H., Wei, M., Chen, X., Zhang, Y., Cao, L., Yuan, P., Wang, F., Yang, G., & Ma, J. (2017). Gold nanoparticle-based miR155 antagonist macrophage delivery restores the cardiac function in ovariectomized diabetic mouse model. *International Journal of Nanomedicine, 12*, 4963–4979. Available from https://doi.org/10.2147/IJN.S138400.

Jia, T., Rao, J., Zou, L., Zhao, S., Yi, Z., Wu, B., Li, L., Yuan, H., Shi, L., Zhang, C., Gao, Y., Liu, S., Xu, H., Liu, H., Liang, S., & Li, G. (2018). Nanoparticle-encapsulated curcumin inhibits diabetic neuropathic pain involving the P2Y12 receptor in the dorsal root ganglia. *Frontiers in Neuroscience, 11* (755). Available from https://doi.org/10.3389/fnins.2017.00755.

Johnson, J. I., Decker, S., Zaharevitz, D., Rubinstein, L. V., Venditti, J. M., Schepartz, S., Kalyandrug, S., Christian, M., Arbuck, S., & Hollingshead, M. (2001). Relationships between drug activity in NCI preclinical in vitro and in vivo models and early clinical trials. *British Journal of Cancer, 84*(10), 1424.

Joshi, R. L., Lamothe, B., Cordonnier, N., Mesbah, K., Monthioux, E., Jami, J., & Bucchini, D. (1996). Targeted disruption of the insulin receptor gene in the mouse results in neonatal lethality. *The EMBO Journal, 15*(7), 1542–1547.

Kandiah, K., Venkatachalam, R., Wang, C., Valiyaveettil, S., & Ganesan, K. (2015). In vitro and preliminary in vivo toxicity screening of high-surface-area TiO_2–chondroitin-4-sulfate nanocomposites for bone regeneration application. *Colloids and Surfaces B: Biointerfaces, 128*, 347–356.

Kang, Y.-F., Li, Y.-H., Fang, Y.-W., Xu, Y., Wei, X.-M., & Yin, X.-B. (2015). Carbon quantum dots for zebrafish fluorescence imaging. *Scientific Reports, 5*, 11835.

Kelley, R. A., Conley, S. M., Makkia, R., Watson, J. N., Han, Z., Cooper, M. J., & Naash, M. I. (2018). DNA nanoparticles are safe and nontoxic in non-human primate eyes. *International Journal of Nanomedicine, 13*, 1361–1379. Available from https://doi.org/10.2147/IJN.S157000.

Khan, L. U., da Silva, G. H., de Medeiros, A. M. Z., Khan, Z. U., Gidlund, M., Brito, H. F., Moscoso-Londoño, O., Muraca, D., Knobel, M., Pérez, C. A., & Martinez, D. S. T. (2019). $Fe_3O_4@SiO_2$ nanoparticles concurrently coated with chitosan and $GdOF:Ce^{3+}$, Tb^{3+} luminophore for bioimaging: Toxicity evaluation in the zebrafish model. *ACS Applied Nano Materials, 2*(6), 3414–3425. Available from https://doi.org/10.1021/acsanm.9b00339.

Khatami, M., Iravani, S., Varma, R. S., Mosazade, F., Darroudi, M., & Borhani, F. (2019). Cockroach wings-promoted safe and greener synthesis of silver nanoparticles and their insecticidal activity. *Bioprocess and Biosystems Engineering, 42*, 2007–2014.

Khlebtsov, N., & Dykman, L. (2011). Biodistribution and toxicity of engineered gold nanoparticles: A review of in vitro and in vivo studies. *Chemical Society Reviews, 40*(3), 1647−1671.

Kim, S.-H., Heo, Y., Choi, S.-J., Kim, Y.-J., Kim, M.-S., Kim, H., Jo, E., Song, C.-W., & Lee, K. (2016). Safety evaluation of zinc oxide nanoparticles in terms of acute dermal toxicity, dermal irritation and corrosion, and skin sensitization. *Molecular & Cellular Toxicology, 12*(1), 93−99.

King, A. J. F. (2012). The use of animal models in diabetes research. *British Journal of Pharmacology, 166*(3), 877−894. Available from https://doi.org/10.1111/j.1476-5381.2012.01911.x.

Kittana, N., Assali, M., Abu-Rass, H., Lutz, S., Hindawi, R., Ghannam, L., Zakarneh, M., & Mousa, A. (2018). Enhancement of wound healing by single-wall/multi-wall carbon nanotubes complexed with chitosan. *International Journal of Nanomedicine, 13*, 7195−7206. Available from https://doi.org/10.2147/IJN.S183342.

Kotb, S., Piraquive, J., Lamberton, F., Lux, F., Verset, M., Di Cataldo, V., Contamin, H., Tillement, O., Canet-Soulas, E., & Sancey, L. (2016). Safety evaluation and imaging properties of gadolinium-based nanoparticles in nonhuman primates. *Scientific Reports, 6*, 35053. Available from https://doi.org/10.1038/srep35053.

Krishnaraj, C., Muthukumaran, P., Ramachandran, R., Balakumaran, M. D., & Kalaichelvan, P. T. (2014). *Acalypha indica* Linn: Biogenic synthesis of silver and gold nanoparticles and their cytotoxic effects against MDA-MB-231, human breast cancer cells. *Biotechnology Reports, 4*, 42−49.

Kteeba, S. M., El-Ghobashy, A. E., El-Adawi, H. I., El-Rayis, O. A., Sreevidya, V. S., Guo, L., & Svoboda, K. R. (2018). Exposure to ZnO nanoparticles alters neuronal and vascular development in zebrafish: Acute and transgenerational effects mitigated with dissolved organic matter. *Environmental Pollution, 242*, 433−448. Available from https://doi.org/10.1016/j.envpol.2018.06.030.

Kubota, N., Tobe, K., Terauchi, Y., Eto, K., Yamauchi, T., Suzuki, R., Tsubamoto, Y., Komeda, K., Nakano, R., & Miki, H. (2000). Disruption of insulin receptor substrate 2 causes type 2 diabetes because of liver insulin resistance and lack of compensatory beta-cell hyperplasia. *Diabetes, 49*(11), 1880−1889.

Labrador-Rached, C. J., Browning, R. T., Braydich-Stolle, L. K., & Comfort, K. K. (2018). Toxicological implications of platinum nanoparticle exposure: Stimulation of intracellular stress, inflammatory response, and Akt signaling in vitro. *Journal of Toxicology, 2018*.

Lajmanovich, R. C., Peltzer, P. M., Martinuzzi, C. S., Attademo, A. M., Colussi, C. L., & Bassó, A. (2018). Acute toxicity of colloidal silicon dioxide nanoparticles on amphibian larvae: Emerging environmental concern. *International Journal of Environmental Research, 12*(3), 269−278.

Lanford, R. E., & Bigger, C. (2002). Advances in model systems for hepatitis C virus research. *Virology, 293*(1), 1−9.

Leahy, J. L., Bonner-Weir, S., & Weir, G. C. (1988). Minimal chronic hyperglycemia is a critical determinant of impaired insulin secretion after an incomplete pancreatectomy. *The Journal of Clinical Investigation, 81*(5), 1407−1414.

Lenzen, S., Tiedge, M., Elsner, M., Lortz, S., Weiss, H., Jörns, A., Klöppel, G., Wedekind, D., Prokop, C. M., & Hedrich, H. J. (2001). The LEW.1AR1/Ztm-iddm rat: A new model of spontaneous insulin-dependent diabetes mellitus. *Diabetologia, 44*(9), 1189−1196.

Leonardi, A., Bucolo, C., Drago, F., Salomone, S., & Pignatello, R. (2015). Cationic solid lipid nanoparticles enhance ocular hypotensive effect of melatonin in rabbit. *International Journal of Pharmaceutics, 478*(1), 180−186.

Leu, J.-G., Chen, S.-A., Chen, H.-M., Wu, W.-M., Hung, C.-F., Yao, Y.-D., Tu, C.-S., & Liang, Y.-J. (2012). The effects of gold nanoparticles in wound healing with antioxidant epigallocatechin gallate and α-lipoic acid. *Nanomedicine: Nanotechnology, Biology and Medicine, 8*(5), 767−775. Available from https://doi.org/10.1016/j.nano.2011.08.013.

Li, J., Ying, G.-G., Jones, K. C., & Martin, F. L. (2015). Real-world carbon nanoparticle exposures induce brain and gonadal alterations in zebrafish (*Danio rerio*) as determined by biospectroscopy techniques. *Analyst, 140*(8), 2687−2695. Available from https://doi.org/10.1039/C4AN02227K.

Li, J., Zhai, D., Lv, F., Yu, Q., Ma, H., Yin, J., Yi, Z., Liu, M., Chang, J., & Wu, C. (2016). Preparation of copper-containing bioactive glass/eggshell membrane nanocomposites for improving angiogenesis, antibacterial activity and wound healing. *Acta Biomaterialia, 36*, 254−266.

Li, X., Zhen, M., Zhou, C., Deng, R., Yu, T., Wu, Y., Shu, C., Wang, C., & Bai, C. (2019). Gadofullerene nanoparticles reverse dysfunctions of pancreas and improve hepatic insulin resistance for type 2 diabetes mellitus treatment. *ACS Nano*. Available from https://doi.org/10.1021/acsnano.9b02050.

Lieschke, G. J., & Currie, P. D. (2007). Animal models of human disease: Zebrafish swim into view. *Nature Reviews Genetics*, *8*(5), 353.

Lin, Z., Monteiro-Riviere, N. A., Kannan, R., & Riviere, J. E. (2016). A computational framework for interspecies pharmacokinetics, exposure and toxicity assessment of gold nanoparticles. *Nanomedicine: Nanotechnology, Biology, and Medicine*, *11*(2), 107−119.

Lushchak, O., Zayachkivska, A., & Vaiserman, A. (2018). Metallic nanoantioxidants as potential therapeutics for type 2 diabetes: A hypothetical background and translational perspectives. *Oxidative Medicine and Cellular Longevity*, *2018*.

Luyts, K., Smulders, S., Napierska, D., Van kerckhoven, S., Poels, K., Scheers, H., Hemmeryckx, B., Nemery, B., Hoylaerts, M. F., & Hoet, P. H. M. (2014). Pulmonary and hemostatic toxicity of multi-walled carbon nanotubes and zinc oxide nanoparticles after pulmonary exposure in Bmal1 knockout mice. *Particle and Fibre Toxicology*, *11*(1), 61. Available from https://doi.org/10.1186/s12989-014-0061-5.

Maity, S., Mukhopadhyay, P., Kundu, P. P., & Chakraborti, A. S. (2017). Alginate coated chitosan core-shell nanoparticles for efficient oral delivery of naringenin in diabetic animals—An in vitro and in vivo approach. *Carbohydrate Polymers*, *170*, 124−132. Available from https://doi.org/10.1016/j.carbpol.2017.04.066.

Mak, I. W. Y., Evaniew, N., & Ghert, M. (2014). Lost in translation: Animal models and clinical trials in cancer treatment. *American Journal of Translational Research*, *6*(2), 114.

Makino, S., Kunimoto, K., Muraoka, Y., Mizushima, Y., Katagiri, K., & Tochino, Y. (1980). Breeding of a non-obese, diabetic strain of mice. *Experimental Animals*, *29*(1), 1−13.

Manshian, B. B., Pokhrel, S., Himmelreich, U., Tämm, K., Sikk, L., Fernández, A., Rallo, R., Tamm, T., Mädler, L., & Soenen, S. J. (2017). In silico design of optimal dissolution kinetics of Fe-doped ZnO nanoparticles results in cancer-specific toxicity in a preclinical rodent model. *Advanced Healthcare Materials*, *6*(9), 1601379.

Mao, Z., Li, Y., Dong, T., Zhang, L., Zhang, Y., Li, S., Hu, H., Sun, C., & Xia, Y. (2019). Exposure to Titanium dioxide nanoparticles during pregnancy changed maternal gut microbiota and increased blood glucose of rat. *Nanoscale Research Letters*, *14*(1), 26. Available from https://doi.org/10.1186/s11671-018-2834-5.

Marins, K., Lazzarotto, L. M. V., Boschetti, G., Bertoncello, K. T., Sachett, A., Schindler, M. S. Z., Chitolina, R., Regginato, A., Zanatta, A. P., & Siebel, A. M. (2019). Iron and manganese present in underground water promote biochemical, genotoxic, and behavioral alterations in zebrafish (*Danio rerio*). *Environmental Science and Pollution Research*, *26*, 23555−23570.

Mathews, C. E., Langley, S. H., & Leiter, E. H. (2002). New mouse model to study islet transplantation in insulin-dependent diabetes mellitus. *Transplantation*, *73*(8), 1333−1336.

Matsuura, Y., Yamashita, A., Zhao, Y., Iwakiri, T., Yamasaki, K., Sugita, C., Koshimoto, C., Kitamura, K., Kawai, K., Tamaki, N., Zhao, S., Kuge, Y., & Asada, Y. (2017). Altered glucose metabolism and hypoxic response in alloxan-induced diabetic atherosclerosis in rabbits. *PLoS One*, *12*(4), e0175976. Available from https://doi.org/10.1371/journal.pone.0175976.

Mears, E. R., Modabber, F., Don, R., & Johnson, G. E. (2015). A review: The current in vivo models for the discovery and utility of new anti-leishmanial drugs targeting cutaneous leishmaniasis. *PLoS Neglected Tropical Diseases*, *9*(9), e0003889.

Mellert, J., Hering, B. J., Liu, X., Brandhorst, D., Brandhorst, H., Brendel, M., Ernst, E., Gramberg, D., Bretzel, R. G., & Hopt, U. T. (1998). Successful islet auto-and allotransplantation in diabetic pigs. *Transplantation*, *66*(2), 200−204.

Merrell, J. G., McLaughlin, S. W., Tie, L., Laurencin, C. T., Chen, A. F., & Nair, L. S. (2009). Curcumin-loaded poly(ε-caprolactone) nanofibres: Diabetic wound dressing with anti-oxidant and anti-inflammatory properties. *Clinical and Experimental Pharmacology and Physiology*, *36*(12), 1149−1156. Available from https://doi.org/10.1111/j.1440-1681.2009.05216.x.

Mesak, C., dos Reis Sampaio, D. M., de Oliveira Ferreira, R., de Oliveira Mendes, B., de Lima Rodrigues, A. S., & Malafaia, G. (2018). The effects of predicted environmentally relevant concentrations of ZnO nanoparticles on the behavior of *Gallus gallus domesticus* (Phasianidae) chicks. *Environmental Pollution*, 242, 1274−1282. Available from https://doi.org/10.1016/j.envpol.2018.08.004.

Modarresi Chahardehi, A., Arsad, H., & Lim, V. (2020). Zebrafish as a successful animal model for screening toxicity of medicinal plants. *Plants*, 9(10), 1345.

Monteiro-Riviere, N. A., Inman, A. O., & Zhang, L. W. (2009). Limitations and relative utility of screening assays to assess engineered nanoparticle toxicity in a human cell line. *Toxicology and Applied Pharmacology*, 234(2), 222−235.

Morningstar, J. E., Nath, A. K., O'Sullivan, J. F., Zheng, B., Peterson, R. T., & Gerszten, R. E. (2017). DMGV is predictive of future diabetes, and modulates glucose homeostasis in zebrafish. *Circulation*, 136(suppl_1), A19501.

Morris, S. N. S., Coogan, C., Chamseddin, K., Fernandez-Kim, S. O., Kolli, S., Keller, J. N., & Bauer, J. H. (2012). Development of diet-induced insulin resistance in adult *Drosophila melanogaster*. *Biochimica et Biophysica Acta (BBA) − Molecular Basis of Disease*, 1822(8), 1230−1237.

Mukhopadhyay, P., Sarkar, K., Chakraborty, M., Bhattacharya, S., Mishra, R., & Kundu, P. P. (2013). Oral insulin delivery by self-assembled chitosan nanoparticles: In vitro and in vivo studies in diabetic animal model. *Materials Science and Engineering: C*, 33(1), 376−382.

Nabofa, W. E. E., Alashe, O. O., Oyeyemi, O. T., Attah, A. F., Oyagbemi, A. A., Omobowale, T. O., Adedapo, A. A., & Alada, A. R. A. (2018). Cardioprotective effects of curcumin−nisin based poly lactic acid nanoparticle on myocardial infarction in guinea pigs. *Scientific Reports*, 8(1), 1−11.

Nakhooda, A. F., Like, A. A., Chappel, C. I., Murray, F. T., & Marliss, E. B. (1977). The spontaneously diabetic Wistar rat: Metabolic and morphologic studies. *Diabetes*, 26(2), 100−112.

Nazaktabar, A., Lashkenari, M. S., Araghi, A., Ghorbani, M., & Golshahi, H. (2017). In vivo evaluation of toxicity and antiviral activity of polyrhodanine nanoparticles by using the chicken embryo model. *International Journal of Biological Macromolecules*, 103, 379−384. Available from https://doi.org/10.1016/j.ijbiomac.2017.05.069.

Nedzvetskii, V. S., Pryshchepa, I. V., Tykhomyrov, A. A., & Baydas, G. (2016). Inhibition of reactive gliosis in the retina of rats with streptozotocin-induced diabetes under the action of hydrated C 60 fullerene. *Neurophysiology*, 48(2), 130−140.

Noda, K., Melhorn, M. I., Zandi, S., Frimmel, S., Tayyari, F., Hisatomi, T., Almulki, L., Pronczuk, A., Hayes, K. C., & Hafezi-Moghadam, A. (2010). An animal model of spontaneous metabolic syndrome: Nile grass rat. *FASEB Journal*, 24(7), 2443−2453.

Okamoto, N., Tobe, T., Hackett, S. F., Ozaki, H., Vinores, M. A., LaRochelle, W., Zack, D. J., & Campochiaro, P. A. (1997). Transgenic mice with increased expression of vascular endothelial growth factor in the retina: A new model of intraretinal and subretinal neovascularization. *The American Journal of Pathology*, 151(1), 281.

Okazaki, F., Zang, L., Nakayama, H., Chen, Z., Gao, Z.-J., Chiba, H., Hui, S.-P., Aoki, T., Nishimura, N., & Shimada, Y. (2019). Microbiome alteration in type 2 diabetes mellitus model of zebrafish. *Scientific Reports*, 9(1), 867.

Oliveira, C. R., Garcia, T. D., Franco-Belussi, L., Salla, R. F., Souza, B. F. S., de Melo, N. F. S., Irazusta, S. P., Jones-Costa, M., Silva-Zacarin, E. C. M., & Fraceto, L. F. (2019). Pyrethrum extract encapsulated in nanoparticles: Toxicity studies based on genotoxic and hematological effects in bullfrog tadpoles. *Environmental Pollution*, 253, 1009−1020.

Oliveira-Filho, E. C., Sousa Filho, J., Novais, L. A., Peternele, W. S., Azevedo, R. B., & Grisolia, C. K. (2016). Effects of γ-Fe$_2$O$_3$ nanoparticles on the survival and reproduction of *Biomphalaria glabrata* (Say, 1818) and their elimination from this benthic aquatic snail. *Environmental Science and Pollution Research*, 23(18), 18362−18368.

Orlando, P., Chellan, N., Muller, C. J. F., Louw, J., Chapman, C. C., Joubert, E., & Tiano, L. (2017). Green rooibos extract improves plasma lipid profile and oxidative status in diabetic non-human primates. *Free Radical Biology and Medicine*, 108, S97.

Pandit, J., Sultana, Y., & Aqil, M. (2017). Chitosan-coated PLGA nanoparticles of bevacizumab as novel drug delivery to target retina: Optimization, characterization, and in vitro toxicity evaluation. *Artificial Cells, Nanomedicine, and Biotechnology, 45*(7), 1397−1407.

Patnaik, S., Gorain, B., Padhi, S., Choudhury, H., Gabr, G. A., Md, S., Mishra, D. K., & Kesharwani, P. (2021). Recent update of toxicity aspects of nanoparticulate systems for drug delivery. *European Journal of Pharmaceutics and Biopharmaceutics, 161*, 100−119.

Pecoraro, R., D'Angelo, D., Filice, S., Scalese, S., Capparucci, F., Marino, F., Iaria, C., Guerriero, G., Tibullo, D., & Scalisi, E. M. (2018). Toxicity evaluation of graphene oxide and titania loaded nafion membranes in zebrafish. *Frontiers in Physiology, 8*, 1039.

Pham, T. T., Nguyen, T. T., Pathak, S., Regmi, S., Nguyen, H. T., Tran, T. H., Yong, C. S., Kim, J. O., Park, P., Park, M. H., Bae, Y. K., Choi, J. U., Byun, Y., Ahn, C., Yook, S., & Jeong, J. (2018). Tissue adhesive FK506-loaded polymeric nanoparticles for multi-layered nano-shielding of pancreatic islets to enhance xenograft survival in a diabetic mouse model. *Biomaterials, 154*, 182−196. Available from https://doi.org/10.1016/j.biomaterials.2017.10.049.

Pick, L., Graham, P., & Pick, L. (2017). *Chapter thirteen − Drosophila as a model for diabetes and diseases of insulin resistance, . Current topics in developmental biology* (vol. 121, pp. 397−419). Academic Press. Available from https://doi.org/10.1016/bs.ctdb.2016.07.011.

Platania, C. B. M., Maisto, R., Trotta, M. C., D'Amico, M., Rossi, S., Gesualdo, C., D'Amico, G., Balta, C., Herman, H., & Hermenean, A. (2019). Retinal and circulating miRNA expression patterns in diabetic retinopathy: An in silico and in vivo approach. *British Journal of Pharmacology, 176*, 2179−2194.

Portha, B., Levacher, C., Picon, L., & Rosselin, G. (1974). Diabetogenic effect of streptozotocin in the rat during the perinatal period. *Diabetes, 23*(11), 889−895.

Prakash, J., Venkatesan, M., Joy Sebastian Prakash, J., Bharath, G., Anwer, S., Veluswamy, P., Prema, D., Venkataprasanna, K. S., & Venkatasubbu, G. D. (2019). Investigations on the in-vivo toxicity analysis of reduced graphene oxide/TiO$_2$ nanocomposite in zebrafish embryo and larvae (*Danio rerio*). *Applied Surface Science, 481*, 1360−1369. Available from https://doi.org/10.1016/j.apsusc.2019.03.287.

Raics, A. B., & Bajic, V. B. (2016). In silico toxicology: Computational methods for the prediction of chemical toxicity. *Wiley Interdisciplinary Reviews: Computational Molecular Science, 6*(2), 147−172.

Rakeiten, N. (1963). Studies on the diabetogenic action of streptozotocin. *Cancer Chemotherapy Reports. Part 1, 29*, 91−98.

Rathore, P., Mahor, A., Jain, S., Haque, A., & Kesharwani, P. (2020). Formulation development, in vitro and in vivo evaluation of chitosan engineered nanoparticles for ocular delivery of insulin. *RSC Advances, 10*(71), 43629−43639.

Rees, D. A., & Alcolado, J. C. (2005). Animal models of diabetes mellitus. *Diabetic Medicine, 22*(4), 359−370.

Safley, S. A., Kenyon, N. S., Berman, D. M., Barber, G. F., Willman, M., Duncanson, S., Iwakoshi, N., Holdcraft, R., Gazda, L., & Thompson, P. (2018). Microencapsulated adult porcine islets transplanted intraperitoneally in streptozotocin-diabetic non-human primates. *Xenotransplantation, 25*(6), e12450.

Sakata, N., Yoshimatsu, G., Tsuchiya, H., Egawa, S., & Unno, M. (2012). Animal models of diabetes mellitus for islet transplantation. *Experimental Diabetes Research, 2012*, 11. Available from https://doi.org/10.1155/2012/256707.

Samuel, R. R., Annadurai, G., & Rajeshkumar, S. (2019). Characterization and toxicology evaluation of zirconium oxide nanoparticles on the embryonic development of zebrafish, *Danio rerio*. *Drug and Chemical Toxicology, 42*(1), 104−111.

Sangabathuni, S., Murthy, R. V., Chaudhary, P. M., Subramani, B., Toraskar, S., & Kikkeri, R. (2017). Mapping the glyco-gold nanoparticles of different shapes toxicity, biodistribution and sequestration in adult zebrafish. *Scientific Reports, 7*(1), 4239.

Sarmento, B., Ribeiro, A., Veiga, F., Ferreira, D., & Neufeld, R. (2007). Oral bioavailability of insulin contained in polysaccharide nanoparticles. *Biomacromolecules, 8*(10), 3054−3060.

Sayes, C. M., Reed, K. L., & Warheit, D. B. (2007). Assessing toxicity of fine and nanoparticles: Comparing in vitro measurements to in vivo pulmonary toxicity profiles. *Toxicological Sciences, 97*(1), 163−180.

Shah, A., Buabeid, M. A., Arafa, E.-S. A., Hussain, I., Li, L., & Murtaza, G. (2019). The wound healing and antibacterial potential of triple-component nanocomposite (chitosan-silver-sericin) films loaded with moxifloxacin. *International Journal of Pharmaceutics, 564*, 22–38.

Shakya, A., Chaudary, S. K., Garabadu, D., Bhat, H. R., Kakoti, B. B., & Ghosh, S. K. (2020). A comprehensive review on preclinical diabetic models. *Current Diabetes Reviews, 16*(2), 104–116.

Shanker, K., Mohan, G. K., Hussain, M. A., Jayarambabu, N., & Pravallika, P. L. (2017). Green biosynthesis, characterization, in vitro antidiabetic activity, and investigational acute toxicity studies of some herbal-mediated silver nanoparticles on animal models. *Pharmacognosy Magazine, 13*(49), 188–192. Available from https://doi.org/10.4103/0973-1296.197642.

Shao, J., Yu, N., Kolwijck, E., Wang, B., Tan, K. W., Jansen, J. A., Walboomers, X. F., & Yang, F. (2017). Biological evaluation of silver nanoparticles incorporated into chitosan-based membranes. *Nanomedicine: Nanotechnology, Biology, and Medicine, 12*(22), 2771–2785.

Shrestha, N., Araújo, F., Shahbazi, M.-A., Mäkilä, E., Gomes, M. J., Airavaara, M., Kauppinen, E. I., Raula, J., Salonen, J., Hirvonen, J., Sarmento, B., & Santos, H. A. (2016). Oral hypoglycaemic effect of GLP-1 and DPP4 inhibitor based nanocomposites in a diabetic animal model. *Journal of Controlled Release, 232*, 113–119. Available from https://doi.org/10.1016/j.jconrel.2016.04.024.

Simon, R. H., Engelhardt, J. F., Yang, Y., Zepeda, M., Weber-Pendleton, S., Grossman, M., & Wilson, J. M. (1993). Adenovirus-mediated transfer of the CFTR gene to lung of nonhuman primates: Toxicity study. *Human Gene Therapy, 4*(6), 771–780.

Sivaji, K., & Kannan, R. R. (2019). Polysorbate 80 coated gold nanoparticle as a drug carrier for brain targeting in zebrafish model. *Journal of Cluster Science, 30*(4), 897–906. Available from https://doi.org/10.1007/s10876-019-01548-1.

Sligar, A. D., Howe, G., Goldman, J., Felli, P., Karanam, V., Smalling, R. W., & Baker, A. B. (2019). Preclinical model of hind limb ischemia in diabetic rabbits. *Journal of Visualized Experiments: JoVE, 148*, e58964.

Sonaje, K., Chen, Y.-J., Chen, H.-L., Wey, S.-P., Juang, J.-H., Nguyen, H.-N., Hsu, C.-W., Lin, K.-J., & Sung, H.-W. (2010). Enteric-coated capsules filled with freeze-dried chitosan/poly(γ-glutamic acid) nanoparticles for oral insulin delivery. *Biomaterials, 31*(12), 3384–3394. Available from https://doi.org/10.1016/j.biomaterials.2010.01.042.

Song, Q., Zhou, H., Zhen, H., Wang, N., Deng, J., Wang, J., & Pan, X. (2016). Establishment and evaluation of a rhesus monkey model of experimental type 2 diabetes mellitus. *Chinese Journal of Tissue Engineering Research, 20*(40), 6048–6053.

Srinivasan, P., Vijayakumar, S., Kothandaraman, S., & Palani, M. (2018). Anti-diabetic activity of quercetin extracted from *Phyllanthus emblica* L. fruit: In silico and in vivo approaches. *Journal of Pharmaceutical Analysis, 8*(2), 109–118.

Stephens, M. L., Betts, K., Beck, N. B., Cogliano, V., Dickersin, K., Fitzpatrick, S., Freeman, J., Gray, G., Hartung, T., & McPartland, J. (2016). The emergence of systematic review in toxicology. *Toxicological Sciences, 152*(1), 10–16.

Sukhanova, A., Bozrova, S., Sokolov, P., Berestovoy, M., Karaulov, A., & Nabiev, I. (2018). Dependence of nanoparticle toxicity on their physical and chemical properties. *Nanoscale Research Letters, 13*(1), 1–21.

Suter, L., Babiss, L. E., & Wheeldon, E. B. (2004). Toxicogenomics in predictive toxicology in drug development. *Chemistry & Biology, 11*(2), 161–171.

Suzuki, K., Kim, K. S., & Bae, Y. H. (2019). Long-term oral administration of Exendin-4 to control type 2 diabetes in a rat model. *Journal of Controlled Release, 294*, 259–267.

Tamemoto, H., Kadowaki, T., Tobe, K., Yagi, T., Sakura, H., Hayakawa, T., Terauchi, Y., Ueki, K., Kaburagi, Y., & Satoh, S. (1994). Insulin resistance and growth retardation in mice lacking insulin receptor substrate-1. *Nature, 372*(6502), 182.

Tang, P., Han, L., Li, P., Jia, Z., Wang, K., Zhang, H., Tan, H., Guo, T., & Lu, X. (2019). Mussel-inspired electroactive and antioxidative scaffolds with incorporation of polydopamine-reduced graphene oxide for enhancing skin wound healing. *ACS Applied Materials & Interfaces, 11*(8), 7703–7714. Available from https://doi.org/10.1021/acsami.8b18931.

Tashakori Miyanroudi, M., & Arabi, M. (2016). Evaluation the effect of silver nanoparticles on oxidative stress biomarkers in blood serum and liver and kidney tissues. *Nanomedicine Journal, 3*(3), 179−185.

Tee, L. B. G., Penrose, M. A., O'Shea, J. E., Lai, C. M., Rakoczy, E. P., & Dunlop, S. A. (2008). VEGF-induced choroidal damage in a murine model of retinal neovascularisation. *British Journal of Ophthalmology, 92*(6), 832−838.

Teijeiro-Valiño, C., Yebra-Pimentel, E., Guerra-Varela, J., Csaba, N., Alonso, M. J., & Sánchez, L. (2017). Assessment of the permeability and toxicity of polymeric nanocapsules using the zebrafish model. *Nanomedicine: Nanotechnology, Biology, and Medicine, 12*(17), 2069−2082. Available from https://doi.org/10.2217/nnm-2017-0078.

Teleman, A. A., Ratzenböck, I., & Oldham, S. (2012). Drosophila: A model for understanding obesity and diabetic complications. *Experimental and Clinical Endocrinology & Diabetes: Official Journal, German Society of Endocrinology [and] German Diabetes Association, 120*(04), 184−185. Available from https://doi.org/10.1055/s-0032-1304566.

Tian, J., Wong, K. K. Y., Ho, C.-M., Lok, C.-N., Yu, W.-Y., Che, C.-M., Chiu, J.-F., & Tam, P. K. H. (2007). Topical delivery of silver nanoparticles promotes wound healing. *ChemMedChem, 2*(1), 129−136. Available from https://doi.org/10.1002/cmdc.200600171.

Tice, R. R., Agurell, E., Anderson, D., Burlinson, B., Hartmann, A., Kobayashi, H., Miyamae, Y., Rojas, E., Ryu, J., & Sasaki, Y. F. (2000). Single cell gel/comet assay: Guidelines for in vitro and in vivo genetic toxicology testing. *Environmental and Molecular Mutagenesis, 35*(3), 206−221.

Tourrel, C., Bailbé, D., Meile, M.-J., Kergoat, M., & Portha, B. (2001). Glucagon-like peptide-1 and exendin-4 stimulate β-cell neogenesis in streptozotocin-treated newborn rats resulting in persistently improved glucose homeostasis at adult age. *Diabetes, 50*(7), 1562−1570.

Truong, L., Zaikova, T., Baldock, B. L., Balik-Meisner, M., To, K., Reif, D. M., Kennedy, Z. C., Hutchison, J. E., & Tanguay, R. L. (2019). Systematic determination of the relationship between nanoparticle core diameter and toxicity for a series of structurally analogous gold nanoparticles in zebrafish. *Nanotoxicology, 13*(7), 879−893. Available from https://doi.org/10.1080/17435390.2019.1592259.

Turner, C. T., McInnes, S. J. P., Melville, E., Cowin, A. J., & Voelcker, N. H. (2017). Delivery of flightless I neutralizing antibody from porous silicon nanoparticles improves wound healing in diabetic mice. *Advanced Healthcare Materials, 6*(2), 1600707. Available from https://doi.org/10.1002/adhm.201600707.

Varshochian, R., Riazi-Esfahani, M., Jeddi-Tehrani, M., Mahmoudi, A.-R., Aghazadeh, S., Mahbod, M., Movassat, M., Atyabi, F., Sabzevari, A., & Dinarvand, R. (2015). Albuminated PLGA nanoparticles containing bevacizumab intended for ocular neovascularization treatment. *Journal of Biomedical Materials Research. Part A, 103*(10), 3148−3156. Available from https://doi.org/10.1002/jbm.a.35446.

Vibe, C. B., Fenaroli, F., Pires, D., Wilson, S. R., Bogoeva, V., Kalluru, R., Speth, M., Anes, E., Griffiths, G., & Hildahl, J. (2016). Thioridazine in PLGA nanoparticles reduces toxicity and improves rifampicin therapy against mycobacterial infection in zebrafish. *Nanotoxicology, 10*(6), 680−688.

Vong, L. B., Kobayashi, M., & Nagasaki, Y. (2016). Evaluation of the toxicity and antioxidant activity of redox nanoparticles in zebrafish (*Danio rerio*) embryos. *Molecular Pharmaceutics, 13*(9), 3091−3097. Available from https://doi.org/10.1021/acs.molpharmaceut.6b00225.

Vranic, S., Shimada, Y., Ichihara, S., Kimata, M., Wu, W., Tanaka, T., Boland, S., Tran, L., & Ichihara, G. (2019). Toxicological evaluation of SiO_2 nanoparticles by zebrafish embryo toxicity test. *International Journal of Molecular Sciences, 20*(4), 882.

Wahba, N. S., Shaban, S. F., Kattaia, A. A. A., & Kandeel, S. A. (2016). Efficacy of zinc oxide nanoparticles in attenuating pancreatic damage in a rat model of streptozotocin-induced diabetes. *Ultrastructural Pathology, 40*(6), 358−373. Available from https://doi.org/10.1080/01913123.2016.1246499.

Wiggenhauser, L. M., & Kroll, J. (2019). Vascular damage in obesity and diabetes: Highlighting links between endothelial dysfunction and metabolic disease in zebrafish and man. *Current Vascular Pharmacology, 17*, 476−490.

Winner, B., Jappelli, R., Maji, S. K., Desplats, P. A., Boyer, L., Aigner, S., Hetzer, C., Loher, T., Vilar, M., & Campioni, S. (2011). In vivo demonstration that α-synuclein oligomers are toxic. *Proceedings of the National Academy of Sciences of the United States of America, 108*(10), 4194−4199.

Wöhler, F., & Liebig, J. (1838). Untersuchungen über die natur der harnsäure. *Annalen der Pharmacie, 26* (3), 241–336.

Wu, W., Yan, L., Wu, Q., Li, Y., Li, Q., Chen, S., Yang, Y., Gu, Z., Xu, H., & Yin, Z. Q. (2016). Evaluation of the toxicity of graphene oxide exposure to the eye. *Nanotoxicology, 10*(9), 1329–1340.

Xiao, J., Zhu, Y., Huddleston, S., Li, P., Xiao, B., Farha, O. K., & Ameer, G. A. (2018). Copper metal–organic framework nanoparticles stabilized with folic acid improve wound healing in diabetes. *ACS Nano, 12*(2), 1023–1032. Available from https://doi.org/10.1021/acsnano.7b01850.

Ye, L., Yong, K. T., Liu, L., Roy, I., Hu, R., Zhu, J., Cai, H., Law, W. C., Liu, J., Wang, K., Liu, J., Liu, Y., Hu, Y., Zhang, X., Swihart, M. T., & Prasad, P. N. (2012). A pilot study in non-human primates shows no adverse response to intravenous injection of quantum dots. *Nature Nanotechnology, 7* (7), 453–458. Available from https://doi.org/10.1038/nnano.2012.74.

Yen, H.-J., Horng, J.-L., Yu, C.-H., Fang, C.-Y., Yeh, Y.-H., & Lin, L.-Y. (2019). Toxic effects of silver and copper nanoparticles on lateral-line hair cells of zebrafish embryos. *Aquatic Toxicology, 215*, 105273.

Younes, N., Pintus, G., Al-Asmakh, M., Rasool, K., Younes, S., Calzolari, S., Mahmoud, K. A., & Nasrallah, G. K. (2019). "Safe" chitosan/zinc oxide nanocomposite has minimal organ-specific toxicity in early stages of zebrafish development. *ACS Biomaterials Science & Engineering, 6*, 38–47.

Zang, L., Shimada, Y., Nakayama, H., Chen, W., Okamoto, A., Koide, H., Oku, N., Dewa, T., Shiota, M., & Nishimura, N. (2019). Therapeutic silencing of centromere protein X ameliorates hyperglycemia in zebrafish and mouse models of type 2 diabetes mellitus. *Frontiers in Genetics, 10*(693). Available from https://doi.org/10.3389/fgene.2019.00693.

Zgheib, C., Hilton, S. A., Dewberry, L. C., Hodges, M. M., Ghatak, S., Xu, J., Singh, S., Roy, S., Sen, C. K., Seal, S., & Liechty, K. W. (2019). Use of cerium oxide nanoparticles conjugated with microRNA-146a to correct the diabetic wound healing impairment. *Journal of the American College of Surgeons, 228*(1), 107–115. Available from https://doi.org/10.1016/j.jamcollsurg.2018.09.017.

Zhang, J.-H., Niu, A., Li, J., Fu, J.-W., Xu, Q., & Pei, D.-S. (2016). In vivo characterization of hair and skin derived carbon quantum dots with high quantum yield as long-term bioprobes in zebrafish. *Scientific Reports, 6*, 37860.

Zhang, J.-H., Sun, T., Niu, A., Tang, Y.-M., Deng, S., Luo, W., Xu, Q., Wei, D., & Pei, D.-S. (2017). Perturbation effect of reduced graphene oxide quantum dots (rGOQDs) on aryl hydrocarbon receptor (AhR) pathway in zebrafish. *Biomaterials, 133*, 49–59.

Zhao, Q., Li, J., Yan, J., Liu, S., Guo, Y., Chen, D., & Luo, Q. (2016). *Lycium barbarum* polysaccharides ameliorates renal injury and inflammatory reaction in alloxan-induced diabetic nephropathy rabbits. *Life Sciences, 157*, 82–90. Available from https://doi.org/10.1016/j.lfs.2016.05.045.

Zunhammer, M., Ploner, M., Engelbrecht, C., Bock, J., Kessner, S. S., & Bingel, U. (2017). The effects of treatment failure generalize across different routes of drug administration. *Science Translational Medicine, 9*(393), eaal2999.

CHAPTER 9

Preclinical and clinical evaluation of nanodrugs for diabetes treatment

Introduction

Preclinical and clinical evaluations are critical to drug development (Ioannidis et al., 2018). Preclinical studies involve the use of ex vivo, in vitro, and in vivo approaches to evaluate the efficacy, toxicity, and biophysical and biochemical properties of nanomedicines (Gad, 2008; Jiang et al., 2021). Even though lab-scale in silico and in vitro studies helps in screening potential nanomedicines for the treatment of specific diseases, preclinical studies are performed via good laboratory practices (GLP) and good clinical practices (GCP) that are mandatory for approval (Chen & Cai, 2015; Satalkar et al., 2016; Xu et al., 2015). Clinical studies, involving human trials, represent a crucial stage to translate nanomedicines from products to treatments (Jang et al., 2016). However, there are currently no standard preclinical and clinical protocols to facilitate the widespread development of nanomedicines.

Cancer is a premier disease with nanomedicines approved for treatment (Bor et al., 2019). It is noteworthy that most of the approved drugs are nanoformulations of conventional drugs (Shi & Lammers, 2019). For diabetes, few nanoformulations have received approvals from regulatory bodies for controlled delivery of insulin (Bahman et al., 2021; Cao, 2018; Joshi et al., 2019). Standard preclinical and clinical assays are needed to evaluate the antidiabetic properties of nanomedicines as well as their toxic effects (Engel et al., 2010; Rivera-Mancía et al., 2018). This chapter presents an overview of various preclinical assays to evaluate the antidiabetic properties of nanomedicines. It also discusses clinical evaluation of antidiabetic efficacy of nanomedicines to support regulatory approval.

Biochemical and cellular assays for antidiabetic analysis

Biochemical and cellular assays for analyzing antidiabetic efficacy can be grouped into three types: carbohydrate digesting enzyme inhibition assay, in vitro cellular models to evaluate insulin secretion, and in vitro antidiabetic assay. Fig. 9.1 shows different biochemical and cellular assays for the evaluation of antidiabetic properties.

Carbohydrate digesting enzyme inhibition assay

Antidiabetic drugs are designed to mostly target and inhibit three main carbohydrate digesting enzymes such as alpha-amylase, alpha-glucosidase, and sucrase. Numerous

Emerging Nanomedicines for Diabetes Mellitus Theranostics
DOI: https://doi.org/10.1016/B978-0-323-85396-5.00011-7

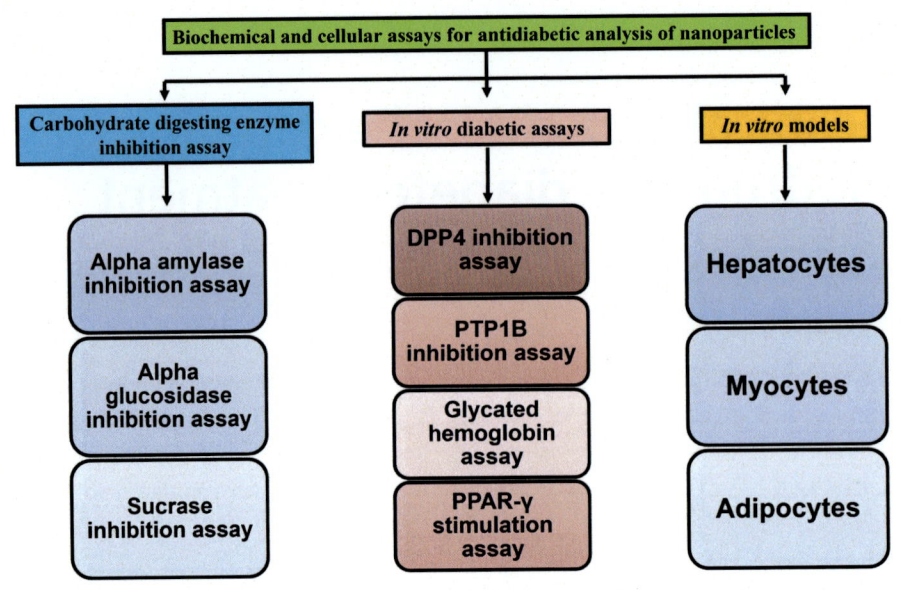

Figure 9.1 Biochemical and cellular assays for antidiabetic evaluation.

research reports have indicated effective inhibition of these carbohydrate digesting enzymes by nanoparticles to reduce glucose generation.

Alpha-amylase inhibition assay

This assay is usually performed in a microtiter plate similar to the standard starch-iodine test with few modifications. The nanoparticles at different concentrations are added to 150 μL of 0.02 M sodium phosphate buffer at pH 6.9 with 6 mM sodium chloride and 75 μL of alpha-amylase solution. This mixture is incubated for 15 min at 37°C. Later, 100 μL of the mixture is aliquoted and mixed with 750 μL of soluble starch and phosphate buffer (500 μL) solution and incubated for 45 min at 37°C. Of note, 25 μL of the incubated mixture is then mixed with 2.5 mL iodine reagent and stirred well. Changes in the color of the mixture are monitored via absorbance at 565 nm using a microplate reader. The presence of starch is indicated by a dark blue color. Partial starch degradation is indicated by a brown color, and complete starch degradation is indicated by a yellow color. If the nanoparticles possess amylase inhibition property, the starch in the mixture will not be degraded, showing a dark blue color (Sudha et al., 2011).

Metal and metal oxides nanoparticles are mostly explored for the inhibition of alpha-amylase enzyme to reduce blood glucose levels. Johnson et al. (2018) reported that silver nanoparticles synthesized using the flower extracts of *Bauhinia variegata* possessed enhanced ability to inhibit alpha-amylase and functioned as an effective antioxidant scavenger (Johnson et al., 2018). The alpha-amylase inhibition assay has been used to show the antidiabetic activity of silver nanoparticles synthesized using *Ocimum basilicum*, *Ocimum sanctum* (Malapermal et al., 2017), *Costus pictus* (Aruna et al., 2014), the leaf extract of *Calophyllum tomentosum* (Govindappa et al., 2018), *Punica granatum* (Saratale et al., 2018), and *Gynura procumbens* encapsulated with fungal chitosan (Sathiyaseelan et al., 2020). Naqvi et al. (2017) reported that copper nanoparticles synthesized using the extracts of *Aspergillus niger* also possessed enhanced antidiabetic activity by inhibiting alpha-amylase carbohydrate digesting enzyme. Dhobale et al. (2008) reported that zinc oxide (ZnO) nanoparticles possessed alpha-amylase inhibition mediated antidiabetic activity. In addition, other studies have reported amylase inhibition activities for ZnO nanoparticles synthesized via mangrove plant extract from *Sonneratia apetala* and *Heritiera fomes* (Thatoi et al., 2016), *Sageretia thea* (Khalil et al., 2017), *Andrographis paniculata* (Rajakumar et al., 2018), *Geranium wallichianum* (Abbasi et al., 2020), and wet chemical method using starch as reducing agent (Aroma et al., 2016). Amylase inhibition assay has also been used to evaluate the antidiabetic activity of polyvinylpyrrolidone and polyethylene glycol (PEG) doped copper oxide (CuO) nanoparticles (Javed et al., 2017). Also, iron oxide nanoparticles synthesized using phytochemicals extracted from the leaves of *Rhamnus virgata* were subjected to in vitro alpha-amylase inhibition assay and were shown to possess enhanced antidiabetic and antioxidant properties (Abbasi, Iqbal, Mahmood, Ahmad, et al., 2019; Abbasi, Iqbal, Mahmood, Qyyum, et al., 2019). Furthermore, gold (Saware et al., 2015), silver-coated silica (Selvan et al., 2020), silver-doped indium oxide (Naik et al., 2016), and nickel oxide nanoparticles (Abbasi, Iqbal, Mahmood, Ahmad, et al., 2019; Abbasi, Iqbal, Mahmood, Qyyum, et al., 2019) have also been shown to possess antidiabetic activities via alpha-amylase inhibition assay.

Several polymer nanoparticles, either free or as a carrier for insulin or conventional antidiabetic drugs, have been shown to have a promising ability to mitigate diabetes complications by inhibiting alpha-amylase (Wang & Zhang, 2012). Al Rashid (2017) reported a novel method to fabricate nanosized ethyl acetate fractions (extracted from *Diospyros melanoxylon* leaves) that are encapsulated with poly(lactic-*co*-glycolic acid) (PLGA) polymer. The alpha-amylase assay revealed that the polymer-encapsulated nanoparticles possessed enhanced antidiabetic activity compared to free nanoparticles (Al Rashid, 2017). Jiang et al. (2018) also reported that spherical and polygonal morphologies of natural starch polymer nanoparticles possessed the ability to inhibit the pancreatic alpha-amylase enzyme in vitro (Jiang et al., 2018). In addition to polymeric nanoparticles, some hybrid and nanocomposites have also proven to possess alpha-amylase inhibition ability. Bakur et al. (2019) demonstrated that a complex nanocomposite prepared with mannosylerythritol lipids (MEL), silver nanoparticles, zinc oxide (ZnO) nanoparticles, and dried gum Arabic exudation extracted from the stems of *Acacia senegal* possessed enhanced ability to inhibit alpha-amylase enzyme compared to MEL−silver and MEL−ZnO nanoparticles (Bakur et al., 2019). Palladium-reduced graphene oxide (Hazarika et al., 2019) and silver-silica in core-shell morphology (Selvan et al., 2019) are some additional nanocomposites with alpha-amylase inhibition abilities.

Alpha-glucosidase inhibition assay

The alpha-glucosidase inhibition assay is performed via dilution of the enzyme solution in phosphate buffer to make a stock solution of 10 mM and pH 6.8. Of note, 10 μL aliquot of the enzyme solution is added to 60 μL of the dispersed sample. The resulting mixture is incubated at 37°C for 20 min. After incubation, 30 μL of 1,4-phenylenediboronic acid is added at room temperature, gently mixed, and incubated for 15 min. The absorbance of the resulting solution is measured at 520 nm to evaluate enzyme inhibition (Zhang et al., 2015). Alpha-glucosidase catalyzes the breakdown of starch and disaccharides into glucose. Thus inhibition of this enzyme reduces glucose generation, lowering blood glucose levels in diabetic patients (Moelands et al., 2018). Generally, acarbose/miglitol and glyset are the main classes of alpha-glucosidase inhibitors, which involve both exercise and diet programs for controlling high blood glucose in type 2 diabetic patients (Coleman et al., 2019). Acarbose is a complex oligosaccharide that reduces the ability to digest ingested carbohydrates by reducing plasma glucose and glycosylated hemoglobin levels in type 2 diabetes conditions, leading to a slower increase in the concentration of blood glucose after meals (Pan et al., 2016). Miglitol also reduces blood glucose level following a similar mechanism as acarbose. However, it is a desoxynojirimycin derivative, which reduces average blood glucose concentration over time (Dash et al., 2018). It should be noted that the treatment of diabetes with alpha-glucosidase inhibitors can lead to severe cardiovascular ailments (Chang et al., 2015) and obesity-related problems (Hao et al., 2017). Similar to alpha-amylase, several nanoparticles possess the ability to inhibit alpha-glucosidase by using either the acarbose or miglitol mechanisms.

Metal nanoparticles are commonly investigated for their alpha-glucosidase inhibition properties. Gold nanoparticles synthesized from the extracts of brown seaweed *Padina boergesenii* (Senthilkumar et al., 2015), procyanidins from *Leucosidea sericea* (Badeggi et al., 2020), and marine algae *Gelidiella acerosa* (Senthilkumar et al., 2019) have been demonstrated to possess alpha-glucosidase inhibition properties. Also, silver nanoparticles synthesized via seagrass *Enhalus acoroides* (Senthilkumar et al., 2016), *Tephrosia tinctoria* (Rajaram et al., 2015), *Allium cepa* (Jini & Sharmila, 2019), *Lonicera japonica* (Balan et al., 2016), sprouted *Zingiberaceae* species (Mohammed et al., 2021), and *Argyreia nervosa* (Saratale et al., 2017) have been shown to possess the ability to inhibit alpha-glucosidase. Chen et al. (2015) showed that unmodified gold nanoparticles can be used in fabricating colorimetric sensors for evaluating alpha-glucosidase activity and inhibition (Chen et al., 2015). Copper nanoparticles fabricated from extracts of *Dioscorea bulbifera* (Ghosh et al., 2015) and *Gnidia glauca* and *Plumbago zeylanica* (Jamdade et al., 2019) have been reported to exhibit alpha-glucosidase inhibition property. Metallic nanorods (Cheng et al., 2019), metal-doped nanomaterials (Naik et al., 2016), silver nanotriangles (Senthilkumar et al., 2016), and bimetallic gold—silver nanoparticles (Malapermal et al., 2015) are additional nanomaterials for detecting alpha-glucosidase levels or inhibit their activity based on morphology and composition.

Some metal oxide nanoparticles possess alpha-glucosidase inhibition properties. ZnO nanoparticles have been shown to possess inhibition properties against alpha-glucosidase (Balogun & Ashafa, 2020). ZnO nanoparticles prepared from the leaf extracts of *Costus igneus* (Vinotha et al., 2019), *A. paniculata* (Rajakumar et al., 2018), and lemon juice extract (Prasad et al., 2019) are reported to possess antidiabetic activity by inhibiting alpha-glucosidase. Silver-doped indium oxide (Naik et al., 2016), ZnO quantum dots (Asok et al., 2015), and superparamagnetic iron oxide (Lunov et al., 2010) are additional metal oxide nanoparticles that have been reported to have antidiabetic properties based on alpha-glucosidase inhibition.

Some carbon-based nanoparticles, including graphene, carbon nanotubes, carbon dots, and fullerenes, have also been used to demonstrate alpha-glucosidase inhibition. Gold—silver—indium—reduced graphene oxide nanocomposites were prepared from the leaf extract of *Piper pedicellatum* and were reported to possess enhanced ability to inhibit alpha-glucosidase with negligible cytotoxic effect toward normal healthy cells (Saikia et al., 2018). Zhang et al. (2018) showed that microreactors fabricated via reduced graphene oxide—iron oxide nanocomposites can serve as a biosensor for screening alpha-glucosidase inhibitors present in *Dioscorea opposita* Thunb. peel (Zhang et al., 2018). Shao et al. (2019) reported that carbon nanoparticles fabricated via carbonization of polysaccharides from the root of *Actium lappa* are beneficial in inducing hypoglycemic effects in diabetic mice by inhibiting alpha-glucosidase enzyme activity. Maltase decorated chiral carbon dots have been used to inhibit alpha-glucosidase to control glucose levels in biological fluids (Zhang et al., 2019). Different studies have also reported the use of fullerenes to inhibit glucosidase and reduce glucose levels (Abellán Flos et al., 2016; Alali et al., 2019; Compain et al., 2010).

Polymer nanoparticles can be used to encapsulate or formulate bioactive materials with alpha-glucosidase inhibition properties. Sechi et al. (2016) reported that PLGA—PEG—carboxyl group and poly-(ε-caprolactone) (PCL)-based polymeric nanoparticles are highly beneficial in formulating dietary flavonoid fisetin to elevate their antioxidant and alpha-glucosidase inhibition activities (Sechi et al., 2016). It has also been shown that F127-poly ethylene glycol micellar system is useful in loading mangiferin to improve water solubility, enhance alpha-glucosidase inhibition activity, and reduce toxicity with low lactate dehydrogenase activity (Bezerra et al., 2019). Of note, 8A dendrimer nanoformulation of phenols from biological sources has been proven to elevate antidiabetic property via inhibition of carbohydrate digesting enzymes, especially alpha-glucosidase (Hoda et al., 2019).

Sucrase inhibition assay

Honda and Hara (1993) used a biochemical assay to analyze sucrase inhibition. Briefly, 10 μL of the enzyme solution is incubated for 10 min at 37°C with 25−200 μg/mL of buffered solubilized sample in maleate buffer at pH 6 to make up a volume of 200 μL. The reaction is initiated by adding 100 μL of 60 mM sucrose solution. The reaction is terminated after 30 min by adding 200 μL of 3,5-dinitrosalysilic acid. The mixture is then incubated for 5 min in a boiling water bath. The absorbance at 540 nm is measured to determine sucrase inhibition activity. Some nanomaterials have demonstrated sucrase inhibition properties. For example, Lu et al. (2019) fabricated carbon dots from charred *Fructus crataegi* and showed their ability to inhibit sucrase activity and reduce in vivo postprandial blood glucose levels (Lu et al., 2019). There are various sucrase inhibition compounds that can be natural sourced (Alam et al., 2019). For example, ethanolic extract from the stem of *Coscinium fenestratum* possesses the ability to inhibit intestinal maltase and sucrase (Adisakwattana et al., 2009). Enzyme inhibition compounds from natural sources can be encapsulated using nanoparticles to enhance their solubility and activity (Jeevanandam et al., 2016).

In vitro diabetic assays

In addition to the inhibition of carbohydrate enzymes, nanomaterials can exhibit antidiabetic activity by inhibiting diabetes markers such as dipeptidyl peptidase IV (DPP4), protein tyrosine phosphatase-1β (PTP1B), and hemoglobin glycation. Nanomaterials can also stimulate peroxisome proliferator−activated receptor-γ, which can be evaluated via specific assays to determine their antidiabetic activity.

Dipeptidyl peptidase IV inhibition assay

DPP4 or adenosine deaminase complexing protein 2 (CD26) is an enzyme that acts as a type 2 transmembrane glycoprotein and exists in the blood plasma and other body fluids (Barnett, 2006). DPP4 inhibitors help to increase the level of incretins by inhibiting glucagon release while increasing the secretion of insulin, decreasing gastric emptying and blood glucose level (Bae, 2016). Several antidiabetic drugs such as sitagliptin, Saxagliptin, and Linagliptin act as DPP4 inhibitors for the treatment of type 2 diabetes (Berger et al., 2018). DPP4 inhibition assay can be performed by four different methods. These are fluorometric assay using glycyl-prolyl-4-methoxy-β-naphthylamide (gly-pro-4-Me-β-NA) and glycyl-prolyl-7-amino-4-methylcoumarin (Gly-Pro-AMC), colorimetric assay using glycyl-prolyl-para-nitroanilide (Gly-Pro-*p*NA), and DPP4-Glo™ protease assay. Gly-pro-4-Me-β-NA and Gly-Pro-AMC are used as fluorogenic substrates for analyzing the activity of DPP4. For tests using blood plasma samples, 10 μL of the plasma sample is mixed in a 96-well plate with 100 μL of 0.5 mM gly-pro-4-Me-β-NA diluted in a 100 mM stock solution of dimethyl sulfoxide (DMSO) at −20°C and 50 mM Tris buffer at pH 8.3 to make a final volume to 110 μL. The activity of the enzyme is determined by incubating the mixture for 10 min at 37°C and measuring the fluorescence at 340 and 430 nm wavelengths. The fluorescent intensity data are used to determine the DPP4 inhibitory activity of the sample (Scharpé et al., 1988). The same method is followed for Gly-Pro-AMC DPP4 fluorometric assay, in which case 0.5 mM Gly-Pro-AMC is used and the fluorescent intensity is measured at 380 and 460 nm (Kim et al., 2012).

The colorimetric assay to determine DPP4 inhibitory activity uses Gly-Pro-pNA as a chromogenic substrate. Briefly, 10 μL of blood plasma sample is mixed in a 96-well plate with 0.5 mM of Gly-Pro-pNA in Tris buffer to make a final volume of 200 μL. The mixture is incubated for 10 min at 37°C and the optical absorbance at 405 nm is measured to determine the concentration of DPP4 (Dubois et al., 2008). The DPP4–Glo protease assay is an assay kit commonly used to measure DPP4 inhibition. Matheeussen et al. (2012) compared all four assays for DPP4 inhibition and concluded that they are all useful in determining DPP4 activity in noninhibiting biological samples. However, they indicated that the fluorometric assay is more suitable for low sample volumes due to the degree of sensitivity. They also suggested that residual in vivo DPP4 activity in the presence of inhibitors can be estimated via elaborated fluorometric methods (Matheeussen et al., 2012).

Various research studies have investigated the DPP4 inhibition capacity of different nanomaterials, nanoformulations, and extracts for nanoparticles generation. Huang et al. (2017) reported that 16-hydroxycleroda-3,13-dine-16,15-olide phytocompound extracted from *Polyalthia longifolia* possesses enhanced antidiabetic activity via DP44 inhibition. The study showed that encapsulation of the phytocompound in mesoporous silicon dioxide nanoparticles enhanced the DPP4 inhibition property and reduced blood glucose levels in diabetic mice (Huang et al., 2017). Also, Sitagliptin drug loaded into PLGA polymeric nanoparticles and conjugated with anti-CD-4 mAB enhanced DPP4 inhibition efficacy to reduce glycemic levels in type 1 diabetes (Thondawada et al., 2018). ZnO nanoparticle with conventional DPP4 inhibitors, such as Vildagliptin, has been reported to enhance antidiabetic activity through DPP4 inhibition compared to the individual nanoparticles and the drug (El-Gharbawy et al., 2016). A multistage dual-drug delivery nanosystem was synthesized using chitosan-modified porous silicon nanoparticles coated with an enteric polymer, hydroxypropylmethylcellulose acetate succinate. The nanosystem showed promise in use as an oral codelivery carrier for incretin glucagon-like peptide-1 hormone and DPP4 inhibitor for the treatment of chronic diabetic complications (Shrestha et al., 2016).

Protein tyrosine phosphatase-1β inhibition assay

PTP1B is an enzyme that acts as a negative regulator of insulin and leptin signaling pathways. The enzyme is associated with diabetes and obesity (Cho, 2013).

The PTP1B inhibition assay is performed using p-nitrophenyl phosphate (pNPP) as a substrate along with a buffer solution containing 25 mM Tris—HCl at pH 7.5, 1 mM ethylenediaminetetraacetic acid, 2 mM β-mercaptoethanol, and 1 mM dithiothreitol. The test uses 10 μL of the sample with 20 μL of 1 μg/mL enzyme and 40 μL of 4 mM pNPP mixed with 130 μL of buffer solution in a 96-well plate for 10 min at 37°C. pNP is produced as a result of pNPP dephosphorylation, and this is monitored at 405 nm (Fang et al., 2014; Song et al., 2017).

A wide range of nanomaterials have been used to deliver PTP1B inhibition and antidiabetic activities. Propionic acid extracted from *Cassia auriculata* can be used to synthesize organic—inorganic hybrid gold nanoparticles with enhanced antidiabetic activity that is associated with the inhibition of PTP1B, along with alpha-glucosidase and alpha-amylase activities (Suganya et al., 2019). It has also been demonstrated that Vicenin-2-coated gold nanoparticles possess the ability to increase glucose uptake in diabetic 3T3-L1 adipocytes by binding and inhibiting enzymes such as PTP1B and adenosine monophosphate—activated protein kinase (Chockalingam et al., 2015).

Glycated hemoglobin inhibition assay

Hemoglobin in the red blood cells serves as an oxygen carrier for transferring oxygen throughout the body. The glucose, which is present in blood, possesses the ability to bind with hemoglobin in case of hypoglycemia and affects oxygen transfer. It is highly essential to inhibit glycated hemoglobin (HbA1C) as a strategy for effective treatment of diabetes condition (Selvin et al., 2010). The assay for glycated hemoglobin inhibition uses a fresh blood sample, which is centrifuge to separate erythrocytes. Of note, 5 mL of erythrocytes is washed with phosphate-buffered saline solution three times to remove impurities. The erythrocytes sample is further separated and centrifuged to obtain hemoglobin from the supernatant. The concentration of hemoglobin is estimated by the Drabkin and Austin method (Drabkin & Austin, 1932; Koenig et al., 1977). The HbA1C assay is performed by adding 60 mg/100 mL hemoglobin solution, 1 mL gentamycin in 0.01 M phosphate buffer at pH 7.4 in the presence of 2 g/100 mL concentration of glucose and incubated for 72 h in the dark at room temperature. The concentration of glycated hemoglobin is measured at 443 nm. The same procedure is used to analyze the HbA1C inhibition ability of the sample and the percentage of glycated hemoglobin is determined.

Different types of nanomaterials have been used as biosensors for the detection of glycated hemoglobin in the biological samples with high sensitivity. Electrochemical sensors can be developed for the detection of HbA1C using gold nanoparticles embedded in nitrogen-doped graphene nanosheets (Jain & Chauhan, 2017). Also, nanosized propolis (Kazemi et al., 2018), zinc oxide (Chawla & Pundir, 2012), lanthanide-doped (Jo et al., 2016), and gold—platinum bimetallic (Jain et al., 2017) particles have been used in developing sensors for HbA1C detection. Various nanomaterials have also been used for the inhibition or reduction of glycated hemoglobin. For example, it has been demonstrated that ZnO nanoparticles possess the ability to reduce the formation of glycated hemoglobin in streptozotocin-induced diabetic rats (Afify et al., 2019). Also, gold nanoparticles that are stabilized using phytochemicals from pomegranate peel extract have been demonstrated to possess the ability to reduce HbA1C of regulating NF-kB and Nrf2 signaling system (Manna et al., 2019). Cerium oxide nanoparticles (Gharebaghi et al., 2019), pelargonidin encapsulated biocompatible PLGA nanoparticles (Roy et al., 2017), gold nanoparticles synthesized from the extracts of *Turbinaria conoide* brown seaweed (Venkatraman et al., 2018), and silver nanoparticles prepared via *Centella asiatica* (Wilson et al., 2015) are additional nanoparticles that have been reported to possess the potential to reduce glycated hemoglobin.

Peroxisome proliferator–activated receptor-γ stimulation assay

Peroxisome proliferator–activated receptor-γ luciferase (PPARγ) belongs to the superfamily of nuclear receptors and plays a crucial role in various biological activities such as lipid metabolism, adipogenesis, and placenta development (Degrelle et al., 2017). PPARγ helps in the regulation of glucose metabolism, adipocyte differentiation, and storage of fatty acids. In the case of diabetes, stimulation of PPARγ helps to reduce serum glucose without elevating the secretion of pancreatic insulin (Semple et al., 2006). In general, a screening assay of transient transfection via PPARγ2 expression construct is performed to measure the inhibition of PPARγ transactivation (Petersen et al., 2012; Kopp et al., 2015). Curcumin-loaded polylactic acid (PLA)-PEG copolymer nanoparticles have a high potential in reducing hepatic PPARγ for the treatment of diabetes (El-Naggar et al., 2019). Gold nanoparticles functionalized with the phytochemicals of fresh *Panax ginseng* leaf extract have been shown to suppress the process of adipogenesis by downregulating PPARγ signaling pathway, thereby reducing glucose level in mature 3T3-L1 adipocytes (Simu et al., 2019). The study demonstrated potential application in the treatment of obesity, hypertension, cancer, and stroke associated with type 2 diabetes.

In vitro models to evaluate insulin secretion

Various cell lines have been used as in vitro models for the evaluation of insulin secretion or other biochemical effects of nanomedicines in diabetic cells (Hattangady & Rajadhyaksha, 2009; Pedersen et al., 2009). Cell lines such as hepatocytes (Thule et al., 2000), myocytes (Lorenzo et al., 2008), and adipocytes (Jeevanandam et al., 2019) are conventional in vitro models that are used to study the effect of nanomedicines on insulin secretion. We have extensively discussed in vitro cell and tissue diabetic models in Chapter 7.

Evaluation of antidiabetic nanoparticles

Clinical evaluation of antidiabetic nanoparticles involves in-depth in vivo determination of blood glucose, glycated proteins, and postprandial blood glucose levels. Clinical evaluation also demands GCP to support regulatory approval. We have discussed in vivo diabetic models extensively in Chapter 8. Here, we focus on important clinical tests that are commonly performed to support diabetes evaluation.

In vivo determination of blood glucose

Mouse models are widely used as in vivo animal subjects to determine the efficacy of nanomedicines in reducing blood glucose levels. Gold nanoparticles synthesized using *Gymnema sylvestre* possess the ability to reduce glucose by exhibiting antiinflammatory effects via modulating tumor necrosis factor-alpha, high-sensitive C-reactive protein, and interleukin-6 in serum, a study performed with alloxan-induced diabetic rats under clinical conditions (Karthick et al., 2014). Polymeric nanoparticles coated with Fc fragments to bind the neonatal Fc receptor in the epithelium of the intestine can act as an oral drug delivery system to reduce glucose levels (Pridgen et al., 2013). Also, streptozotocin-induced diabetic rats have been used to validate the hypoglycemic property of gold nanoparticles synthesized using the aqueous phytochemical extracts of *Cassia fistula* stem bark (Daisy & Saipriya, 2012).

In vivo analysis of glycated proteins

Apart from blood glucose, glycated proteins are a major indicator of the severity of diabetic conditions and help to evaluate the efficacy of nanomedicines. It has been demonstrated that cerium chloride loaded mesoporous silicon dioxide nanoparticles has the potential to alleviate the development and progression of diabetic cataract conditions, possessing the ability to reduce oxidative stress by modulating glycated proteins (Yang et al., 2017). Also, alloxan-induced diabetic male Wistar Albino rats have been used to show the antidiabetic efficacy of gold nanoparticles synthesized using the algal extracts of *Sargassum swartzii*. The study showed that the antidiabetic effect of the gold nanoparticles was due to reduction of fasting blood glucose levels and glycosylated hemoglobin (Dhas et al., 2016). An in vivo study using diabetic ob/ob mice showed the pharmacodynamics of exenatide nanoparticles embedded in gastro-resistance particles administered via the oral route. The study showed that the antidiabetic activity of the nanoparticles was associated with decreasing blood glucose level and glycated hemoglobin (Soudry-Kochavi et al., 2018).

In vivo analysis of postprandial blood glucose

Postprandial glucose is the quantity of glucose present in diabetic patients after a meal, which generally increases due to increase in carbohydrate intake as food, compared to fasting (before meal) conditions. There are numerous studies using animal models that demonstrate the efficacy of nanomedicines to reduce the concentration of postprandial glucose in blood. Miglitol nanoparticles exhibited antidiabetic properties by inhibiting alpha-glucosidase enzyme in the small intestines of streptozotocin-induced diabetic Wistar rats. The nanoparticles helped in reducing both fasting and postprandial plasma glucose (Karuppusamy et al., 2017). Type 1 diabetic Swiss mice have been used to demonstrate that hydroxy-aluminum phthalocyanine with nanostructured copolymeric Pluronic and Carbopol hydrogel possesses the ability to reduce postprandial plasma glucose level. The study showed that the hydrogel had the ability to treat cutaneous ulcers in diabetic conditions (Melo et al., 2019). Gold nanoparticles synthesized using *Streptomyces coelicoflavus* have the ability to reduce postprandial hyperglycemic condition via in vivo studies conducted using streptozotocin-induced diabetic male albino Wister rats (Sathish Kumar & Bhaskara Rao, 2016).

FDA approved polymeric antidiabetic nanomedicines

Some nanomedicines have received the United States Food and Drug Administration (USFDA) approval to be marketed as drug materials for the treatment of diabetes. PEGylated liposomal nanoparticles system was one of the early FDA-approved nano-formulations used to encapsulate the cancer drug, doxorubicin (Barenholz, 2012). This approval catalyzed the immediate filing of about 20 nanoformulations for clinical investigation (Bertrand et al., 2014). There are about 50 nanopharmaceuticals that have been approved by FDA in the time period between 1995 and 2021. Nanomedicines fail the FDA approval process due to challenges associated with successful clinical demonstrations of efficacy and safety. Recent FDA guidelines have several amendments that can promote the development of nanoformulations for clinical applications (Leong et al., 2019). PLGA is a novel copolymer with biodegradable properties that was approved by FDA to be used in therapeutic devices, sutures, and nanomedicine formulation (Bernhardt, 2014). Biodegradable PLGA nanoparticles have been approved for use as a next-generation tolerogenic vaccine against several autoimmune diseases such as type 1 diabetes, multiple sclerosis, and rheumatoid arthritis (Cappellano et al., 2019).

Chitosan is a type of natural polymer approved by FDA for biomedical applications, due to its biocompatibility, bioavailability, and antimicrobial property. Chitosan in its nanoform (nanochitosan) can be used to synthesize novel nanoparticles-encapsulated wound dressings for controlled drug release to regenerate wound-affected body parts (Lin et al., 2017). HemCon is a novel chitosan-based dressing approved by FDA as a hemostatic agent used as a bandage to control heavy bleeding wounds (Wedmore et al., 2006). In addition, other synthetic polymers such as PLA, polyglycolide, and PCL are approved by FDA and European Medicines Agency

(EMA) for the fabrication of surgical devices, implants, and suture threads (Tyler et al., 2016). However, the nanosized versions of these polymers are mostly under clinical trials. Polymeric nanoparticles can serve as effective carriers for conventional wound healing drugs to enhance diabetes-associated wound regeneration via controlled delivery (Berthet et al., 2017). Several nanodrugs and nanomedicines for the treatment of various diseases are in the pipeline to obtain FDA approval or under research investigations. Inorganic nanoparticles such as iron dextran, Ferumoxytol, sodium ferric gluconate complex in sucrose, and iron sucrose are approved by FDA to treat iron deficiency in chronic kidney disease associated with diabetes (Ventola, 2017). Hence, a widespread use of nanomedicines for the treatment of diabetes is achievable with improve clinical validation.

Limitations to the use of nanoparticles in disease treatments

FDA and EMA are the major regulatory bodies that control the use of nanomedicines for the treatment of diseases. There are regulations much stricter than conventional drugs for nanoparticles-based medicines due to the potential to cause severe life-threatening side-effects if not properly validated clinically (Nie et al., 2020). A comprehensive efficacy and toxicological assessment are needed to support the approval of nanoparticles. Nanoparticles require a proper documentation based on their specific toxicity and safety levels. Numerous data on the efficacy of nanoformulations such as liposomes, micelles, and polymers have been reported. However, it is important to document the data based on size, morphology, encapsulation efficiency, and surface charge to create a standardized repository for drug nanoencapsulation toward specific applications (Havel et al., 2016). The next significant factor that affects nanoparticles is toxicity. Full toxicological analyses of nanomedicines are important to support regulatory approvals. The toxicity of nanomedicines can differ depending on the administration route, thus, clinical studies should also focus on evaluating administration route-mediated toxicity (Wolfram et al., 2015). Another major concern for nanomedicines is the lack of specific regulatory guidelines. FDA is the primary regulatory body, which approves nanomedicines. Initially, preclinical studies with animal models are used to evaluate the safety, efficacy, and dosage. Later, clinical trials are carried out for the determination of efficacy and safety in humans. The clinical trials are classified into Phases I to III. Typically, in Phase I, dose, excretion, and toxicity in healthy subjects are evaluated, and several assays are used to assess the cytotoxicity, hemolysis, hemagglutination, thrombogenicity, cardiotoxicity, endothelial permeability, inflammation, phagocytosis, and complement activation (Saha et al., 2019). Safety and efficacy in disease-affected subjects are evaluated in Phase II, whereas randomized, multicenter, placebo-controlled trials are performed in Phase III. After Phase II, a novel drug application is filed with FDA for approval. Phase IV trial is conducted after receiving approval, and this is performed upon request from healthcare professionals or FDA. Several researchers have proposed that nanomedicines should have specific regulatory guidelines due to their unique characteristics (Ventola, 2012). In 2014, FDA issued guidelines for the utilization of nanotechnology in products that are regulated by FDA. It is anticipated that with time FDA will release a specific guideline to facilitate the use of nanomedicines and nanobiologics (Ventola, 2017).

Recommendations and future perspective

Nanomedicines continue to emerge as novel therapeutics for the treatment of diseases, including diabetes. Even after, several preclinical and clinical studies, there are significant hurdles to address before bringing nanomedicines to the market. The limitations are still prevailing and even more so when nanomedicines are developed for diabetes as diabetic patients take drugs more routinely and this can lead to exacerbated toxic effects. Predictive toxicity profiling of nanomedicines that leverages advanced data analytics approaches is important to facilitate the development and application of nanomedicines. In silico methods are also under extensive research investigations to determine the toxicity of nanomedicines, allowing extrapolation to live animal models or humans with high precision. Furthermore, it is essential to establish more specific regulatory guidelines to validate and control biomedical products and therapeutics generated from nanotechnology.

Conclusion

This chapter presented an overview of various inhibition assays of diabetes biochemical markers, in vitro studies, and antidiabetic assays that are commonly used to evaluate the antidiabetic properties of nanomedicines. Furthermore, in vivo studies used to evaluate the antidiabetic efficacy of some nanomedicines are discussed. Nanomedicines that are approved by FDA and other regulatory bodies are also discussed in relation to current the regulatory framework for FDA approval. Despite the large numbers of preclinical and clinical studies on nanomedicines for the formulation of conventional drugs and as direct therapeutics for diabetes, nano-based medicines are less on the market due to the various aforementioned limitations. More comprehensive and high-precision biomedical research approaches are required to address these limitations to support the validation and approval of nanodrugs for diabetes treatment applications.

References

Abbasi, B. A., Iqbal, J., Ahmad, R., Zia, L., Kanwal, S., Mahmood, T., Wang, C., & Chen, J.-T. (2020). Bioactivities of *Geranium wallichianum* leaf extracts conjugated with zinc oxide nanoparticles. *Biomolecules*, *10*(1), 38.

Abbasi, B. A., Iqbal, J., Mahmood, T., Ahmad, R., Kanwal, S., & Afridi, S. (2019). Plant-mediated synthesis of nickel oxide nanoparticles (NiO) via *Geranium wallichianum*: Characterization and different biological applications. *Materials Research Express, 6*(8), 0850a7.

Abbasi, B. A., Iqbal, J., Mahmood, T., Qyyum, A., & Kanwal, S. (2019). Biofabrication of iron oxide nanoparticles by leaf extract of *Rhamnus virgata*: Characterization and evaluation of cytotoxic, antimicrobial and antioxidant potentials. *Applied Organometallic Chemistry, 33*(7), e4947.

Abellán Flos, M., García Moreno, M. I., Ortiz Mellet, C., Garcia Fernandez, J. M., Nierengarten, J., & Vincent, S. P. (2016). Potent glycosidase inhibition with heterovalent fullerenes: Unveiling the binding modes triggering multivalent inhibition. *Chemistry—A European Journal, 22*(32), 11450−11460.

Adisakwattana, S., Chantarasinlapin, P., Thammarat, H., & Yibchok-Anun, S. (2009). A series of cinnamic acid derivatives and their inhibitory activity on intestinal α-glucosidase. *Journal of Enzyme Inhibition and Medicinal Chemistry, 24*(5), 1194−1200.

Afify, M., Samy, N., Hafez, N. A., Alazzouni, A. S., Mahdy, E. S., El Mezayen, H. A. E.-M., & Kelany, M. M. (2019). Evaluation of zinc-oxide nanoparticles effect on treatment of diabetes in streptozotocin-induced diabetic rats. *Egyptian Journal of Chemistry, 62*(10), 1771−1783. Available from https://doi.org/10.21608/ejchem.2019.11350.1735.

Al Rashid, H. (2017). Preparation and characterization of PLGA loaded nanoparticles obtained from *D. melanoxylon* Roxb. leaves for their antiproliferative and antidiabetic activity. *International Journal of Green Pharmacy (IJGP), 11*(03).

Alali, U., Vallin, A., Bil, A., Khanchouche, T., Mathiron, D., Przybylski, C., Beaulieu, R., Kovensky, J., Benazza, M., & Bonnet, V. (2019). The uncommon strong inhibition of α-glucosidase by multivalent glycoclusters based on cyclodextrin scaffolds. *Organic & Biomolecular Chemistry, 17*(30), 7228−7237.

Alam, F., Shafique, Z., Amjad, S. T., & Bin Asad, M. H. H. (2019). Enzymes inhibitors from natural sources with antidiabetic activity: A review. *Phytotherapy Research, 33*(1), 41−54.

Aroma, S. R., Namasivayam, S. K. R., & Thyagarajan, R. (2016). An investigation of chemogenic zinc oxide nanoparticles mediated enzyme activity inhibition under in vitro condition. *Pharma Letter, 8,* 425−431.

Aruna, A., Nandhini, R., Karthikeyan, V., Bose, P., & Vijayalakshmi, K. (2014). Comparative antidiabetic effect of methanolic extract of insulin plant (*Costus pictus*) leaves and its silver nanoparticle. *Indo American Journal of Pharmaceutical Research, 4*(7), 3217−3230.

Asok, A., Ghosh, S., More, P. A., Chopade, B. A., Gandhi, M. N., & Kulkarni, A. R. (2015). Surface defect rich ZnO quantum dots as antioxidants inhibiting α-amylase and α-glucosidase: A potential anti-diabetic nanomedicine. *Journal of Materials Chemistry B, 3*(22), 4597−4606. Available from https://doi.org/10.1039/C5TB00407A.

Badeggi, U. M., Ismail, E., Adeloye, A. O., Botha, S., Badmus, J. A., Marnewick, J. L., Cupido, C. N., & Hussein, A. A. (2020). Green synthesis of gold nanoparticles capped with procyanidins from *Leucosidea sericea* as potential antidiabetic and antioxidant agents. *Biomolecules, 10*(3), 452.

Bae, E. J. (2016). DPP-4 inhibitors in diabetic complications: Role of DPP-4 beyond glucose control. *Archives of Pharmacal Research, 39*(8), 1114−1128.

Bahman, F., Greish, K., & Taurin, S. (2021). *Insulin nanoformulations for nonparenteral administration in diabetic patients. Theory and applications of nonparenteral nanomedicines* (pp. 409−443). Elsevier.

Bakur, A., Elshaarani, T., Niu, Y., & Chen, Q. (2019). Comparative study of antidiabetic, bactericidal, and antitumor activities of MEL@AgNPs, MEL@ZnONPs, and Ag−ZnO/MEL/GA nanocomposites prepared by using MEL and gum arabic. *RSC Advances, 9*(17), 9745−9754. Available from https://doi.org/10.1039/C9RA00344D.

Balan, K., Qing, W., Wang, Y., Liu, X., Palvannan, T., Wang, Y., Ma, F., & Zhang, Y. (2016). Antidiabetic activity of silver nanoparticles from green synthesis using *Lonicera japonica* leaf extract. *Rsc Advances, 6*(46), 40162−40168.

Balogun, F. O., & Ashafa, A. O. T. (2020). Green-synthesized zinc oxide nanoparticles from aqueous root extract of *Dicoma anomala* (Sond.) mitigates free radicals and diabetes-linked enzymes. *Nanoscience & Nanotechnology-Asia, 10*(6), 918−929.

Barenholz, Y. C. (2012). Doxil®—The first FDA-approved nano-drug: Lessons learned. *Journal of Controlled Release, 160*(2), 117−134.

Barnett, A. (2006). DPP-4 inhibitors and their potential role in the management of type 2 diabetes. *International Journal of Clinical Practice, 60*(11), 1454−1470.

Berger, J. P., SinhaRoy, R., Pocai, A., Kelly, T. M., Scapin, G., Gao, Y., Pryor, K. A. D., Wu, J. K., Eiermann, G. J., & Xu, S. S. (2018). A comparative study of the binding properties, dipeptidyl peptidase-4 (DPP-4) inhibitory activity and glucose-lowering efficacy of the DPP-4 inhibitors alogliptin, linagliptin, saxagliptin, sitagliptin and vildagliptin in mice. *Endocrinology, Diabetes & Metabolism, 1*(1), e00002.

Bernhardt, P. (2014). Self-monitoring blood glucose test systems for over-the-counter use. Draft Guidance for Industry and Food and Drug Administration.

Berthet, M., Gauthier, Y., Lacroix, C., Verrier, B., & Monge, C. (2017). Nanoparticle-based dressing: The future of wound treatment? *Trends in Biotechnology, 35*(8), 770−784. Available from https://doi.org/10.1016/j.tibtech.2017.05.005.

Bertrand, N., Wu, J., Xu, X., Kamaly, N., & Farokhzad, O. C. (2014). Cancer nanotechnology: The impact of passive and active targeting in the era of modern cancer biology. *Advanced Drug Delivery Reviews, 66*, 2−25.

Bezerra, F. W. A., Fechine, L. M. U. D., Lopes, K. P. S., de Sousa, A. F., do Nascimento, G. O., Amaral, H. H., de, S., Leal, L. K. A. M., Trevisan, M. T. S., Ribeiro, M. E. N. P., & Ricardo, N. M. P. S. (2019). α-Glucosidase inhibitory activity of mangiferin-loaded F127/PEG micellar system. *Materials Letters, 255*, 126522. Available from https://doi.org/10.1016/j.matlet.2019.126522.

Bor, G., Mat Azmi, I. D., & Yaghmur, A. (2019). Nanomedicines for cancer therapy: Current status, challenges and future prospects. *Therapeutic Delivery, 10*(2), 113−132.

Cao, R. (2018). Diffusion of PLGA nanoparticle from alginate microcapsule to control immune response after implantation for type I diabetes.

Cappellano, G., Comi, C., Chiocchetti, A., & Dianzani, U. (2019). Exploiting PLGA-based biocompatible nanoparticles for next-generation tolerogenic vaccines against autoimmune disease. *International Journal of Molecular Sciences, 20*(1). Available from https://doi.org/10.3390/ijms20010204.

Chang, C.-H., Chang, Y.-C., Lin, J.-W., Chen, S.-T., Chuang, L.-M., & Lai, M.-S. (2015). Cardiovascular risk associated with acarbose vs metformin as the first-line treatment in patients with type 2 diabetes: A nationwide cohort study. *The Journal of Clinical Endocrinology & Metabolism, 100*(3), 1121−1129.

Chawla, S., & Pundir, C. S. (2012). An amperometric hemoglobin A1c biosensor based on immobilization of fructosyl amino acid oxidase onto zinc oxide nanoparticles−polypyrrole film. *Analytical Biochemistry, 430*(2), 156−162.

Chen, F., & Cai, W. (2015). Nanomedicine for targeted photothermal cancer therapy: Where are we now? *Nanomedicine: Nanotechnology, Biology, and Medicine, 10*(1), 1−3.

Chen, H., Zhang, J., Wu, H., Koh, K., & Yin, Y. (2015). Sensitive colorimetric assays for α-glucosidase activity and inhibitor screening based on unmodified gold nanoparticles. *Analytica Chimica Acta, 875*, 92−98.

Cheng, X., Huang, Y., Yuan, C., Dai, K., Jiang, H., & Ma, J. (2019). Colorimetric detection of α-glucosidase activity based on the etching of gold nanorods and its application to screen antidiabetic drugs. *Sensors and Actuators B: Chemical, 282*, 838−843.

Cho, H. (2013). Protein tyrosine phosphatase 1B (PTP1B) and obesity. In G. Litwack (Ed.), *Vitamins & hormones* (Vol. 91). Academic Press. Available from https://doi.org/10.1016/B978-0-12-407766-9.00017-1.

Chockalingam, S., Thada, R., Dhandapani, R. K., & Panchamoorthy, R. (2015). Biogenesis, characterization, and the effect of vicenin-gold nanoparticles on glucose utilization in 3T3-L1 adipocytes: A bioinformatic approach to illuminate its interaction with PTP 1B and AMPK. *Biotechnology Progress, 31*(4), 1096−1106. Available from https://doi.org/10.1002/btpr.2112.

Coleman, R. L., Scott, C. A. B., Lang, Z., Bethel, M. A., Tuomilehto, J., & Holman, R. R. (2019). *Meta*-analysis of the impact of alpha-glucosidase inhibitors on incident diabetes and cardiovascular outcomes. *Cardiovascular Diabetology, 18*(1), 135.

Compain, P., Decrooq, C., Iehl, J., Holler, M., Hazelard, D., Mena Barragan, T., Ortiz Mellet, C., & Nierengarten, J. (2010). Glycosidase inhibition with fullerene iminosugar balls: A dramatic multivalent effect. *Angewandte Chemie International Edition, 49*(33), 5753−5756.

Daisy, P., & Saipriya, K. (2012). Biochemical analysis of *Cassia fistula* aqueous extract and phytochemically synthesized gold nanoparticles as hypoglycemic treatment for diabetes mellitus. *International Journal of Nanomedicine, 7,* 1189−1202. Available from https://doi.org/10.2147/IJN.S26650.

Dash, R. P., Babu, R. J., & Srinivas, N. R. (2018). Reappraisal and perspectives of clinical drug−drug interaction potential of α-glucosidase inhibitors such as acarbose, voglibose and miglitol in the treatment of type 2 diabetes mellitus. *Xenobiotica; The Fate of Foreign Compounds in Biological Systems, 48*(1), 89−108.

Degrelle, S. A., Shoaito, H., & Fournier, T. (2017). New transcriptional reporters to quantify and monitor PPARγ activity. *PPAR Research, 2017,* 7. Available from https://doi.org/10.1155/2017/6139107.

Dhas, T. S., Kumar, V. G., Karthick, V., Vasanth, K., Singaravelu, G., & Govindaraju, K. (2016). Effect of biosynthesized gold nanoparticles by *Sargassum swartzii* in alloxan induced diabetic rats. *Enzyme and Microbial Technology, 95,* 100−106. Available from https://doi.org/10.1016/j.enzmictec.2016.09.003.

Dhobale, S., Thite, T., Laware, S. L., Rode, C. V., Koppikar, S. J., Ghanekar, R.-K., & Kale, S. N. (2008). Zinc oxide nanoparticles as novel alpha-amylase inhibitors. *Journal of Applied Physics, 104*(9), 094907.

Drabkin, D. L., & Austin, J. H. (1932). Spectrophotometric studies I. Spectrophotometric constants for common hemoglobin derivatives in human, dog, and rabbit blood. *Journal of Biological Chemistry, 98* (2), 719−733.

Dubois, V., Lambeir, A.-M., Van der Veken, P., Augustyns, K., Creemers, J., Chen, X., Scharpe, S., & De Meester, I. (2008). Purification and characterization of a dipeptidyl peptidase 9-like enzyme from bovine testes. *Frontiers in Bioscience: A Journal and Virtual Library, 13,* 3558−3568.

El-Gharbawy, R. M., Emara, A. M., & Abu-Risha, S. E.-S. (2016). Zinc oxide nanoparticles and a standard antidiabetic drug restore the function and structure of beta cells in Type-2 diabetes. *Biomedicine & Pharmacotherapy, 84,* 810−820. Available from https://doi.org/10.1016/j.biopha.2016.09.068.

El-Naggar, M. E., Al-Joufi, F., Anwar, M., Attia, M. F., & El-Bana, M. A. (2019). Curcumin-loaded PLA-PEG copolymer nanoparticles for treatment of liver inflammation in streptozotocin-induced diabetic rats. *Colloids and Surfaces B: Biointerfaces, 177,* 389−398. Available from https://doi.org/10.1016/j.colsurfb.2019.02.024.

Engel, S. S., Williams-Herman, D. E., Golm, G. T., Clay, R. J., Machotka, S. V., Kaufman, K. D., & Goldstein, B. J. (2010). Sitagliptin: Review of preclinical and clinical data regarding incidence of pancreatitis. *International Journal of Clinical Practice, 64*(7), 984−990.

Fang, L., Cao, J., Duan, L., Tang, Y., & Zhao, Y. (2014). Protein tyrosine phosphatase 1B (PTP1B) and α-glucosidase inhibitory activities of *Schisandra chinensis* (Turcz.) Baill. *Journal of Functional Foods, 9,* 264−270.

Gad, S. C. (2008). *Preclinical development handbook: Toxicology* (Vol. 4). Wiley.

Gharebaghi, A., Ranjbar, A., Artimani, T., Mirzaeiseresht, B., & Asl, S. S. (2019). The effect of cerium oxide nanoparticles on memory impairment and antioxidant capacity in streptozotocin-induced diabetic rats. *Physiology & Pharmacology, 23*(3).

Ghosh, S., More, P., Nitnavare, R., Jagtap, S., Chippalkatti, R., Derle, A., Kitture, R., Asok, A., Kale, S., & Singh, S. (2015). Antidiabetic and antioxidant properties of copper nanoparticles synthesized by medicinal plant *Dioscorea bulbifera. Journal of Nanomedicine & Nanotechnology, S6,* 1.

Govindappa, M., Hemashekhar, B., Arthikala, M.-K., Ravishankar Rai, V., & Ramachandra, Y. L. (2018). Characterization, antibacterial, antioxidant, antidiabetic, anti-inflammatory and antityrosinase activity of green synthesized silver nanoparticles using *Calophyllum tomentosum* leaves extract. *Results in Physics, 9,* 400−408. Available from https://doi.org/10.1016/j.rinp.2018.02.049.

Hao, L., Schlussel, Y., Fieselmann, K., Schneider, S., & Shapses, S. (2017). Appetite and gut hormones response to a putative α-glucosidase inhibitor, *Salacia chinensis,* in overweight/obese adults: A double blind randomized controlled trial. *Nutrients, 9*(8), 869.

Hattangady, N. G., & Rajadhyaksha, M. S. (2009). A brief review of in vitro models of diabetic neuropathy. *International Journal of Diabetes in Developing Countries, 29*(4), 143.

Havel, H., Finch, G., Strode, P., Wolfgang, M., Zale, S., Bobe, I., Youssoufian, H., Peterson, M., & Liu, M. (2016). Nanomedicines: From bench to bedside and beyond. *The AAPS Journal, 18*(6), 1373−1378.

Hazarika, M., Boruah, P. K., Pal, M., Das, M. R., & Tamuly, C. (2019). Synthesis of Pd-rGO nanocomposite for the evaluation of in vitro anticancer and antidiabetic activities. *ChemistrySelect*, *4*(4), 1244−1250. Available from https://doi.org/10.1002/slct.201802789.

Hoda, M., Hemaiswarya, S., & Doble, M. (2019). *Polyphenol nanoformulations with potential antidiabetic properties. Role of phenolic phytochemicals in diabetes management* (pp. 145−157). Springer.

Honda, M., & Hara, Y. (1993). Inhibition of rat small intestinal sucrase and α-glucosidase activities by tea polyphenols. *Bioscience, Biotechnology, and Biochemistry*, *57*(1), 123−124.

Huang, P.-K., Lin, S.-X., Tsai, M.-J., Leong, K. M., Lin, S.-R., Kankala, K. R., Lee, C.-H., & Weng, C.-F. (2017). Encapsulation of 16-hydroxycleroda-3,13-dine-16,15-olide in mesoporous silica nanoparticles as a natural dipeptidyl peptidase-4 inhibitor potentiated hypoglycemia in diabetic mice. *Nanomaterials*, *7*(5). Available from https://doi.org/10.3390/nano7050112.

Ioannidis, J. P. A., Kim, B. Y. S., & Trounson, A. (2018). How to design preclinical studies in nanomedicine and cell therapy to maximize the prospects of clinical translation. *Nature Biomedical Engineering*, *2*(11), 797.

Jain, U., & Chauhan, N. (2017). Glycated hemoglobin detection with electrochemical sensing amplified by gold nanoparticles embedded N-doped graphene nanosheet. *Biosensors and Bioelectronics*, *89*, 578−584.

Jain, U., Gupta, S., & Chauhan, N. (2017). Construction of an amperometric glycated hemoglobin biosensor based on Au−Pt bimetallic nanoparticles and poly (indole-5-carboxylic acid) modified Au electrode. *International Journal of Biological Macromolecules*, *105*, 549−555.

Jamdade, D. A., Rajpali, D., Joshi, K. A., Kitture, R., Kulkarni, A. S., Shinde, V. S., Bellare, J., Babiya, K. R., & Ghosh, S. (2019). *Gnidia glauca*-and *Plumbago zeylanica*-mediated synthesis of novel copper nanoparticles as promising antidiabetic agents. *Advances in Pharmacological Sciences*, *2019*.

Jang, H. L., Zhang, Y. S., & Khademhosseini, A. (2016). Boosting clinical translation of nanomedicine. *Future Medicine*.

Javed, R., Ahmed, M., ul Haq, I., Nisa, S., & Zia, M. (2017). PVP and PEG doped CuO nanoparticles are more biologically active: Antibacterial, antioxidant, antidiabetic and cytotoxic perspective. *Materials Science and Engineering: C*, *79*, 108−115.

Jeevanandam, J., Chan, Y. S., Danquah, M. K., & Law, M. C. (2019). Cytotoxicity analysis of morphologically different sol-gel-synthesized MgO nanoparticles and their in vitro insulin resistance reversal ability in adipose cells. *Applied Biochemistry and Biotechnology*. Available from https://doi.org/10.1007/s12010-019-03166-z.

Jeevanandam, J., San Chan, Y., & Danquah, M. K. (2016). Nano-formulations of drugs: Recent developments, impact and challenges. *Biochimie*, *128*, 99−112.

Jiang, S., Li, M., Chang, R., Xiong, L., & Sun, Q. (2018). In vitro inhibition of pancreatic α-amylase by spherical and polygonal starch nanoparticles. *Food & Function*, *9*(1), 355−363.

Jiang, W., Wang, Y., Wargo, J. A., Lang, F. F., & Kim, B. Y. S. (2021). Considerations for designing preclinical cancer immune nanomedicine studies. *Nature Nanotechnology*, *16*(1), 6−15.

Jini, D., & Sharmila, S. (2019). Green synthesis of silver nanoparticles from *Allium cepa* and its in vitro antidiabetic activity. *Materials Today: Proceedings*, *22*, 432−438.

Jo, E.-J., Mun, H., & Kim, M.-G. (2016). Homogeneous immunosensor based on luminescence resonance energy transfer for glycated hemoglobin detection using upconversion nanoparticles. *Analytical Chemistry*, *88*(5), 2742−2746.

Johnson, P., Krishnan, V., Loganathan, C., Govindhan, K., Raji, V., Sakayanathan, P., Vijayan, S., Sathish Kumar, P., & Palvannan, T. (2018). Rapid biosynthesis of *Bauhinia variegata* flower extract-mediated silver nanoparticles: An effective antioxidant scavenger and α-amylase inhibitor. *Artificial Cells, Nanomedicine, and Biotechnology*, *46*(7), 1488−1494. Available from https://doi.org/10.1080/21691401.2017.1374283.

Joshi, M., Shankar, R., & Pathak, K. (2019). Bioadhesive nanoformulations—Concepts and preclinical studies: A critical review. *Reviews of Adhesion and Adhesives*, *7*(3), 295−329.

Karthick, V., Kumar, V. G., Dhas, T. S., Singaravelu, G., Sadiq, A. M., & Govindaraju, K. (2014). Effect of biologically synthesized gold nanoparticles on alloxan-induced diabetic rats—An in vivo approach. *Colloids and Surfaces B: Biointerfaces*, *122*, 505−511. Available from https://doi.org/10.1016/j.colsurfb.2014.07.022.

Karuppusamy, C., Venkatesan, P., & Kalaiselvan, R. (2017). Evaluation of antidiabetic activity of miglitol nanoparticles in streptozotocin induced diabetic rats. *International Journal of Research in Pharmaceutical Sciences*, *8*(1), 103–108.

Kazemi, F., Divsalar, A., & Saboury, A. A. (2018). Structural analysis of the interaction between free, glycated and fructated hemoglobin with propolis nanoparticles: A spectroscopic study. *International Journal of Biological Macromolecules*, *109*, 1329–1337.

Khalil, A. T., Ovais, M., Ullah, I., Ali, M., Shinwari, Z. K., Khamlich, S., & Maaza, M. (2017). *Sageretia thea* (Osbeck.) mediated synthesis of zinc oxide nanoparticles and its biological applications. *Nanomedicine: Nanotechnology, Biology, and Medicine*, *12*(15), 1767–1789.

Kim, M.-K., Chae, Y. N., Kim, H. D., Yang, E. K., Cho, E. J., Choi, S., Cheong, Y.-H., Kim, H.-S., Kim, H. J., & Jo, Y. W. (2012). DA-1229, a novel and potent DPP4 inhibitor, improves insulin resistance and delays the onset of diabetes. *Life Sciences*, *90*(1–2), 21–29.

Koenig, R. J., Blobstein, S. H., & Cerami, A. (1977). Structure of carbohydrate of hemoglobin AIc. *Journal of Biological Chemistry*, *252*(9), 2992–2997.

Kopp, T. I., Lundqvist, J., Petersen, R. K., Oskarsson, A., Kristiansen, K., Nellemann, C., & Vogel, U. (2015). In vitro screening of inhibition of PPAR-gamma activity as a first step in identification of potential breast carcinogens. *Human & Experimental Toxicology*, *34*(11), 1106–1118. Available from https://doi.org/10.1177/0960327115569811.

Leong, H. S., Butler, K. S., Brinker, C. J., Azzawi, M., Conlan, S., Dufès, C., Owen, A., Rannard, S., Scott, C., & Chen, C. (2019). On the issue of transparency and reproducibility in nanomedicine. *Nature Nanotechnology*, *14*(7), 629–635.

Lin, Y., Lin, J., & Hong, Y. (2017). Development of chitosan/poly-γ-glutamic acid/pluronic/curcumin nanoparticles in chitosan dressings for wound regeneration. *Journal of Biomedical Materials Research, Part B: Applied Biomaterials*, *105*(1), 81–90.

Lorenzo, M., Fernández-Veledo, S., Vila-Bedmar, R., Garcia-Guerra, L., De Alvaro, C., & Nieto-Vazquez, I. (2008). Insulin resistance induced by tumor necrosis factor-α in myocytes and brown adipocytes. *Journal of Animal Science*, *86*(suppl_14), E94–E104.

Lu, F., Zhang, Y., Cheng, J., Zhang, M., Luo, J., Qu, H., Zhao, Y., & Wang, Q. (2019). Maltase and sucrase inhibitory activities and hypoglycemic effects of carbon dots derived from charred *Fructus crataegi*. *Materials Research Express*, *6*(12), 125005. Available from https://doi.org/10.1088/2053-1591/ab4fd8.

Lunov, O., Syrovets, T., Röcker, C., Tron, K., Nienhaus, G. U., Rasche, V., Mailänder, V., Landfester, K., & Simmet, T. (2010). Lysosomal degradation of the carboxydextran shell of coated superparamagnetic iron oxide nanoparticles and the fate of professional phagocytes. *Biomaterials*, *31*(34), 9015–9022.

Malapermal, V., Botha, I., Krishna, S. B. N., & Mbatha, J. N. (2017). Enhancing antidiabetic and antimicrobial performance of *Ocimum basilicum*, and *Ocimum sanctum* (L.) using silver nanoparticles. *Saudi Journal of Biological Sciences*, *24*(6), 1294–1305. Available from https://doi.org/10.1016/j.sjbs.2015.06.026.

Malapermal, V., Mbatha, J. N., Gengan, R. M., & Anand, K. (2015). Biosynthesis of bimetallic Au-Ag nanoparticles using *Ocimum basilicum* (L.) with antidiabetic and antimicrobial properties. *Advanced Materials Letter*, *6*, 1050–1057.

Manna, K., Mishra, S., Saha, M., Mahapatra, S., Saha, C., Yenge, G., Gaikwad, N., Pal, R., Oulkar, D., Banerjee, K., & Das Saha, K. (2019). Amelioration of diabetic nephropathy using pomegranate peel extract-stabilized gold nanoparticles: Assessment of NF-κB and Nrf2 signaling system. *International Journal of Nanomedicine*, *14*, 1753–1777. Available from https://doi.org/10.2147/IJN.S176013.

Matheeussen, V., Lambeir, A.-M., Jungraithmayr, W., Gomez, N., Mc Entee, K., Van der Veken, P., Scharpé, S., & De Meester, I. (2012). Method comparison of dipeptidyl peptidase IV activity assays and their application in biological samples containing reversible inhibitors. *Clinica Chimica Acta*, *413*(3), 456–462. Available from https://doi.org/10.1016/j.cca.2011.10.031.

Melo, M. A. B., Caetano, W., Oliveira, E. L., Barbosa, P. M., Rando, A. L. B., Pedrosa, M. M. D., & Godoi, V. A. F. (2019). Effects of nanoparticles of hydroxy-aluminum phthalocyanine on markers of liver injury and glucose metabolism in diabetic mice. *Brazilian Journal of Medical and Biological*

Research,   52.   Available   from   http://www.scielo.br/scielo.php?script = sci_arttext&pid = S0100-879X2019000100609&nrm = iso.

Moelands, S. V. L., Lucassen, P. L. B. J., Akkermans, R. P., De Grauw, W. J. C., & Van de Laar, F. A. (2018). Alpha-glucosidase inhibitors for prevention or delay of type 2 diabetes mellitus and its associated complications in people at increased risk of developing type 2 diabetes mellitus. *Cochrane Database of Systematic Reviews, 12.*

Mohammed, S. S. S., Lawrance, A. V., Sampath, S., Sunderam, V., & Madhavan, Y. (2021). *Facile green synthesis of silver nanoparticles from sprouted* Zingiberaceae *species: Spectral characterisation and its potential biological applications. Materials technology* (pp. 1—14). Taylor & Francis.

Naik, M. Z., Meena, S. N., Ghadi, S. C., Naik, M. M., & Salker, A. V. (2016). Evaluation of silver-doped indium oxide nanoparticles as in vitro α-amylase and α-glucosidase inhibitors. *Medicinal Chemistry Research, 25*(3), 381—389.

Naqvi, S. T. Q., Shah, Z., Fatima, N., Qadir, M. I., Ali, A., & Muhammad, S. A. (2017). Characterization and biological studies of copper nanoparticles synthesized by *Aspergillus niger. Journal of Bionanoscience, 11*(2), 136—140.

Nie, X., Chen, Z., Pang, L., Wang, L., Jiang, H., Chen, Y., Zhang, Z., Fu, C., Ren, B., & Zhang, J. (2020). Oral nano drug delivery systems for the treatment of type 2 diabetes mellitus: An available administration strategy for antidiabetic phytocompounds. *International Journal of Nanomedicine, 15*, 10215.

Pan, Q., Xu, Y., Yang, N., Gao, X., Liu, J., Yang, W., & Wang, G. (2016). Comparison of acarbose and metformin on albumin excretion in patients with newly diagnosed type 2 diabetes: A randomized controlled trial. *Medicine, 95*(14).

Pedersen, M. G., Toffolo, G. M., & Cobelli, C. (2009). Cellular modeling: Insight into oral minimal models of insulin secretion. *American Journal of Physiology-Endocrinology and Metabolism, 298*(3), E597—E601.

Petersen, R. K., Larsen, S. B., Jensen, D. M., Christensen, J., Olsen, A., Loft, S., Nellemann, C., Overvad, K., Kristiansen, K., & Tjønneland, A. (2012). PPARgamma-PGC-1alpha activity is determinant of alcohol related breast cancer. *Cancer Letters, 315*(1), 59—68.

Prasad, A. R., Basheer, S. M., Williams, L., & Joseph, A. (2019). Highly selective inhibition of α-glucosidase by green synthesised ZnO nanoparticles—In-vitro screening and in-silico docking studies. *International Journal of Biological Macromolecules, 139*, 712—718. Available from https://doi.org/10.1016/j.ijbiomac.2019.08.033.

Pridgen, E. M., Alexis, F., Kuo, T. T., Levy-Nissenbaum, E., Karnik, R., Blumberg, R. S., Langer, R., & Farokhzad, O. C. (2013). Transepithelial transport of Fc-targeted nanoparticles by the neonatal Fc receptor for oral delivery. *Science Translational Medicine, 5*(213). Available from https://doi.org/10.1126/scitranslmed.3007049, 213ra167.

Rajakumar, G., Thiruvengadam, M., Mydhili, G., Gomathi, T., & Chung, I.-M. (2018). Green approach for synthesis of zinc oxide nanoparticles from *Andrographis paniculata* leaf extract and evaluation of their antioxidant, anti-diabetic, and anti-inflammatory activities. *Bioprocess and Biosystems Engineering, 41*(1), 21—30.

Rajaram, K., Aiswarya, D. C., & Sureshkumar, P. (2015). Green synthesis of silver nanoparticle using *Tephrosia tinctoria* and its antidiabetic activity. *Materials Letters, 138*, 251—254.

Rivera-Mancía, S., Trujillo, J., & Chaverri, J. P. (2018). Utility of curcumin for the treatment of diabetes mellitus: Evidence from preclinical and clinical studies. *Journal of Nutrition & Intermediary Metabolism, 14*, 29—41.

Roy, M., Pal, R., & Chakraborti, A. S. (2017). Pelargonidin-PLGA nanoparticles: Fabrication, characterization, and their effect on streptozotocin induced diabetic rats. *Indian Jounal of Experimental Biology, 55*, 819—830.

Saha, A. K., Zhen, M.-Y. S., Erogbogbo, F., & Ramasubramanian, A. K. (2019). *Design considerations and assays for hemocompatibility of FDA-approved nanoparticles.* Thieme Medical Publishers.

Saikia, I., Hazarika, M., Yunus, S., Pal, M., Das, M. R., Borah, J. C., & Tamuly, C. (2018). Green synthesis of Au-Ag-In-rGO nanocomposites and its α-glucosidase inhibition and cytotoxicity effects. *Materials Letters, 211*, 48—50. Available from https://doi.org/10.1016/j.matlet.2017.09.084.

Saratale, G. D., Saratale, R. G., Benelli, G., Kumar, G., Pugazhendhi, A., Kim, D.-S., & Shin, H.-S. (2017). Anti-diabetic potential of silver nanoparticles synthesized with *Argyreia nervosa* leaf extract high synergistic antibacterial activity with standard antibiotics against foodborne bacteria. *Journal of Cluster Science*, 28(3), 1709−1727.

Saratale, R. G., Shin, H. S., Kumar, G., Benelli, G., Kim, D.-S., & Saratale, G. D. (2018). Exploiting antidiabetic activity of silver nanoparticles synthesized using *Punica granatum* leaves and anticancer potential against human liver cancer cells (HepG2). *Artificial Cells, Nanomedicine, and Biotechnology*, 46 (1), 211−222. Available from https://doi.org/10.1080/21691401.2017.1337031.

Satalkar, P., Elger, B. S., & Shaw, D. M. (2016). Stakeholder views on participant selection for first-in-human trials in cancer nanomedicine. *Current Oncology*, 23(6), e530.

Sathish Kumar, S. R., & Bhaskara Rao, K. V. (2016). Postprandial anti-hyperglycemic activity of marine *Streptomyces coelicoflavus* SRBVIT13 mediated gold nanoparticles in streptozotocin induced diabetic male albino Wister rats. *IET Nanobiotechnology*, 10(5), 308−314. Available from https://doi.org/10.1049/iet-nbt.2015.0094.

Sathiyaseelan, A., Saravanakumar, K., Mariadoss, A. V. A., & Wang, M.-H. (2020). Biocompatible fungal chitosan encapsulated phytogenic silver nanoparticles enhanced antidiabetic, antioxidant and antibacterial activity. *International Journal of Biological Macromolecules*, 153, 63−71. Available from https://doi.org/10.1016/j.ijbiomac.2020.02.291.

Saware, K., Aurade, R. M., Kamala Jayanthi, P. D., & Abbaraju, V. (2015). Modulatory effect of citrate reduced gold and biosynthesized silver nanoparticles on α-amylase activity. *Journal of Nanoparticles*, 2015.

Scharpé, S., De Meester, I., Vanhoof, G., Hendriks, D., Van Sande, M., Van Camp, K., & Yaron, A. (1988). Assay of dipeptidyl peptidase IV in serum by fluorometry of 4-methoxy-2-naphthylamine. *Clinical Chemistry*, 34(11), 2299−2301.

Sechi, M., Syed, D. N., Pala, N., Mariani, A., Marceddu, S., Brunetti, A., Mukhtar, H., & Sanna, V. (2016). Nanoencapsulation of dietary flavonoid fisetin: Formulation and in vitro antioxidant and α-glucosidase inhibition activities. *Materials Science and Engineering: C*, 68, 594−602. Available from https://doi.org/10.1016/j.msec.2016.06.042.

Selvan, D. S. A., Shobana, S., Thiruvasagam, P., Murugesan, S., & Rahiman, A. K. (2019). Evaluation of antimicrobial and antidiabetic activities of Ag@SiO$_2$ core−shell nanoparticles synthesized with diverse shell thicknesses. *Journal of Cluster Science*. Available from https://doi.org/10.1007/s10876-019-01682-w.

Selvan, D. S. A., Shobana, S., Thiruvasagam, P., Murugesan, S., & Rahiman, A. K. (2020). Evaluation of antimicrobial and antidiabetic activities of Ag@ SiO$_2$ core−shell nanoparticles synthesized with diverse shell thicknesses. *Journal of Cluster Science*, 31, 1−11.

Selvin, E., Steffes, M. W., Zhu, H., Matsushita, K., Wagenknecht, L., Pankow, J., Coresh, J., & Brancati, F. L. (2010). Glycated hemoglobin, diabetes, and cardiovascular risk in nondiabetic adults. *New England Journal of Medicine*, 362(9), 800−811.

Semple, R. K., Chatterjee, V. K. K., & O'Rahilly, S. (2006). PPARγ and human metabolic disease. *The Journal of Clinical Investigation*, 116(3), 581−589.

Senthilkumar, P., Kumar, D. S. R. S., Sudhagar, B., Vanthana, M., Parveen, M. H., Sarathkumar, S., Thomas, J. C., Mary, A. S., & Kannan, C. (2016). Seagrass-mediated silver nanoparticles synthesis by *Enhalus acoroides* and its α-glucosidase inhibitory activity from the Gulf of Mannar. *Journal of Nanostructure in Chemistry*, 6(3), 275−280.

Senthilkumar, P., Priya, L., Kumar, R. S., & Bhuvaneshwari, D. S. (2015). Potent α-glucosidase inhibitory activity of green synthesized gold nanoparticles from the brown seaweed *Padina boergesenii*. *International Journal of Recent Advances in Multidisciplinary Research*, 2(11), 917−923.

Senthilkumar, P., Surendran, L., Sudhagar, B., & Ranjith Santhosh Kumar, D. S. (2019). Facile green synthesis of gold nanoparticles from marine algae *Gelidiella acerosa* and evaluation of its biological potential. *SN Applied Sciences*, 1(4), 284. Available from https://doi.org/10.1007/s42452-019-0284-z.

Shao, T., Yuan, P., Zhu, L., Xu, H., Li, X., He, S., Li, P., Wang, G., & Chen, K. (2019). Carbon nanoparticles inhibit A-glucosidase activity and induce a hypoglycemic effect in diabetic mice. *Molecules (Basel, Switzerland)*, 24(18). Available from https://doi.org/10.3390/molecules24183257.

Shi, Y., & Lammers, T. (2019). Combining nanomedicine and immunotherapy. *Accounts of Chemical Research*, 52, 1543−1554.

Shrestha, N., Araújo, F., Shahbazi, M.-A., Mäkilä, E., Gomes, M. J., Airavaara, M., Kauppinen, E. I., Raula, J., Salonen, J., Hirvonen, J., Sarmento, B., & Santos, H. A. (2016). Oral hypoglycaemic effect of GLP-1 and DPP4 inhibitor based nanocomposites in a diabetic animal model. *Journal of Controlled Release, 232*, 113−119. Available from https://doi.org/10.1016/j.jconrel.2016.04.024.

Simu, S. Y., Ahn, S., Castro-Aceituno, V., Singh, P., Mathiyalagan, R., Jiménez-Pérez, Z. E., Hurh, J., Oi, L. Z., Hun, N. J., Kim, Y.-J., & Yang, D.-C. (2019). Gold nanoparticles synthesized with fresh *Panax ginseng* leaf extract suppress adipogenesis by downregulating PPARγ/CEBPα signaling in 3T3-L1 mature adipocytes. *Journal of Nanoscience and Nanotechnology, 19*(2), 701−708. Available from https://doi.org/10.1166/jnn.2019.15753.

Song, Y. H., Uddin, Z., Jin, Y. M., Li, Z., Curtis-Long, M. J., Kim, K. D., Cho, J. K., & Park, K. H. (2017). Inhibition of protein tyrosine phosphatase (PTP1B) and α-glucosidase by geranylated flavonoids from *Paulownia tomentosa*. *Journal of Enzyme Inhibition and Medicinal Chemistry, 32*(1), 1195−1202. Available from https://doi.org/10.1080/14756366.2017.1368502.

Soudry-Kochavi, L., Naraykin, N., Di Paola, R., Gugliandolo, E., Peritore, A., Cuzzocrea, S., Ziv, E., Nassar, T., & Benita, S. (2018). Pharmacodynamical effects of orally administered exenatide nanoparticles embedded in gastro-resistant microparticles. *European Journal of Pharmaceutics and Biopharmaceutics, 133*, 214−223. Available from https://doi.org/10.1016/j.ejpb.2018.10.013.

Sudha, P., Zinjarde, S. S., Bhargava, S. Y., & Kumar, A. R. (2011). Potent α-amylase inhibitory activity of Indian Ayurvedic medicinal plants. *BMC Complementary and Alternative Medicine, 11*(1), 5.

Suganya, K. S. U., Govindaraju, K., Vani, C. V., Premanathan, M., & Kumar, V. K. G. (2019). *In vitro biological evaluation of anti-diabetic activity of organic−inorganic hybrid gold nanoparticles*. IET nanobiotechnology (13, pp. 226−229). Institution of Engineering and Technology 2. Available from https://digital-library.theiet.org/content/journals/10.1049/iet-nbt.2018.5139.

Thatoi, P., Kerry, R. G., Gouda, S., Das, G., Pramanik, K., Thatoi, H., & Patra, J. K. (2016). Photo-mediated green synthesis of silver and zinc oxide nanoparticles using aqueous extracts of two mangrove plant species, *Heritiera fomes* and *Sonneratia apetala* and investigation of their biomedical applications. *Journal of Photochemistry and Photobiology B: Biology, 163*, 311−318.

Thondawada, M., Wadhwani, A. D. S., Palanisamy, D., Rathore, H. S., Gupta, R. C., Chintamaneni, P. K., Samanta, M. K., Dubala, A., Varma, S., Krishnamurthy, P. T., & Gowthamarajan, K. (2018). An effective treatment approach of DPP-IV inhibitor encapsulated polymeric nanoparticles conjugated with anti-CD-4 mAb for type 1 diabetes. *Drug Development and Industrial Pharmacy, 44*(7), 1120−1129. Available from https://doi.org/10.1080/03639045.2018.1438460.

Thule, P. M., Liu, J., & Phillips, L. S. (2000). Glucose regulated production of human insulin in rat hepatocytes. *Gene Therapy, 7*(3), 205.

Tyler, B., Gullotti, D., Mangraviti, A., Utsuki, T., & Brem, H. (2016). Polylactic acid (PLA) controlled delivery carriers for biomedical applications. *Advanced Drug Delivery Reviews, 107*, 163−175.

Venkatraman, A., Yahoob, S. A. M., Nagarajan, Y., Harikrishnan, S., & Vasudevan, S. (2018). Pharmacological activity of biosynthesized gold nanoparticles from brown algae-seaweed *Turbinaria conoide*. *NanoWorld Journal, 4*(1), 17−22.

Ventola, C. L. (2012). The nanomedicine revolution: Part 3: Regulatory and safety challenges. *P & T: A Peer-Reviewed Journal for Formulary Management, 37*(11), 631−639. Available from https://www.ncbi.nlm.nih.gov/pubmed/23204818.

Ventola, C. L. (2017). Progress in nanomedicine: Approved and investigational nanodrugs. *P & T: A Peer-Reviewed Journal for Formulary Management, 42*(12), 742−755. Available from https://www.ncbi.nlm.nih.gov/pubmed/29234213.

Vinotha, V., Iswarya, A., Thaya, R., Govindarajan, M., Alharbi, N. S., Kadaikunnan, S., Khaled, J. M., Al-Anbr, M. N., & Vaseeharan, B. (2019). Synthesis of ZnO nanoparticles using insulin-rich leaf extract: Anti-diabetic, antibiofilm and anti-oxidant properties. *Journal of Photochemistry and Photobiology B: Biology, 197*, 111541. Available from https://doi.org/10.1016/j.jphotobiol.2019.111541.

Wang, X.-Q., & Zhang, Q. (2012). pH-sensitive polymeric nanoparticles to improve oral bioavailability of peptide/protein drugs and poorly water-soluble drugs. *European Journal of Pharmaceutics and Biopharmaceutics, 82*(2), 219−229. Available from https://doi.org/10.1016/j.ejpb.2012.07.014.

Wedmore, I., McManus, J. G., Pusateri, A. E., & Holcomb, J. B. (2006). A special report on the chitosan–based hemostatic dressing: Experience in current combat operations. *Journal of Trauma and Acute Care Surgery, 60*(3), 655−658.

Wilson, S., Cholan, S., Vishnu, U., Sannan, M., Jananiya, R., Vinodhini, S., Manimegalai, S., & Rajeswari, D. V. (2015). In vitro assessment of the efficacy of free-standing silver nanoparticles isolated from *Centella asiatica* against oxidative stress and its antidiabetic activity. *Der Pharmacia Lettre*, 7 (12), 194–205.

Wolfram, J., Zhu, M., Yang, Y., Shen, J., Gentile, E., Paolino, D., Fresta, M., Nie, G., Chen, C., & Shen, H. (2015). Safety of nanoparticles in medicine. *Current Drug Targets*, 16(14), 1671–1681.

Xu, X., Ho, W., Zhang, X., Bertrand, N., & Farokhzad, O. (2015). Cancer nanomedicine: From targeted delivery to combination therapy. *Trends in Molecular Medicine*, 21(4), 223–232.

Yang, J., Gong, X., Fang, L., Fan, Q., Cai, L., Qiu, X., Zhang, B., Chang, J., & Lu, Y. (2017). Potential of $CeCl_3$@$mSiO_2$ nanoparticles in alleviating diabetic cataract development and progression. *Nanomedicine: Nanotechnology, Biology and Medicine*, 13(3), 1147–1155. Available from https://doi.org/10.1016/j.nano.2016.12.021.

Zhang, J., Liu, Y., Lv, J., & Li, G. (2015). A colorimetric method for α-glucosidase activity assay and its inhibitor screening based on aggregation of gold nanoparticles induced by specific recognition between phenylenediboronic acid and 4-aminophenyl-α-d-glucopyranoside. *Nano Research*, 8(3), 920–930.

Zhang, M., Wang, H., Wang, B., Ma, Y., Huang, H., Liu, Y., Shao, M., Yao, B., & Kang, Z. (2019). Maltase decorated by chiral carbon dots with inhibited enzyme activity for glucose level control. *Small*, 15, e1901512.

Zhang, S., Qiu, B., Zhu, J., Hu, W., Ma, F., Khan, M. Z. H., & Liu, X. (2018). Rapidly screening of α-glucosidase inhibitors from *Dioscorea opposita* Thunb. peel based on rGO@ Fe_3O_4 nanocomposites microreactor. *Journal of Enzyme Inhibition and Medicinal Chemistry*, 33(1), 1335–1342.

CHAPTER 10

Future of nanoparticles, nanomaterials, and nanomedicines in diabetes treatment

Introduction

Nanosized particles, materials, and medicines are gaining significant attention in the biomedical field due to their exclusive biological properties and the ability to cure diseases at the molecular level (Khan et al., 2020; Pelaz et al., 2017). Numerous nanosized materials have been introduced in recent times. These include nanosized ribonucleic acid particles (Jasinski et al., 2019), polymers (Bordat et al., 2019), micelles (Tambe et al., 2019), carbon dots (Ghosal & Ghosh, 2019), carbon nanotubes (Pastorin, 2019), superparamagnetic particles (Xiao & Du, 2019), metal (Azharuddin et al., 2019), metal oxides (Andra et al., 2019), quantum dots (Wagner et al., 2019), nanoclusters (Wang et al., 2019), and upconversion particles (Chen & Wang, 2020). Some of these nanomaterials are extensively under research investigations for the diagnosis and treatment of various diseases (Anderson et al., 2019; Eom et al., 2020; Zhang et al., 2019; Zor et al., 2019), as implants (Teh & Lai, 2019), and in bioimaging (Key & Leary, 2014), biosensing (Vallabani et al., 2019), transplantation (Janjic & Gorantla, 2020), and regenerative surgery (Amin et al., 2019; Janjic & Gorantla, 2020). Nanomedicines have the potential to address some of the drawbacks of conventional drugs, offering unique biomedical and biophysical properties that can enhance therapeutic performance indices. Despite research advancements in the development of nanosized particles for biomedical applications, significant challenges to the translation of such nanotechnologies to commercial scales persist (Lin-Ping et al., 2020; Patel et al., 2021). There is a need for more aggressive and comprehensive research pursuits to address these limitations.

Toxicological effects of nanomedicines through their cellular and molecular level interactions with organisms are a major concern (Benchimol et al., 2019; Prasad et al., 2017). Green and biosynthesized nanomedicines offer an alternative approach to address some of the toxicity-related challenges of chemically fabricated nanomedicines (Jeevanandam et al., 2016). However, issues with stability and large-scale production represent major limitations of green and biosynthesized nanomedicines (Jahangirian et al., 2017).

Emerging Nanomedicines for Diabetes Mellitus Theranostics
DOI: https://doi.org/10.1016/B978-0-323-85396-5.00013-0

The emergence of nanomedicines has resulted in new opportunities for develop personalized treatment modalities for diabetic patients (Mohsen, 2019). It has been projected that the rapid research advancements in the field of nanomedicines have the potential to catalyze their use as alternatives to conventional diabetes drugs (Sabahi et al., 2019), thus technological efforts toward large-scale production of nanodiabetic drugs are paramount (Hassan et al., 2016; Rehman et al., 2020). This chapter discusses some of the limitations of nanomedicines toward commercialization, emphasizing on limitations relating to diabetes, and some potential solutions. In addition, the future of nanomedicines in diagnosis and treatment applications of diabetes are discussed.

Limitations of nanomedicines

While nanomedicines have the potential to offer significant theranostic benefits for various diseases, there are significant limitations to their full-scale application. The difficulties in transforming nanomedicines from lab-scale production to clinical applications require a balance between efficacy and toxicity as well as promoting applications as nanocarriers. The combination of drugs in nanocarriers has the potential to be key to the long-term clinical success of nanomedicines. However, the lack of reliable techniques to evaluate the stability of nanocarriers in human fluids and the utilization of nonpredictive animal models for preclinical studies are major concerns. The existence of organic solvent residues in nanoparticles preparation, stability of nanoformulations, level of toxicity, sterility, and consistency of batch-to-batch production are some additional challenges to commercialization (Landesman-Milo & Peer, 2016). Some specific limitations of nanomedicines are discussed here. Bari et al. (2018) validated a GMP-compliant process for pilot-scale production of freeze-dried mesenchymal stem or stromal secretome for cell-free regenerative nanomedicine development. The study led to a successful isolation and lyophilization process for product development without altering the integrity and morphology of the extracellular vesicles. However, technologies for large-scale production of secretome-based nanomedicines require more research efforts (Bari et al., 2018). Bregoli et al. (2016) discussed various perspectives of applying nanomedicine (liposomes) to translational oncology for effective treatment of cancer. Several limitations of nanomedicines were indicated, including controlled release challenges, stability, drug localization at the site of the tumor, dosage, pharmacokinetic drawbacks, and difficulties in predicting biodistribution (Bregoli et al., 2016). Nanomedicines have been recommended for oral administration of drugs to bypass physiological barriers via diverse lipid-based nanoformulations, polymer-based nanoplatforms, and nanosuspensions. Limitations to such applications relate to toxicity, economics of production, and technical challenges to the development of stable nanosized medicines (Sim et al., 2016). Nanovesicles, including liposomes, niosomes, and ethosomes can be used to develop nanomedicines for drug delivery applications. However, some limitations demonstrated by these nanovesicles in drug formulation include issues of stability, rapid clearance and leakage of drugs, reduced bioavailability, hydrolysis, and limited shelf-life (Weissig et al., 2019). Agrahari and Agrahari (2018) discussed several challenges associated with the translation of nanosized medicines to clinical products. Lack of reproducibility in manufacturing and scale-up challenges, inadequate cutting-edge technologies to probe cellular interactions with nanomedicines, concerns relating to the safety of patients and the environment, barriers related to regulations, and inadequate knowledge about biological and physicochemical interactions of nanoformulations are some of the challenges that hurdle clinical translation of nanomedicines (Agrahari & Agrahari, 2018). Virus and virus-like nanoparticles are promising candidates for the development of nanomedicine in gene therapy. However, they have been reported to show limitations including the expression of transgene over a short duration, cytotoxicity, and metabolism (Yan et al., 2019). In addition, concerns associated with stability, intrinsic immunogenicity, and limitations in the fusion of antigens are other drawbacks of using virus nanoparticles as nanomedicines for gene therapy (Jeevanandam et al., 2019). Hybrids of carbon nanotubes and hydrogels have been explored in the development of nanomedicines. Challenges such as low conductivity and instability of hydrogels and the toxicity of carbon nanotubes hinder their immediate introduction in biomedical applications as nanomedicines (Vashist et al., 2018). Problems relating to the permeability and retention of hybrid nanomedicines, drug targeting in biological systems, intracellular trafficking, loss of drug cargo routes, nonspecific interactions between serum proteins, and economic barriers also impact the clinical translation of hybrid nanomedicines (Ashok Bohara et al., 2019). It is important that these limitations are addressed through research advancements to support the clinical translation of nanomedicines. Table 10.1 provides a summary of reported limitations of nanomedicines toward clinical translation and commercialization.

Future of nanobased technologies in diabetes diagnosis

Various nanomaterials and nanotechnologies are under extensive research for the development of novel personalized diabetes diagnostics. Glucose nanosensors as discussed in Chapter 3 and nanotattoos as discussed in Chapter 4 are promising approaches with the potential to herald the future of diabetes diagnostics. Kuralay (2019) discussed various nanomaterial-based enzyme biosensors for diabetes and reported that carbon-based nanomaterials such as carbon nanotubes and graphene are nanosized materials commonly explored for enzymatic electrochemical-mediated biosensing for diabetes. In addition, nanosized particles such as gold, palladium, silver, and platinum; and nanosized metal oxide particles, such as iron oxide, possess the ability to bind with enzymes and electrochemically detect glucose for monitoring of diabetes. The article emphasized the potential of nanocomposites to address some of the limitations of nanosized particles (Ozkan et al., 2019). An affinity nanosensor has been

Table 10.1 List of nanomedicines and their possible limitations in clinical translation and commercialization.

Nanomedicines	Limitations	Reference
Large-scale nanomedicine production	Presence of residual organic solvent, challenges of formulation stability, endotoxin level, sterility, and batch-to-batch production reproducibility.	Landesman-Milo and Peer (2016)
Cell-free regenerative nanomedicine	Limitations in large-scale production of secretome-based nanomedicines.	Bari et al. (2018)
Immunotherapeutic polymeric nanomedicines	Noninteractive behavior, low possibility of Pickering emulsion to load drug, and low dissolution of liposomes.	Sun et al. (2019)
Nanomedicines in translational oncology	Stability and drug localization at target sites. Dosing and pharmacokinetics drawbacks of nanosized cholesterol-based liposomes. Difficulties in predicting the biodistribution of nanocarrier.	Bregoli et al. (2016)
Orally administered nanomedicines	Toxicity, economic drawbacks, and biopharmaceutical formulation strategy.	Sim et al. (2016)
Nanovesicles	Issues of stability, rapid clearance, leakage of drugs, low bioavailability, and hydrolysis	Wadhwa et al. (2019)
Translation nanomedicines	Lack of reproducibility in manufacturing and scale-up processes. Inadequate characterization tools to analyze cellular interactions with nanomedicine. Concerns of safety among patients and the environment. Insufficient knowledge about biological and physicochemical interactions of nanoformulations.	Agrahari and Agrahari (2018)
Virus and virus-like nanoparticles	Expression of transgene for a short duration. Cytotoxicity concerns about metabolism and elimination of viral nanoparticles.	Yan et al. (2019)
Nanoformulation-based gene therapy	Instability of particles, intrinsic immunogenicity, and limitations in the fusion of antigens.	Jeevanandam et al. (2019)
Hybrids of carbon nanotubes and hydrogels as nanomedicine	Low conductivity and instability of hydrogels. Toxicity of carbon nanotubes	Vashist et al. (2018)
Translation of hybrid nanomedicines	Glitches in permeability and retention. Intracellular trafficking, and loss of drug cargo routes. Nonspecific interaction between serum proteins. Economic barriers.	Thorat et al. (2019)

fabricated using graphene nanostructures for the detection of low-molecular weight and low-charge molecules, such as glucose. The results showed that the nanosensor possessed the ability to measure the concentration of glucose in the range of $2\,\mu M$ to 25 mM, and can be fabricated as a noninvasive tool to monitor glucose (Zhu et al., 2016). A theoretical system prototype capable of alerting diabetic patients of altering blood glucose levels has been proposed. The alerts will be transmitted using a biological network of nanosensors, providing a platform to control diabetes in the early stage (Suarez et al., 2019).

Last few years, a wide range of novel nanosensors have been developed in an attempt to address the limitations of conventional diabetes diagnostics. Shafiee et al. (2019) showed that nanosensors can be used to monitor the efficacy of drugs for the treatment of diabetes. They proposed that nanosensors are beneficial for islet transplantation and can be incorporated into follow-up technologies to monitor recipients of islet and point-of-care devices (Shafiee et al., 2019). Furthermore, a quencher-based viologen fluorescent dye system incorporated into poly(2-hydroethyl methacrylate) hydrogels was reported to perform as effective near-infrared optical nanosensors for continuous detection of glucose. The incorporation of this system with nanosized silica particles facilitated efficient real-time monitoring of glucose even at low concentrations (Le et al., 2019). Nanosized gold particles with star morphology can be applied in colorimetric fructosyl valine detection for use as a future point-of-care biosensor to monitor glycated hemoglobin with the potential to be monitored with the naked eye (Mulder et al., 2019).

Nanoparticle-based smart tattoos are gaining significant attention among young diabetic patients to monitor glucose levels and/or manage insulin administration without any sophisticated equipment. Nanoparticle-incorporated smart tattoos can be a highly beneficial in monitoring glucose among diabetes patients in the future. The emergence of smart tattoos will lead to innovative, personalized diagnostic tools for diabetic patients (Meetoo et al., 2019). Functional artistic tattoos can be a platform to inject colorimetric biosensors directly into the skin to quantify metabolite biomarkers in the skin interstitial fluid with observable color changes in response to alterations in the pH as well as biomarker concentration (Graziano, 2019). Smart tattoos can serve as an efficient continuous glucose monitoring system and are beneficial for patients with nocturnal hypoglycemia (Mian et al., 2019).

Aptamer-based nanosensors are gaining significance as a next generation biosensor for diabetes, particularly for the detection of low concentration diabetes biomarkers. For example, nanosized gold particles, zero-dimensional semiconductor quantum dots, and deoxyribonucleic acid aptamer can be used to fabricate optical nanosensors for the detection of glycated albumin to diagnose diabetes (Ghosh et al., 2017). Also, a novel approach to fabricate a graphene-based aptameric field-effect transistor nanosensor for real-time detection of insulin has been demonstrated. The device was used for label-free insulin monitoring and rapid prediction of specific insulin dosage (Hao et al., 2017). Aptamers modified with thiol molecules and immobilized over nanosized gold particles have been used to develop label-free electrochemical biosensor array for high-sensitivity monitoring of total and glycated hemoglobin in human whole blood samples with detection limits of 0.34 and 0.2 ng mL for total and glycated hemoglobin, respectively, offering a simple, stable, easy-to-use, and low-cost array platform for monitoring diabetes (Eissa & Zourob, 2017). It is evident that the emergence of nanotechnology will enhance the efficacy of conventional diagnostic tools for monitoring glucose, insulin level, and other biomarkers in diabetes patients.

Future of nanomedicines and nanodevices in diabetes therapy

Various types of nanomedicines have been developed for the treatment of diabetes. Glucose-responsive microneedle patches with advanced nanotechnological approaches are widely utilized for transdermal delivery of insulin and other antidiabetic drugs via hypodermic administration. This approach can offer a closed-loop drug delivery system for self-regulated administration of glucose, insulin, and antidiabetic drug (Chen et al., 2018). Also, nanobased approaches to promote insulinotropic, proangiogenic effects, and immune modulation are being explored in cell replacement therapies for type 1 diabetes. Nanosized thin films made up of polymers have been introduced as a potential coating agent of isles to protect islet modality and overcome the challenges in islet transplantation. Nanosized inorganic, lipid, and polymer particles, self-assembled peptides as a nanosized antidiabetic agent carrier, nanotechnology-based microneedle patches, and nanoparticle hybrid therapies are highly useful in the treatment different types diabetes conditions such as obesity (Tsou et al., 2019). Nanocarriers can be used as an alternative to conventional insulin pen, pump, and syringes due to their capacity to overcome limitations such as rapid insulin degradation in gastric fluid and low bioavailability. They can be engineered to acquire a superior performance in delivering insulin and managing diabetic conditions compared to enzyme inhibitors, mucoadhesive polymers, adsorption enhancers, and chemically modified, receptor-mediated absorbers (Bahman et al., 2019). Nanoparticles synthesized via plant extracts have the potential to address secondary diabetes conditions by inhibiting the generation of deleterious advanced glycation end products. Nanoparticles can be a carrier for insulin administration via oral or nasal routes. Natural polymeric nanoparticles of chitosan, alginate, and dextran, nanosized synthetic polymers such as poly(lactic-co-glycolic acid), polylactic acid, polyallylamine, and other nanosized polymers with cell-penetrating peptides, niosomes, poly(amidoamine) dendrimers, polymeric micelles, inorganic nanoparticles, nanosized lipid-based carriers, including nanosized solid lipid particles, liposomes, nanoemulsions, and nanosuspensions can be used to deliver antidiabetic drugs through oral and nasal routes

(Souto et al., 2019). Nanoparticles have also been employed in the treatment of other diseases associated with diabetes. These include cardiomyopathy (Mao et al., 2020), wound healing (Lee et al., 2020), retinopathy (Gupta & Kanna, 2019), foot ulcer (Liu et al., 2020), and nephropathy (Desai et al., 2021; Mohamed, 2019).

Research advancements and recommendations

There are various scientific concepts and ideas that are being investigated as potential nanotechnologies for diabetes theranostics as shown in Fig. 10.1. Nanorobots are used in the field of nanotechnology to describe nanosized systems that can detect impairments in cellular activities and possess the ability to deliver drugs accordingly to revert cell damage without any external monitoring systems (Gupta & Singh, 2021; Sujatha et al., 2010). Nanorobots can be used in diabetes theranostics. For example, conventionally, blood glucose level is monitored via pricks (e.g., finger pricking), and insulin is injected or orally administered along with antidiabetic drugs to control hyperglycemia. This may lead to hypoglycemic conditions in certain cases and can be normalized using glucose monitoring systems and hypoglycemic drugs (Jeevanandam et al., 2015). Nanorobots are effective for real-time monitoring of fluctuations in blood glucose level and as well as the administration of insulin and antidiabetic drugs accordingly until the glucose level is back to normal. Nanorobots or nanobots can be designed to combat diabetes effectively to avoid side effects and future emergence of complications (Rifat et al., 2019). Nanobot-based drug delivery systems with smart tattoos are expected to be more relevant in the pharmaceutical market as an effective diabetes treatment modality in future (Siwach et al., 2019).

One of the most recent developments in the field of biomedicine is the application of data analytics and machine learning approaches for high-precision disease diagnosis.

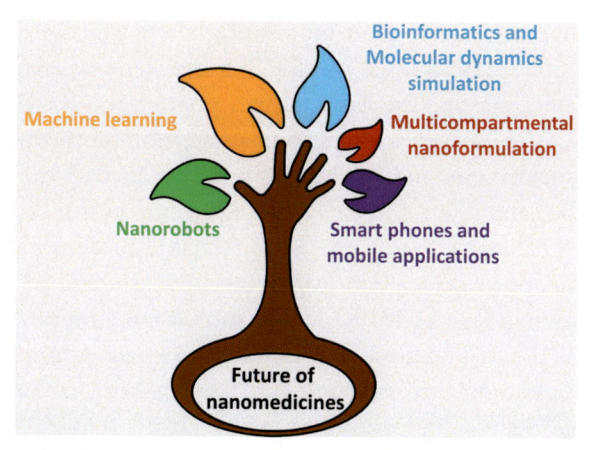

Figure 10.1 Emerging developments in the field of nanomedicines for diabetes treatment.

These computational approaches can be applied in developing advanced theranostics for diabetes. Nanoparticles or nanomaterials are employed in the bioimaging of diabetic cells (Hussain et al., 2021; Urrejola et al., 2018) and early detection of diabetes (Zhao et al., 2020). The combination of nanobioimaging and machine learning approaches can be used to probe alterations in the physicochemical properties of diabetic cells in response to insulin drug delivery, providing useful insights in determining specific insulin and drug formulations to reduce blood glucose level (Ko et al., 2017). Machine learning algorithms can also be used to predict the toxicity of nanomedicines (Furxhi et al., 2019). Bioinformatics and molecular dynamics simulation approaches can be used to identify novel biomarkers, bioaffinity, and functional attributes to create new theranostics for diabetes. The ability to design molecular targets in diabetic cells and the application of nanomedicines as drugs will expedite the screening of novel drugs and the cost of drug development processes.

Nanoformulation of nanosized conventional medicines or nanosized particles is another approach that is being explored for diabetes treatment (Ling et al., 2019). Such nanoformulations can be developed via encapsulation using dendrimers to generate a multicompartmental nanoformulation to harbor drug ingredients, insulin, and peptides in a single formulation for all-in-one diabetes treatment (Jeevanandam et al., 2017). These nanocompartments can also be decorated with bioaffinity ligands such as aptamers to target diabetic cells and can be engineered for drug release based on fluctuations in blood glucose and insulin levels in diabetes patients (Tan et al., 2020). Several attempts have been made to develop multicompartmental nanoformulations using dendrimers. However, incorporating more than two drug entities has been challenging based on available synthesis approaches. There are research opportunities to improve the synthesis approach for these nanoformulations.

Smartphones and mobile applications will continue to play a crucial role in delivering effective, remote, and personalized monitoring of glucose and insulin levels in diabetic patients. Smartphones are used in monitoring blood glucose levels and there are presently several mobile applications for the management of diabetes (Tran et al., 2012). An image-based colorimetric assay using a smartphone RGB (red, green, blue) camera for a point-of-care diabetes monitoring application has been developed (Wang et al., 2020). Also, smartphone and wearable fitness tracker data can be used to monitor diabetes conditions and retain the data as an electronic health record system (Wang et al., 2018). 922-P:Diabits has been developed for rapid monitoring of blood glucose levels and to predict outcomes after drug intake by artificial intelligence-powered smartphone application (Hayeri, 2019). Other mobile applications, such as Dexcom G5, which is one of the first United States Food and Drug Administration (USFDA) approved mobile application, are currently used by diabetic patients for continuous monitoring of blood glucose (Perez-Guzman et al., 2021). Contour plus one, Apollo smart glucometer, Omron elite, Eversense, and GlucoMe are additional mobile applications that can be integrated with smartphones, smartwatches, and other wearable devices for continuous monitoring of blood glucose (Brophy et al., 2021).

Conclusion

This chapter presents an overview of the limitations of nanomedicines that hurdle their potential as alternatives to conventional diabetes drugs and treatment modalities, as well as ongoing research approaches to address the challenges. Furthermore, the future outlook of nanomedicines and nanobased technologies in the diagnosis and treatment of diabetes is discussed. More research investment is required to address the limitations of nanomedicines and nanodevices to advance their development for widespread use. Computational design and development of nanodrugs via bioinformatics and molecular simulations, nanoparticle-mediated bioimaging, machine learning algorithms, smartphone-mediated monitoring, and nanorobot-mediated targeted delivery of antidiabetic drugs are some of the advanced research approaches that are implemented to develop enhanced diabetes theranostic nanotechnologies to meet future market demands.

References

Agrahari, V., & Agrahari, V. (2018). Facilitating the translation of nanomedicines to a clinical product: Challenges and opportunities. *Drug Discovery Today*, 23(5), 974–991.

Amin, K., Moscalu, R., Imere, A., Murphy, R., Barr, S., Tan, Y., Wong, R., Sorooshian, P., Zhang, F., & Stone, J. (2019). The future application of nanomedicine and biomimicry in plastic and reconstructive surgery. *Nanomedicine: Nanotechnology, Biology, and Medicine*, 14(20), 2679–2696.

Anderson, C. F., Grimmett, M. E., Domalewski, C. J., & Cui, H. (2019). Inhalable nanotherapeutics to improve treatment efficacy for common lung diseases. *Wiley Interdisciplinary Reviews: Nanomedicine and Nanobiotechnology*, 12(1), e1586.

Andra, S., Balu, S. K., Jeevanandham, J., Muthalagu, M., Vidyavathy, M., San Chan, Y., & Danquah, M. K. (2019). Phytosynthesized metal oxide nanoparticles for pharmaceutical applications. *Naunyn-Schmiedeberg's Archives of Pharmacology*, 392(7), 755–771.

Ashok Bohara, R., Thorat, N., Thorat, N.D., Brennan, G., Bauer, J., Silien, C., & Tofail, S.A.M. (2019). Chapter 12: Strengths and limitations of translating the hybrid nanostructures to the clinic. In *Hybrid nanostructures for cancer theranostics* (pp. 229–254). Elsevier. Available from https://doi.org/10.1016/B978-0-12-813906-6.00012-3.

Azharuddin, M., Zhu, G.H., Das, D., Ozgur, E., Uzun, L., Turner, A.P.F., & Patra, H.K. (2019). A repertoire of biomedical applications of noble metal nanoparticles. *Chemical Communications*.

Bahman, F., Greish, K., & Taurin, S. (2019). Nanotechnology in insulin delivery for management of diabetes. *Pharmaceutical Nanotechnology*, 7(2), 113–128. Available from https://doi.org/10.2174/2211738507666190321110721.

Bari, E., Perteghella, S., Di Silvestre, D., Sorlini, M., Catenacci, L., Sorrenti, M., Marrubini, G., Rossi, R., Tripodo, G., Mauri, P., Marazzi, M., & Torre, L. M. (2018). Pilot production of mesenchymal stem/stromal freeze-dried secretome for cell-free regenerative nanomedicine: A validated GMP-compliant process. *Cells*, 7(11). Available from https://doi.org/10.3390/cells7110190.

Benchimol, M. J., Bourne, D., Moghimi, S. M., & Simberg, D. (2019). Pharmacokinetic analysis reveals limitations and opportunities for nanomedicine targeting of endothelial and extravascular compartments of tumours. *Journal of Drug Targeting, 27*(5–6), 690–698. Available from https://doi.org/10.1080/1061186X.2019.1566339.

Bordat, A., Boissenot, T., Nicolas, J., & Tsapis, N. (2019). Thermoresponsive polymer nanocarriers for biomedical applications. *Advanced Drug Delivery Reviews, 138,* 167–192.

Bregoli, L., Movia, D., Gavigan-Imedio, J. D., Lysaght, J., Reynolds, J., & Prina-Mello, A. (2016). Nanomedicine applied to translational oncology: A future perspective on cancer treatment. *Nanomedicine: Nanotechnology, Biology and Medicine, 12*(1), 81–103. Available from https://doi.org/10.1016/j.nano.2015.08.006.

Brophy, K., Davies, S., Olenik, S., Çotur, Y., Ming, D., VanZalk, N., O'Hare, D., Guder, F., & Yetisen, A.K. (2021). The future of wearable technologies.

Chen, B., & Wang, F. (2020). Emerging frontiers of upconversion nanoparticles. *Trends in Chemistry, 2* (5), 427–439.

Chen, G., Yu, J., & Gu, Z. (2018). Glucose-responsive microneedle patches for diabetes treatment. *Journal of Diabetes Science and Technology, 13*(1), 41–48. Available from https://doi.org/10.1177/1932296818778607.

Desai, N., Koppisetti, H., Pande, S., Shukla, H., Sirsat, B., Ditani, A. S., Mallick, P. P., Kathar, U., Kalia, K., & Tekade, R. K. (2021). Nanomedicine in the treatment of diabetic nephropathy. *Future Medicinal Chemistry, 13*(07), 663–686.

Eissa, S., & Zourob, M. (2017). Aptamer-based label-free electrochemical biosensor array for the detection of total and glycated hemoglobin in human whole blood. *Scientific Reports, 7*(1), 1016. Available from https://doi.org/10.1038/s41598-017-01226-0.

Eom, S., Choi, G., Nakamura, H., & Choy, J.-H. (2020). 2-Dimensional nanomaterials with imaging and diagnostic functions for nanomedicine: A review. *Bulletin of the Chemical Society of Japan, 93*(1). Available from https://doi.org/10.1246/bcsj.20190270.

Furxhi, I., Murphy, F., Mullins, M., & Poland, C. A. (2019). Machine learning prediction of nanoparticle in vitro toxicity: A comparative study of classifiers and ensemble-classifiers using the Copeland Index. *Toxicology Letters, 312,* 157–166. Available from https://doi.org/10.1016/j.toxlet.2019.05.016.

Ghosal, K., & Ghosh, A. (2019). Carbon dots: The next generation platform for biomedical applications. *Materials Science and Engineering: C, 96,* 887–903.

Ghosh, S., Datta, D., Cheema, M., Dutta, M., & Stroscio, M. A. (2017). Aptasensor based optical detection of glycated albumin for diabetes mellitus diagnosis. *Nanotechnology, 28*(43), 435505. Available from https://doi.org/10.1088/1361-6528/aa893a.

Graziano, G. (2019). Functional tattoos. *Nature Reviews Chemistry, 3*(8), 463. Available from https://doi.org/10.1038/s41570-019-0119-x.

Gupta, A., & Singh, L. (2021). A review on emerging trend of medical armour-nanorobot. *Current Nanomaterials, 6*(1), 58–65.

Gupta, V., & Kanna, S. (2019). The current trends and treatments in diabetic retinopathy. *Asian Journal of Pharmaceutical and Clinical Research, 12*(7), 27–33.

Hao, Z., Zhu, Y., Wang, X., Rotti, P. G., DiMarco, C., Tyler, S. R., Zhao, X., Engelhardt, J. F., Hone, J., & Lin, Q. (2017). Real-time monitoring of insulin using a graphene field-effect transistor aptameric nanosensor. *ACS Applied Materials & Interfaces, 9*(33), 27504–27511. Available from https://doi.org/10.1021/acsami.7b07684.

Hassan, S., Bhat, A., Bhonde, R. R., & Lone, M. A. (2016). Fighting diabetes: Lessons from xenotransplantation and nanomedicine. *Current Pharmaceutical Design, 22*(11), 1494–1505.

Hayeri, A. (2019). 922-P: Diabits—An AI-powered smartphone application for blood glucose monitoring and predictions. *Diabetes, 68*(Suppl. 1). Available from https://doi.org/10.2337/db19-922-P.

Hussain, Z., Rahim, M. A., Jan, N., Shah, H., Rawas-Qalaji, M., Khan, S., Sohail, M., Thu, H. E., Ramli, N. A., & Sarfraz, R. M. (2021). Cell membrane cloaked nanomedicines for bio-imaging and immunotherapy of cancer: Improved pharmacokinetics, cell internalization and anticancer efficacy. *Journal of Controlled Release, 335,* 130–157.

Jahangirian, H., Lemraski, E. G., Webster, T. J., Rafiee-Moghaddam, R., & Abdollahi, Y. (2017). A review of drug delivery systems based on nanotechnology and green chemistry: Green nanomedicine. *International Journal of Nanomedicine*, *12*, 2957.

Janjic, J. M., & Gorantla, V. S. (2020). Nanomedicine in organ transplantation and regenerative surgery: An interview with Vijay Gorantla and Jelena Janjic. *Nanomedicine: Nanotechnology, Biology, and Medicine*, *15*(3), 215−218.

Jasinski, D. L., Binzel, D. W., & Guo, P. (2019). One-pot production of RNA nanoparticles via automated processing and self-assembly. *ACS Nano*, *13*(4), 4603−4612.

Jeevanandam, J., Aing, Y. S., Chan, Y. S., Pan, S., & Danquah, M. K. (2017). Nanoformulation and application of phytochemicals as antimicrobial agents. *Antimicrobial nanoarchitectonics: From synthesis to applications* (Vol. 1, pp. 62−82). Elsevier.

Jeevanandam, J., Chan, Y. S., & Danquah, M. K. (2016). Biosynthesis of metal and metal oxide nanoparticles. *ChemBioEng Reviews*, *3*(2), 55−67.

Jeevanandam, J., Danquah, M. K., Debnath, S., Meka, V. S., & Chan, Y. S. (2015). Opportunities for nano-formulations in type 2 diabetes mellitus treatments. *Current Pharmaceutical Biotechnology*, *16*(10), 853−870.

Jeevanandam, J., Pal, K., & Danquah, M. K. (2019). Virus-like nanoparticles as a novel delivery tool in gene therapy. *Biochimie*, *157*, 38−47. Available from https://doi.org/10.1016/j.biochi.2018.11.001.

Key, J., & Leary, J. F. (2014). Nanoparticles for multimodal in vivo imaging in nanomedicine. *International Journal of Nanomedicine*, *9*, 711.

Khan, A. U., Khan, M., Cho, M. H., & Khan, M. M. (2020). Selected nanotechnologies and nanostructures for drug delivery, nanomedicine and cure. *Bioprocess and Biosystems Engineering*, *43*(8), 1339−1357.

Ko, J., Bhagwat, N., Yee, S. S., Ortiz, N., Sahmoud, A., Black, T., Aiello, N. M., McKenzie, L., O'Hara, M., & Redlinger, C. (2017). Combining machine learning and nanofluidic technology to diagnose pancreatic cancer using exosomes. *ACS Nano*, *11*(11), 11182−11193.

Kuralay, F. (2019). Nanomaterials-based enzyme biosensors for electrochemical applications: Recent trends and future prospects. *New Developments in Nanosensors for Pharmaceutical Analysis*, 381−408.

Landesman-Milo, D., & Peer, D. (2016). Transforming nanomedicines from lab scale production to novel clinical modality. *Bioconjugate Chemistry*, *27*(4), 855−862. Available from https://doi.org/10.1021/acs.bioconjchem.5b00607.

Le, L. V., Chendke, G. S., Gamsey, S., Wisniewski, N., & Desai, T. A. (2019). Near-infrared optical nanosensors for continuous detection of glucose. *Journal of Diabetes Science and Technology*. Available from https://doi.org/10.1177/1932296819886928.

Lee, C.-H., Hung, K.-C., Hsieh, M.-J., Chang, S.-H., Juang, J.-H., Hsieh, I. C., Wen, M.-S., & Liu, S.-J. (2020). Core-shell insulin-loaded nanofibrous scaffolds for repairing diabetic wounds. *Nanomedicine: Nanotechnology, Biology and Medicine*, *24*, 102123.

Ling, J. K. U., Hii, Y. S., Jeevanandam, J., Chan, Y. S., & Danquah, M. K. (2019). Nanoencapsulation of phytochemicals and in-vitro applications. *Phytochemistry: An in-silico and in-vitro update* (Vol. 1, pp. 315−330). Singapore: Springer.

Lin-Ping, W. U., Wang, D., & Li, Z. (2020). Grand challenges in nanomedicine. *Materials Science and Engineering: C*, *106*, 110302.

Liu, D., Liu, L., Yao, L., Peng, X., Li, Y., Jiang, T., & Kuang, H. (2020). Synthesis of ZnO nanoparticles using radish root extract for effective wound dressing agents for diabetic foot ulcers in nursing care. *Journal of Drug Delivery Science and Technology*, *55*, 101364.

Mao, Y., Hu, Y., Feng, W., Yu, L., Li, P., Cai, B., Li, C., & Guan, H. (2020). Effects and mechanisms of PSS-loaded nanoparticles on coronary microcirculation dysfunction in streptozotocin-induced diabetic cardiomyopathy rats. *Biomedicine & Pharmacotherapy*, *121*, 109280.

Meetoo, D., Wong, L., & Ochieng, B. (2019). Smart tattoo: Technology for monitoring blood glucose in the future. *British Journal of Nursing*, *28*(2), 110−115. Available from https://doi.org/10.12968/bjon.2019.28.2.110.

Mian, Z., Hermayer, K. L., & Jenkins, A. (2019). Continuous glucose monitoring: Review of an innovation in diabetes management. *The American Journal of the Medical Sciences, 358*(5), 332–339. Available from https://doi.org/10.1016/j.amjms.2019.07.003.

Mohamed, S. M. (2019). *Nanoparticles based approach for direct detection of urinary miRNA in diabetic nephropathy* (CU theses). Mansoura University.

Mohsen, A. M. (2019). Nanotechnology advanced strategies for the management of diabetes mellitus. *Current Drug Targets, 20*(10), 995–1007.

Mulder, W. D., Phiri, M. M., & Vorster, C. B. (2019). Gold nanostar colorimetric detection of fructosyl valine as a potential future point of care biosensor candidate for glycated haemoglobin detection. *Biosensors, 9*(3). Available from https://doi.org/10.3390/bios9030100.

Ozkan, S.A., Shah, A., & Kuralay, F. (2019). Chapter 12: Nanomaterials-based enzyme biosensors for electrochemical applications: Recent trends and future prospects. In *New developments in nanosensors for pharmaceutical analysis* (pp. 381–408). Academic Press. Available from https://doi.org/10.1016/B978-0-12-816144-9.00012-2.

Pastorin, G. (2019). *Carbon nanotubes: From bench chemistry to promising biomedical applications.* CRC Press.

Patel, D. M., Patel, N. N., & Patel, J. K. (2021). Nanomedicine scale-up technologies: Feasibilies and challenges. *Emerging technologies for nanoparticle manufacturing* (pp. 511–539). Springer.

Pelaz, B., Alexiou, C., Alvarez-Puebla, R. A., Alves, F., Andrews, A. M., Ashraf, S., Balogh, L. P., Ballerini, L., Bestetti, A., Brendel, C., Bosi, S., Carril, M., Chan, W. C. W., Chen, C., Chen, X., Chen, X., Cheng, Z., Cui, D., Du, J., & Parak, W. J. (2017). Diverse applications of nanomedicine. *ACS Nano, 11*(3), 2313–2381. Available from https://doi.org/10.1021/acsnano.6b06040.

Perez-Guzman, M. C., Shang, T., Zhang, J. Y., Jornsay, D., & Klonoff, D. C. (2021). Continuous glucose monitoring in the hospital. *Endocrinology and Metabolism, 36*(2), 240.

Prasad, P., Sukumaran, S., Shaji, N., Yadunath, V.K., Jose, J., Kalarikkal, N., & Thomas, S. (2017). Toxicity of nanomaterials used in nanomedicine. *Recent Trends in Nanomedicine and Tissue Engineering,* 365.

Rehman, A., Jafari, S. M., Tong, Q., Riaz, T., Assadpour, E., Aadil, R. M., Niazi, S., Khan, I. M., Shehzad, Q., & Ali, A. (2020). Drug nanodelivery systems based on natural polysaccharides against different diseases. *Advances in Colloid and Interface Science, 284,* 102251.

Rifat, T., Hossain, M. S., Alam, M. M., & Rouf, A. S. S. (2019). A review on applications of nanobots in combating complex diseases. *Bangladesh Pharmaceutical Journal, 22*(1), 99–108.

Sabahi, M., Ahmadi, S. A., Mahjub, R., & Ranjbar, A. (2019). Oxidative toxicity in diabetes mellitus: Role of nanoparticles and future therapeutic strategies. *Precision Nanomedicine, 2*(4), 382–392.

Shafiee, A., Ghadiri, E., Kassis, J., & Atala, A. (2019). Nanosensors for therapeutic drug monitoring: Implications for transplantation. *Nanomedicine: Nanotechnology, Biology, and Medicine, 14*(20), 2735–2747. Available from https://doi.org/10.2217/nnm-2019-0150.

Sim, T., Lim, C., Hoang, N. H., Joo, H., Lee, J. W., Kim, D., Lee, E. S., Youn, Y. S., Kim, J. O., & Oh, K. T. (2016). Nanomedicines for oral administration based on diverse nanoplatform. *Journal of Pharmaceutical Investigation, 46*(4), 351–362. Available from https://doi.org/10.1007/s40005-016-0255-y.

Siwach, R., Pandey, P., Chawla, V., & Dureja, H. (2019). Role of nanotechnology in diabetic management. *Recent Patents on Nanotechnology, 13*(1), 28–37.

Souto, B. E., Souto, B. S., Campos, R. J., Severino, P., Pashirova, N. T., Zakharova, Y. L., Silva, M. A., Durazzo, A., Lucarini, M., Izzo, A. A., & Santini, A. (2019). Nanoparticle delivery systems in the treatment of diabetes complications. *Molecules (Basel, Switzerland), 24*(23). Available from https://doi.org/10.3390/molecules24234209.

Suarez, E. A. Q., Acosta, H. F. H., & Montiel, G. A. C. (2019). Biological nanosensors on network for diabetes control with alert emission for users. *Ingeniería Solidaria, 15,* 29.

Sujatha, V., Suresh, M., & Mahalaxmi, S. (2010). Nanorobotics: A futuristic approach. *Nano Digest, 3,* 34–37.

Sun, Y., Ma, W., Yang, Y., He, M., Li, A., Bai, L., ... Yu, Z. (2019). Cancer nanotechnology: Enhancing tumor cell response to chemotherapy for hepatocellular carcinoma therapy. *Asian journal of pharmaceutical sciences, 14*(6), 581–594.

Tambe, P., Kumar, P., Paknikar, K. M., & Gajbhiye, V. (2019). Smart triblock dendritic unimolecular micelles as pioneering nanomaterials: Advancement pertaining to architecture and biomedical applications. *Journal of Controlled Release, 299*, 64–89.

Tan, K. X., Jeevanandam, J., Pan, S., Yon, L. S., & Danquah, M. K. (2020). Aptamer-navigated copolymeric drug carrier system for in vitro delivery of MgO nanoparticles as insulin resistance reversal drug candidate in Type 2 diabetes. *Journal of Drug Delivery Science and Technology, 57*, 101764.

Teh, S. J., & Lai, C. W. (2019). Carbon nanotubes for dental implants. *Applications of nanocomposite materials in dentistry* (pp. 93–105). Elsevier.

Thorat, N. D., Townley, H. E., Patil, R. M., Tofail, S. A., & Bauer, J (2019). Comprehensive approach of hybrid nanoplatforms in drug delivery and theranostics to combat cancer. *Drug discovery today, 25*(7), 1245–1252.

Tran, J., Tran, R., & White, J. R. (2012). Smartphone-based glucose monitors and applications in the management of diabetes: An overview of 10 salient "apps" and a novel smartphone-connected blood glucose monitor. *Clinical Diabetes, 30*(4), 173–178.

Tsou, Y.-H., Wang, B., Ho, W., Hu, B., Tang, P., Sweet, S., Zhang, X.-Q., & Xu, X. (2019). Nanotechnology-mediated drug delivery for the treatment of obesity and its related comorbidities. *Advanced Healthcare Materials, 8*(12), 1801184. Available from https://doi.org/10.1002/adhm.201801184.

Urrejola, M. C., Soto, L. V., Zumaran, C. C., Pablo Penaloza, J., Alvarez, B., Fuentevilla, I., & Haidar, Z. S. (2018). Polymer nanoparticle systems: From biodetection and glucose monitoring in diabetes to bioimaging, nano-oncology, gene therapy, tissue engineering/regeneration to nano-dentistry. *International Journal of Morphology, 36*(4), 1490–1499.

Vallabani, N. V. S., Singh, S., & Karakoti, A. (2019). Magnetic nanoparticles: Current trends and future aspects in diagnostics and nanomedicine. *Current Drug Metabolism, 20*(6), 457–472.

Vashist, A., Kaushik, A., Vashist, A., Sagar, V., Ghosal, A., Gupta, Y. K., Ahmad, S., & Nair, M. (2018). Advances in carbon nanotubes–hydrogel hybrids in nanomedicine for therapeutics. *Advanced Healthcare Materials, 7*(9), 1701213. Available from https://doi.org/10.1002/adhm.201701213.

Wadhwa, S., Garg, V., Gulati, M., Kapoor, B., Singh, S. K., & Mittal, N. (2019). Nanovesicles for Nanomedicine: Theory and Practices. *Pharmaceutical Nanotechnology*, 1–7.

Wagner, A. M., Knipe, J. M., Orive, G., & Peppas, N. A. (2019). Quantum dots in biomedical applications. *Acta Biomaterialia, 94*, 44–63.

Wang, J., Coleman, D. C., Kanter, J., Ummer, B., & Siminerio, L. (2018). Connecting smartphone and wearable fitness tracker data with a nationally used electronic health record system for diabetes education to facilitate behavioral goal monitoring in diabetes care: Protocol for a Pragmatic Multi-Site Randomized Trial. *JMIR Research Protocols, 7*(4), e10009.

Wang, Q., Wang, S., Hu, X., Li, F., & Ling, D. (2019). Controlled synthesis and assembly of ultra-small nanoclusters for biomedical applications. *Biomaterials Science, 7*(2), 480–489.

Wang, T.-T., kit Lio, C., Huang, H., Wang, R.-Y., Zhou, H., Luo, P., & Qing, L.-S. (2020). A feasible image-based colorimetric assay using a smartphone RGB camera for point-of-care monitoring of diabetes. *Talanta, 206*, 120211.

Weissig, V., Elbayoumi, T., Wadhwa, S., Garg, V., Gulati, M., Kapoor, B., Singh, S.K., & Mittal, N. (2019). Nanovesicles for nanomedicine: Theory and practices. In Pharmaceutical nanotechnology: Basic protocols (pp. 1–17). Springer New York. Available from https://doi.org/10.1007/978-1-4939-9516-5_1.

Xiao, Y., & Du, J. (2019). Superparamagnetic nanoparticles for biomedical applications. *Journal of Materials Chemistry B, 8*(3). Available from https://doi.org/10.1039/C9TB01955C.

Yan, C., Quan, X.-J., & Feng, Y.-M. (2019). Nanomedicine for gene delivery for the treatment of cardiovascular diseases. *Current Gene Therapy, 19*(1), 20–30. Available from https://doi.org/10.2174/1566523218666181003125308.

Zhang, N., Ming-Yuan Wei, M., & Ma, Q. (2019). Nanomedicines: A potential treatment for blood disorder diseases. *Frontiers in Bioengineering and Biotechnology, 7*, 369.

Zhao, Y., Yang, J., Shan, G., Liu, Z., Cui, A., Wang, A., Chen, Y., & Liu, Y. (2020). Photothermal-enhanced tandem enzyme-like activity of Ag2-xCuxS nanoparticles for one-step colorimetric glucose detection in unprocessed human urine. *Sensors and Actuators B: Chemical, 305,* 127420.

Zhu, Y., Hao, Y., Adogla, E. A., Yan, J., Li, D., Xu, K., Wang, Q., Hone, J., & Lin, Q. (2016). A graphene-based affinity nanosensor for detection of low-charge and low-molecular-weight molecules. *Nanoscale, 8*(11), 5815—5819. Available from https://doi.org/10.1039/C5NR08866F.

Zor, F., Selek, F. N., Orlando, G., & Williams, D. F. (2019). Biocompatibility in regenerative nanomedicine. *Nanomedicine: Nanotechnology, Biology, and Medicine, 14*(20), 2763—2775.

Index

Printed in the United States
by Baker & Taylor Publisher Services